*The Science of
Christian Economy
and other prison writings*

IN DEFENSE OF COMMON SENSE

PROJECT A

THE SCIENCE OF CHRISTIAN ECONOMY

by Lyndon H. LaRouche, Jr.

SCHILLER INSTITUTE, INC.   Washington, D.C.   1991

Library of Congress catalogue number: 91-062722

Front cover photo: Scala/Art Resource, New York
Back cover photo: Nancy B. Spannaus
Cover design: Virginia Baier
Book design: Alan Yue
Project editors: Marianna Wertz, Christina Huth

Please direct all inquiries to the publisher:
Schiller Institute, Inc.
P.O. Box 66082
Washington, D.C. 20035-6082

SIB 91-003

Contents

Project A

LaRouche Discusses His Solution to the 'Riddle of
the Ages,' the Parmenides Paradox of Plato

The Science of Christian Economy

Appendices

Illustrations follow pages 89 and 203.

Author's Preface

Since January 1989, I have found myself often appearing momentarily like a Homeric Cassandra, watching the ruin of your doomed world, watching as if from the crumbling ramparts of extinct, ancient Troy.

By your next door neighbor's consent, long-Kissinger-tormented Lebanon has at last been murdered. By means of a kindred culpability, dying Uganda is perhaps now in the process of vanishing from the biological map of Africa. The years 1990 and 1991 have come, and with them, the onset of that spiral of wars of which I spoke in launching my 1990 congressional campaign.

As I forewarned you this was imminent on a nationwide TV broadcast back in autumn 1988, the civil war in Yugoslavia has now erupted; and this, as I also forewarned you, is an integral part of turning the entire region of the eastern Mediterannean into a vast hecatomb of bloody particularist dismemberments.

Your proud financial and monetary system, which had become the principal article of your religious faith, collapsed during October 1987—as I had foretold earlier—and is now bleeding internally of mortal, self-inflicted wounds.

Had you understood even the ABCs of history, this terrible thing could not have happened. You would not have permitted your next door neighbor to bring all of this ruin upon us by his behaving as he has done.

I am no Cassandra. This poor but precious civilization of ours could yet be rescued from what may appear to many, more and more often, the accelerating onrush of apocalyptic doom.

This civilization could be saved—if we *earn* that.

If we are not all to drown, your neighbor too, must learn now to swim.

What therefore did you urgently need to know, which I had either neglected to tell you, or, perhaps, had not said clearly enough? What did you require most urgently, that you might rescue us from your neighbor's folly? A grander strategic perspective, a more alluring set of programs of economic reconstruction? I thought that was not where my omission lay.

What your neighbor required, most urgently, was not instruction on what to think, but remedial assistance in the matter of how to think. One must never make apology for saying even unpleasant things which are needed, most urgently, to be said. One need not apologize for saying that as well as possible—if no one else were saying it better.

I wish devoutly it were better; but nonetheless, it had been better said than not.

Now, my friends have elected, very kindly, to reissue these three published philosophical writings together, in a single volume. May it enrich you and so give you pleasure. I can do no better but share with you something slightly better than that which I have to give.

August 8, 1991

In Defense of
Common Sense

Ibykus Revisited[1]

It were an ill wind which had not sent some Plato dialogues into my cell. From the prison in which the politician's career expires, the influence of the statesman is raised toward the summits of his life's providential mission. Since Solon (638–558 B.C.), the Socratic method has become the mark of the great Western statesman. Without the reemergence of that leadership, our imperiled civilization will not survive this century's waning years.

The available translations of Plato (427–347 B.C.) were in the tradition set by Benjamin Jowett's hoaxes: a few titles from a recently cooked Penguin Classics series. Robin Waterford's edition of the *Theaetetus* is referenced, to introduce a crucial point.[2]

On page 23, Waterford renders a passage:

Theaetetus: ". . . Any numbers which can be the product of multiplying some number by itself, we called 'square and equal-sided,' on the model of the geometrical square."

Let us make good use of the folly embedded in a footnote which Waterford appends to that passage:

One of Theaetetus's terms, ["square and equal-sided"], is redundant [sic!]. This and other slight awkwardnesses in his account have been taken to suggest he is being portrayed as a schoolboy rather than a master mathematician.

The relevant fact here is Waterford's mathematics blunder. This shows that the editor of this translation could not have understood the crucial thread of the argument throughout that portion of the dialogue.

The construction of a preliminary, geometrical definition of irrational magnitudes, illustrates the existence of cases in which a critical treatment of an English translation, for example, needs no reference to the Greek original. We shall indicate, that that can be said also of *Platonic ideas* generally, and of the notion of *Socratic method* itself.

In his essay, which accompanies his translation, Waterford devotes his effort chiefly to attempting to discredit Socratic method and Platonic ideas. The mask of the translator is dropped. So, more generally, in today's English-speaking world, British empiricism and the progeny of Marburg neo-Kantianism have secured monopolies of putative authority in academic matters bearing upon translation and interpretation of Plato's work.

In its own right, the original Greek text has a specific, irreplaceable function. It is a leading part of the living history of civilization during that entire century and following. It is an integral and central element in the whole of the historical existence of classical Greece; it is a central feature of the recent 2,500 years of universal history; it plays a crucial part in the development of classical Indo-European philology. Nonetheless, here we focus on conceptions which are susceptible of fully intelligible representation in terms of reference to English editions, almost as if Greek texts no longer existed.

The referenced case, Theaetetus's geometrical definition of simple irrationals, is a rudimentary illustration of the point.

In every case Plato uses words to describe a conception he intends should also be represented by methods of *constructive geometry,* constructive geometry affords us a rigorous basis for criticism of an English edition's treatment of that aspect of the dialogue.[3]

It happens to be the case, that all Platonic ideas fit into the class of geometrically constructed *intelligible representations.*[4] That is also true for *Socratic method* as such. The following pages illustrate the point; these pages are not a restatement of *Theaetetus;* they are supplied to complement it, in ways whose significance should be clear from a direct comparison of the two.

I | What Is Common Sense?

Perhaps, there is some truth in the old saw, that the grass is always greener in the other fellow's yard. During my 66-odd years, the people who praised common sense the longest and loudest, turned out to be fellows who possessed very little of that particular *virtù*. The rest among us, watching those performances, were ashamed to defend common sense by its true name.

Now, it appears sometimes, that common sense has about the same prospects as did the extinct dodo. This occasion appears to be our last opportunity to save it, before our nation, too, might become extinct. Let us now, at last, defend it openly by its true name. To that end, let us construct agreement to a proper definition.

We begin with two simple decisions. We should have no reluctance at throwing out all connotations of "horse sense"; the image of cleaning our living-room carpets with a pitchfork, improves our resolution on that account. It should be obvious, that we must reject all definitions premised axiomatically upon the notion of an *individual* person "with the brains to survive." Societies may or may not be immortal; no individual is.

We require a definition which points not only toward the preconditions for successful survival of entire societies. This must be a definition which includes a proof, that some quality, possessed by at least some persons within the deliberative processes of that society, is indispensable to make possible the successful survival of that society as a whole.

Is survival itself not success? There exist states of social existence, which must appear to be *momentary* survival, but which are also a step upward along a pathway which leads but to ruin over the longer term.

Whatever "common sense" should mean, the included term, "common," signifies that a society were unlikely to attain *successful survival*, unless "common sense" were more or less commonplace. That is a view not inconsistent with an opinion expressed by some notable founders of our federal republic, that a certain quality of universal public education were indispensable, to ensure future generations capable of preserving the freedoms which had then been won recently at such great peril and expense.

Such are the broadest generalizations, the guidelines which must circumscribe our continuing search for an adequate definition of true common sense. First, we must focus upon the subject of the successful survival of an entire society. Within that frame of reference, we must define that essential quality of the individual, without the which the society as a whole could not survive.

II | Successful Survival

A swimmer is rescued from the sea, moments before his strength had ebbed fatally. This is *momentary* survival. Is this a paradigm for the successful survival of an entire society? How must we distinguish between merely *momentary* and *durable* survival?

What has gone before now generates what follows.

Select some moment of *a continuing process of survival* of the society as a whole. This process is the aggregation and interaction of the deaths, maturations, births, and conceptions, which, together, define a collection of individuals functionally as an indivisible whole.

That selected moment serves as the time of departure for a continuing journey into the future. Assign the label "A" to that moment. To each of a series of selected, subsequent moments of that continuing journey, assign, respectively, the labels B, C, D, and so on.

Does the swimmer's survival at moment B portend a likelihood for his continuing survival, through C, D, and so on? Perhaps, to foster a successful posterity? Such is the raw distinction between *momentary* and *successful* survival.

Regard that journey into the future as if it were a kind of mathematical physicist's *continuous function*. *Continuously*, cause generates effect and effect has a causal relationship to the subsequent effects. This function is expressed in terms of

8

increase or decrease of a magnitude termed *potential population-density*.

This is no ordinary or "linear" sort of continuous function; it is "nonlinear," but not the less an *efficiently continuous function* in experimental terms of reference. We shall come to that in due course here, after we have examined *population-density* and *potential population-density*.

Population-density signifies what the very term must appear to imply. It signifies, most loosely, the *number of persons* inhabiting *an average square kilometer of land-area*. On closer inspection, this simple ratio becomes more and more interesting.

Some anthropologists teach that, once upon a time, when the surface of our planet was still a utopian wilderness, all mankind lived in a state of infantile innocence and harmony with nature, called "a simple hunting-and-gathering society." There is no surviving evidence to support the alleged existence of such a primitive society, although numerous cases are known in which failed societies have been collapsed into a condition resembling the utopian ideal.

Whether such states of society's development occur only as the next-to-lowest degree of degeneration of a culture, or might also have come into existence as a moment in an upward trend of society's development, is not a relevant issue, beyond our mere obligation to note that as fact. For such a state of society, the demographic and related economic characteristics are known within fairly good degrees of estimation.

One estimate of population-density, is 10 square kilometers of average wilderness land-area per person. Otherwise, the utopian "hunting-and-gathering society" resembles a "tribe" of baboons or "yahoos." The post-infancy life-expectancy is significantly less than an average of 20 years. One estimate for the upper limit to the Earth's population, in such a utopia, would be about 10 million persons.

In contrast, today, there are more than five billion persons. Had the world as a whole employed, to the degree it might have done so, the levels of technology already available by 1970, not only would the planet's population be signficantly larger than it is; the average standard of living per capita would approximate that of North America 1970, and the potential population, as distinct from the actual, would be approaching 25 billion.

The increase of both the actual and the potential population-densities, during the recent thousands of years, is the outcome of the continued and interdependent *generation, transmission, and efficient assimilation of scientific and technological progress.*

This signifies, that the transformation of land and productive powers of labor, which reflect directly the causal impact of scientific progress *per se,* has the following qualities of impact upon potential population-density:

1) The fertility of the land, for production and for human habitation, is increased;

2) The physical productivity of labor is increased;

3) The per capita physical standard of living of labor's households (market-basket) is improved.

The amount of average land-area required to sustain an average person is reduced, while the required, and realized standard of physical consumption is increased.

We discover also, that the following changes occur, as well. With the rise in the potential physical productivity of labor and the correlated rise in required per capita consumption, the following trends prevail.

4) The amount of power consumed, per capita and per hectare, increases.

5) The amount of power concentrated upon a cross-section of work-area, tends to be increased.

Also, the following social changes correlate with increases in physical productivity.

6) The ratio of urban to rural population increases toward what seems to be an asymptotic upper limit.

7) The ratio of labor employed in production of produc-

er's goods, relative to labor employed in production of households' and related goods, rises.

These changes proceed with the restriction, that the per hectare and per capita output of rural and of households' and related goods rises, and never declines. These preconditions are associated with other demographic changes:

8) The age of maturation (e.g., full entry into the regular labor-force) rises toward, apparently, an asymptotic upper limit.

9) There is a general trend of shift from labor-intensive to power-intensive modes of employment of operatives engaged in physical output.

To survive successfully signifies satisfying all these preconditions. These preconditions determine the causal connection between advances in technology and increases in *potential* population-density.

Consider the series, A, B, C, D, . . . , Z, in this light. Assign to each term a specific moment of human development, and also assign to that term the appropriate estimate of *potential population-density.*

To console the anthropologist among us, let us assign to "A" a time in imaginary pre-history—once upon a time—where we find his "simple hunting-and-gathering" utopia. We have somewhere in later parts of the series, real history of real societies, such moments as Athens, 599 B.C.; Western Europe in the moment of Charlemagne's (A.D. 742–814) census; Italy, A.D. 1439; England, A.D. 1588; France, A.D. 1672; France, A.D. 1776; U.S.A., A.D. 1866; U.S.A., A.D. 1946 and so on. Each term of that series represents, *for humanity as a whole,* a relatively higher value of *implicit* potential population-density than any preceding term.

The question, implicitly, is: Does an advance in potential at B *cause* a further advance appearing at C, D, and so on? If there is some *causal necessity* linking the occurrence of a relatively higher potential at moment M to a still greater potential at later moment N, and this is also true of the causal bearing of N upon later moment Q, Q upon P, and so forth, then the increase of potential at M implicitly causes the increases at N, Q, P, and so forth.

That is a crude paradigm for *successful survival* of an entire society.

This is no ordinary sort of algebraic function. Now, as we turn, next, to our first relevant glimpse into the deep interior of the individual human mind, we shall encounter a proof, why only *non-algebraic geometric numbers* can enable us to reflect *potential population-density* as we have represented that magnitude thus far.

III | Potential Population-Density

So far, until the onset of the 1970s, virtually no sane, literate member of our Western European cultural heritage would have required an explanation of the manner we have referenced the causal correlation between scientific and technological progress, as proximate cause, and increase of *momentary* potential population-density, as effect.

We can not leave matters so. The manner in which we defined *potential population-density,* not as merely *momentary,* but *successful* survival, has changed everything.

Before we define the function for successful survival, there are two sub-topics which must be examined. The first, is the deeper meaning of *scientific and technological progress.* After that, we examine the fact that all true *natural law* must be coherent with the central feature of non-algebraic, geometric functions, *isochronicity.*

The simple fact respecting technological progress, is that no species of beast can effect such behavior. *Scientific and technological progress reflects a quality of the human individual which sets mankind apart from and above all other living creatures.* In all beasts, the range of behavioral change which can be generated *on behalf of the species from within the species itself* is delimited in effective range, as if by fixed instinct. This is to the effect, that the potential population-density of the species may be altered either locally, or generally, by changes introduced to the beast's environment; but the beast can not significantly improve that potential *willfully*.

Since we are examining this unique quality of the human species's behavior from the standpoint of a continuous function in mathematical physics, it is appropriate that we define scientific progress as a class of mathematical event. At the instant we resolve to do that, the fun begins.

No formalist's system of mathematics, that is, no mathematics in the spirit of deductive logic, can represent even the simplest kind of act of scientific discovery (see Appendix IV). Therefore, *deductive logic* is useless as a way of representing human mental life; however, by showing what deductive logic can not do, we are able to see more clearly what occurs in every genuine scientific or analogous creative act of the human mind.

At first glance, *population-density* is a simple ratio, *a countable number of persons* to *a countable number of square kilometers* of inhabited land-area. This might be admired as a way of representing some useful ideas bearing upon *momentary survival*, at moment B; but, it is useless as a term, when our attention is turned to *successful survival*, the imputed values of each and all of B, C, D, etc., at moment B.

In a related example, John Von Neumann's systems of linear inequalities might appear to be applied successfully to estimating potential population-densities in *animal ecologies;* those mathematical methods are worse than useless for human "ecologies."

The paramount difficulty, in both instances, is human mental behavior: the role of scientific and technological progress, *as a causal principle*, in *generating increase of potential population-density* as an *effect*. In effect, *a direction of scientific and technological progress* causes a trend of *increase of potential population-density*. We examine the construction of an intelligible representation of this continuous function.

This function, whose construction we are to examine in broad terms here, is characteristic of all human behavior which is associated with *successful survival.* It is not operative in any form of animal behavior.

The general form of the function is given thus:

1) Technological progress fosters a rise in potential population-density.

2) A rise in potential population-density, so caused, fosters the conditions for an advance in the technological level of generalized practice.

3) That, *if realized,* fosters a further increase of potential population-density.

The foregoing presumes a fulfillment of the correlated preconditions specified earlier.

In the cases such a function is continuous in effect, a direction of scientific and technological progress contained in an impulse exerted over interval A to B, chooses a "track" of alternating advances in *technology* and *population-density potential,* which represents a continuous mathematical function in terms of *rate of increase of the rate of increase* of potential population-density. Hence, a value for this function at moment B, expresses also an imputed value for C, for D, etc.

Now, let us construct an intelligible representation of this continuous function. We do this by refuting the central thesis of Immanuel Kant's (1724–1804) *Critiques.*

What we shall now refute is Kant's false conceit on the subject of the possibility of intelligible representation of those forms of creative mental activity expressed by a valid fundamental discovery in mathematical physics. Kant's neo-Aristotelian theorem has the perverse relative merit of being the most exhaustively rigorous among the modern proponents of the same opinion on this point: e.g., Francis Bacon, Hobbes, Galileo, Descartes, Locke, Hume, LaPlace, Cauchy, Kelvin, Maxwell, Kronecker, B. Russell et al. So, if we refute Kant's dogma on this point, we also expose the falseness, the absurdity of the anti-Platonic dogmas generally.

The central feature of Kant's fallacious dogma, like that

of his most important predecessor in this view, Descartes (1596–1650), is the dogma's blind faith in deductive logic.

As we shall show, deductive logic, the root of today's accepted classroom mathematics, is incapable of permitting its user to construct an intelligible representation of a true nonlinear function; therefore, deductive logic and any mathematics enslaved to its formalism, is incapable of comprehension of any truly elementary functions of the physical universe.

To the immediate point, Kant's relative merit was that he recognized the existence of valid fundamental discoveries in physics. However, insofar as he recognized this phenomenon, he also recognized that the process by which such discoveries are generated in the human mind, can not itself be supplied an intelligible representation within the scope of deductive method.

Deductive method says that creative mental processes have no valid existence in our universe. The best deduction can do, is to recognize that such phenomena do exist. We shall now summarize the relevant kernel of Kant's argument from a modern standpoint.

Any ideal system of deductive logic begins with *a set of axioms and postulates*.

Axioms are *theorems* adopted without proof, so adopted on the basis of the assertion that their truth is "self-evident." *Postulates* are *theorems* adopted with no other proof than the indication that such assumptions must be added to the set of axioms for the purpose of eliminating, arbitrarily, ambiguities in the system derived from a combined set of axioms and postulates.

A first layer of theorems is derived from an initial set of axioms and postulates. In turn, a second layer of theorems is derived from the combination of the first layer of theorems, plus the original set of axioms and postulates. So, the edifice of theorems grows, layer by layer, upon the foundation of the initial axioms and postulates.

The result is known as a (deductive) *theorem-lattice*. Every possible theorem in this "lattice" is *deductively consistent* with each and all of the underlying set of axioms and postulates. The corollary is: No theorem of a truly consistent, deductive theorem-lattice claims anything which was not already claimed implicitly by the underlying set of axioms and postulates. This latter "property" is known as "the hereditary property" of all consistent deductive theorem-lattices.

Against that background, consider the implications of a single valid, crucial physical experiment which disproves a single theorem of a body of mathematical-physics dogma.

A formal-deductive form of mathematics dominates the classroom and mathematical physics today. Although, in practice, no body of taught mathematical physics is as consistent as the deductive "hereditary principle" implies it ought to become, it is the implicit intent of formalist mathematical-physics teaching and practice, that such perfect consistency ought to prevail. In that degree, with such qualification, the overwhelming majority of mathematical-physics teaching today is, at its least worst, Kantian.

In that sense, that body of current practice is obliged to meet any standard of proof of consistency implicit in the Kantian deductive method.

If a single, crucial physical experiment shows any strongly defended, present theorem of a school of mathematical physics to be wrong, this physical evidence implies, by virtue of the deductive "hereditary principle," that the entire set of axioms and postulates, upon which the entirety of that mathematical physics rests, is in error.

In practice, when presented with such crucial experimental evidence, we must not plunge immediately to the assumption that the hereditary principle applies with such seismic force. No existing mathematical-physics dogma (excepting perhaps some very obscure, abstract ones) is deductively consistent. That inconsistency, as it bears upon any popular theorem, must be examined, first, before proceeding boldly toward scientific-revolutionary conclusions.

Assume we have taken all of the implied precautions. Assume, therefore, that we have a proven case, a crucial experiment which disproves at least one among the assumptions embedded in the underlying set of axioms and postulates. In that case, the following steps of correction are implicit, using computer "language":

1) The flawed axiomatic assumption is replaced by means' of a relevant hypothesis.

2) The entirety of the existing theorem-lattice is revised, as if theorem by theorem, to reflect this change in the "hereditary basis."

3) Each and all of the revised theorems are subjected to an appropriate crucial experiment, at least implicitly so.

4) If this does not yield a valid result, a new, corrective hypothesis is required, and the process repeated.

5) If the reconstruction is successful, by standard of crucial experiment, the corresponding "scientific revolution" may be considered a success.

Such a "scientific revolution" is used now as typifying those kinds of *transformations,* as from A to B, associated with the causal impact of *scientific and technological progress* relative to a resulting *increase of potential population-density.*

Define Kant's error from the illustrative standpoint just outlined.

Assume a series of such scientific/technological transformations of a society: A→B; B→C; C→D; D→ . . .; etc. Let each such transformation be defined "axiomatically" by a crucial-experimental case of the type illustrated. In this representation, A, B, C, D, etc. are defined as in the hypothetical case of the swimmer, above.

Permit the following ruse. Assume that each of the moments A, B, C, D, etc. corresponds to a specific set of formal-deductive axioms and postulates, such that no set associated with any one term of that series is deductively consistent

III. Potential Population-Density 19

with the set associated with any other among the designated moments.

The arrangement chosen for this ruse will fail us; that failure is our proximate goal. We shall learn more from experiencing the failure than had we not hazarded it.

Assume that the set of axioms and postulates attributed, in each case, is in some form of correspondence to the level of science-technology "state of the art" at that moment. That science-technology level might be represented formally in a different way than the correlated choice of set of axioms and postulates or not. We assume only that the chosen set maps the actual level of practice as well as is possible from the standpoint of deductive theorem-lattices.

The event which is of interest to us, is the occurrence of a crucial-experimental transformation which compels us to replace the set of axioms and postulates α_0 by α_1, α_1 by α_2, and so forth. For our purpose, let us choose A, B, C, D, etc., such that A$\rightarrow$B represents an interval during which α_0 is implicitly replaced by α_1, and so forth and so on. Hence A$\equiv$[set]α_0, B$\equiv\alpha_1$, C$\equiv\alpha_2$, and so on.

The *event* which becomes the elementary subject ("variable term") of our *historically continuous function,* is *the process of transformation* depicted by the arrow linking each pair of successive terms (moments). Let us, then, compare α_0 and α_1. By this means, let us seek to define the nature of the event upon which the notion of this continuous function depends.

No theorem of theorem-lattice A is consistent with any theorem of theorem-lattice B. There is no *deductive continuity* between the two lattices. Between the two, there is a logical gulf, a chasm which no deductive method might bridge.

In the language of formal mathematics, commonplace classroom mathematics, that logical gulf is known as a simple *mathematical discontinuity.*

This *discontinuity* is the *substance* (e.g., the "matter," *"material* substance") of that *action* which defines that *contin-*

uous function, whose intelligible representation we are constructing: *the rate of increase of potential population-density.* This function, as it is key to defining *successful survival,* is indispensable, in turn, for constructing a true definition of common sense.

IV Continuous Transformation

This *action* is the transformation of Lattice A into Lattice B, and so on. So, we are obliged to represent the matter from the viewpoint of deductive lattice-theory. All valid, fundamental, crucial-experimental discovery in physical science, for example, is of this form, at least as this form is to be recognized from that deductive standpoint.

This *action*, even if considered only abstractly, by itself, is a quality which sets mankind absolutely apart from and absolutely above the beasts. Consider, once again, the process-phase so abstracted in terms of these steps:

1) Scientific and technological progress, effected by generation of the indicated discontinuity, *causes* an increase in society's per capita power over nature as a whole, an effect which is expressed by an increase of potential population-density.

2) So, *a mental activity per se,* creative thought, is a powerful *physical cause* of a physical effect.

3) The transformation of knowledge and practice, *caused* by scientific and technological progress's upward transformation of potential population-density, now *conditions,* causally, a new, upward surge in generation, transmission, and efficient assimilation of scientific and technological progress.

4) This, in turn, generates implicitly a further increase of potential population-density.

And, so on.

If we had traced that sequence of cause and effect only as far as completion of the second of the foregoing steps, it would appear that the result of scientific and technological progress were merely *momentary survival*. If we continue through the third and fourth steps, we show a complete cycle, in which the result is measured in terms of the concept *successful survival*.

In contrast to the relatively fixed range of *potential (relative) population-density*[1] of each species and variety of beast, in which the range of adoptive variation appears to be as if fixed by "instinct," man transforms his species's "instinct" at will. Although our species can not change its "instincts" arbitrarily, societies can accomplish such an effect along lawful pathways, by lawful means.

What we have considered on this account, thus far, indicates that any moment A, we may choose, implicitly, among a number of different journeys into the future:

$$A_0 \rightarrow B_1 \rightarrow C_1 \rightarrow D_1 \rightarrow$$
$$\downarrow \rightarrow B_2 \rightarrow C_2 \rightarrow D_2 \rightarrow$$
$$\downarrow \rightarrow B_3 \rightarrow C_3 \rightarrow D_3 \rightarrow$$
$$\downarrow \rightarrow \dots \dots \dots$$

and so on. Also, any among the B's, C's, D's, and so forth, in that array, of any choice of series, may become a new starting-point, A_i, after which any among a number of B_j's might determine a branching, new pathway into the future.

This signifies, that we are viewing the transformation $(A \rightarrow B)$ as relatively crucial, and that we are viewing this transformation in the light of the four-step cycle implicitly defining an action of *successful survival*. It is the *causal* conditioning of the transformation $(B \rightarrow C)$ by the effect of transformation $(A \rightarrow B)$, which defines implicitly the quality of *continuing*, hence *continuous transformation* which we are examining.

Such functions can not be comprehended by any mathematical physics which is consistent with principles of any deductive logic. We summarize the case, in terms of the relevant facts noted up to this point.

The *action* which expresses the transformation of Lattice A into Lattice B, is located uniquely in the mathematical discontinuity separating the domain of Lattice B absolutely from the deductive domain of Lattice A. That discontinuity is the division which defines the exact discreteness of B with respect to A. Yet, that *action which is the transformation* can not be represented intelligibly in any form of deductive logic.

To ensure that what should be obvious, is obvious, we add this. *To be comprehended* in any deductive schema, the chosen subject must be defined in the form of a consistent theorem of a deductive theorem-lattice. By definition—*by construction*—there exists no possible deductive theorem-lattice which could subsume the discontinuity as a consistent theorem-statement, except to assert, as did Kant, that discontinuity's deductive incomprehensibility.

It is the central feature of Kant's *Critiques,* that the kind of creative processes which generate valid, fundamental discoveries in physical science, are intrinsically incomprehensible *as processes*. He shows that this is true for all deductive analysis, and denies the existence of any mode of constructing *intelligible representations* but deduction.

In the first line of argument to this effect, he conceded the fact of creative discovery's occurrence. He recognized such discovery in the physical sciences as a rationally comprehensible fact, but only after the fact of its occurrence—*a posteriori*. Therefore, he would concede, the occurrence of Lattice B—*a posteriori*—as the successor to A. He would be self-obliged to recognize B as, analytically, distinct from A. Similarly, he would be self-obliged to concede the viewpoint of deductive lattice-theory, that B is everywhere inconsistent (*deductively*) with A.

Similarly, the Kantian method would be obliged to concede the existence of a discontinuity, as the characteristic of that transformation which generated B out of A. However, although the Kantian method can recognize the existence of the *discontinuity* as such, it can recognize it only as a bottomless chasm in the expanses of deductive-logical space, a chasm which one may not bridge, but merely leap if one were able to do so. Kantian method could not construct an intelligible representation of that apparent chasm.

Put the latter point in another frame of reference.

At most, deductive analysis may recognize the existence of the discontinuity as to *form,* but not as to *substance.* It may recognize the *existence* of the discontinuity in the sense one recognizes a line of indefinite length as separating two general areas of a plane surface, or a surface as enclosing a volume of space, *as preventing the space from within from finding any continuity with the volume of space without.* To the degree deduction limits the notion of "matter" to *discrete (singularity-bounded)* particles in empty space and empty time, it can not tolerate the attribution of the qualities of matter, excepting the quality of discreteness (form) itself, to a discontinuity.

Axiomatically, analysis recognizes points, lines, surfaces, solids, and hypersolids as *singularities,* and recognizes that *discontinuity* is an essential property of singularity. Yet, axiomatically, *ponderability* is not allowed for the point, line, or surface itself, etc.

Yet, as we have indicated, the *singularity* signed by the *discontinuity* of (A→B) is the *efficient action,* by means of which Lattice A is transformed into Lattice B. That transformation, so depicted, is but the deductive sign for *the action of scientific-technological progress.* The *effect* of this action as cause, is an increase of the potential population-density of society. The latter is a very ponderable effect; scientific and technological progress is a very ponderable "force"; the transformation which generates the force of progress, is a physically efficient "force."

From this point, everything which follows hangs upon our understanding of the kind of mathematical discontinuity we have identified. We must resolve, step by step, the difficult questions bearing upon the nature of its *existence,* and of the possibility of the *intelligible representation of that existence.*

A *discontinuity* might suggest something which has no *existence,* as we think ordinarily of existence.

For example, the idea that two objects differ, is not itself an *existence.* Similarly, the mere idea that a finite object ceases in extent at the outer limit of its finiteness, is not a distinct existence. To attribute existence to such objects is like saying, "I am taking my not-wife out for a walk, so that I may walk alone"; such sentences make good sense to a truly thorough logician, but not to a person with true common sense.

However, if a limit has a distinct *form,* that *form* represents a kind of existence for that limit. This form may be *negative,* in the sense of an externally imposed limitation, or *positive,* as a self-generated limit.

Then, there is the notion of *material existence.* It would be absurd to attribute the quality of *matter* to any *expressed property* but *efficient causal agency.* That suggests that, since the discontinuity *causes* the transformation of Lattice A into B, the discontinuity is *an efficient agency,* and therefore *material.*

Before we can do much with these observations on the subject of *existence,* we must address a second stratum of question.

When we refer, here, to *a distinct form,* we are begging a question: How do we define that form as "distinct"? Our ideas about "distinct forms" fall into two sub-classes. The first sub-class is those forms which we can replicate by use of a coherent principle of construction. That *construction* of an observed form is an *intelligible representation* of that form. *Distinctness* demonstrated by means of such intelligible representation, generates one general sub-class of distinct forms.

The second sub-class of distinct forms comprises those instances which appear to defy an adequate form of such intelligible representation. Bernhard Riemann (1826–66) has referred to such forms as expressing an *arbitrary function.*[2]

However, forms of the second class are implicitly comprehensible nonetheless. If we attempt to replicate them by construction, we create *intelligible representations* which can be termed "near misses." The difference between such a "near miss" and the apparently arbitrary "curve" can be quantified as a distinct number of *singularities* (discontinuities) of each of an observed assortment of classes. This latter, supplemental sort of intelligibility of margins of difference, is a special kind of intelligible representation, sufficient to facilitate *recognition*.

In the case that the apparently arbitrary form is found in the universe apart from our own conjectures, we know that, sooner or later, an intelligible representation will be constructed.

The second aspect of existence is that we associate with "matter," *substance*. It is the issue of the hypothetical *substantiality* of the discontinuities (*singularities*) in *transformation-functions* implicit in $(A \rightarrow B)$, $(B \rightarrow C)$, ..., which bears in a most crucial way on the strict definition of successful survival and, hence, of true common sense. We turn to that issue of *substantiality* next.

V | Circular Action and Maximum-Minimum

We must make a detour at this juncture. The moment we argue, as we shall do here, that the role of scientific and technological progress, as *efficient cause*, defines the transformation-function (A→B) as *existent* in respect to both *form* and *substantiality*, we set all the guard-dogs in the deductive formalists' kennels to howling, yapping, snarling, and snapping. That obliges us to bring into view a page or two from the geometrical history of modern mathematical physics. (See Appendix I.)

What prompts the detour, is specifically this.

To qualify not only for *formal existence*, but also *ontological existence*, a singularity must satisfy certain elementary requirements. Primarily, its existence must be an efficient cause of a change of state in a physical process.

Take the simplest aspect of our outline-function for increase of potential population-density; the local, elementary transformation T(A→B), in which:

    1) T: designates, arbitrarily, a *transformation-function,* implicitly a *continuous function* of some kind.

    2) A: represents the rate of increase of potential population-density upon which the transformation acts.

    3) →: represents the transformation.

    4) B: represents the rate of increase of potential population-density effected by the transformation of A.

$\rightarrow$ is represented, from the formal (deductive) standpoint as a mathematical (logical) *discontinuity,* which we shall reference henceforth, primarily, as a more general term than discontinuity, *singularity.* This singularity ($\rightarrow$) is the *efficient cause* of the transformation in rate of increase of population-density $B-A$.

So far, so good. So it appears, until we go beyond what we have said in reference to this immediate context. The instant we attempt to fit what we have said respecting the inherent *substantiality* of our singularity, we must discard implicitly the Galileo, Descartes, Newton et al. picture of matter, space, and time, to replace that with the Kepler-Leibniz-Riemann notion of *physical space-time.* (See Appendix XI.)

From the inside of schoolbook Euclidean geometry or (Kronecker's) deductive arithmetic, the correction appears to be impossible. We are obliged to consider a different geometry, a so-called *constructive,* or true *non-Euclidean* geometry. To put the elements into order for critical examination, the following, brief historical summary is supplied.

The period of European history from the death of the Emperor Friedrich II in A.D. 1250, to the onset of the Black Death, was a hundred years of the steepest descent into a Dark Age in the known existence of that continent until today. The population was halved and the productivity and cultural and moral level reached abysmal depths relative to the first half of the thirteenth century.

The Augustine heritage, centered around the program of Dante Alighieri (1265–1321), led the re-creation of Christian Church and secular life upward into the fifteenth-century Golden Renaissance. It is to that Renaissance that we owe the establishment and initial rich development of modern physical science.

Although Filippo Brunelleschi (1379–1446), Paulo Toscanelli (1397–1482), Pietro Francesco Alberti (1584–1638), Leonardo da Vinci (1452–1519), and others made indispensable contributions, in varying degree, it was the fundamental contribution of Cardinal Nicolaus of Cusa (1401–64), beginning his *De Docta Ignorantia,* on which the possibility of a comprehensible mathematical physics depends.

During the first 200 years of the emergence of mathematical physics, deductive method played virtually no role. The elaboration of a comprehensive mathematical physics had been nearly completed by Johannes Kepler (1571–1630), before the first significant deductionist, Galileo Galilei (1564–1642), became known. (See Appendix VI). Virtually nothing salvageable was contributed by deductionists such as Galileo, Descartes, Newton, and Boyle, which was not better done earlier by such as Kepler, Gilbert, Desargues, Fermat, Pascal, Huygens, or Leibniz.

So, Newton (1642–1727) presented a failed pseudo-calculus approximately 11 years after Leibniz (1646–1716) had presented the first detailed exposition of a differential calculus to his Paris printer (1676). So, Newton's gravitation was a mere parody of Kepler's earlier discovery, turning Kepler's copied formulation for it inside-out. (See Appendix V.) As for the famed Francis Bacon (1561–1626), his work is trash.

The crucial feature was Cusa's discovery of a "Maximum-Minimum" principle (*De Docta Ignorantia*), from which, for example, the isoperimetric principle of topology is derived, and also Leibniz's *principle of least-action.* This Maximum-Minimum principle freed geometry and physics from all axioms, postulates, and the deductive method. That is the only meaningful definition of a *non-Euclidean geometry.* (See Appendix II.)

Maximum-Minimum, applied to cause-effect, signifies—possibly—the minimum (perimetric) *action* generating the relatively maximum area, volume, as the *work* accomplished (*generated*) by that *action.* Rather than saying, "That works out to show that the circle is the only self-evident form in topology," let us be content to let circular action *express* minimum *action,* and circular area *express* (positively) maximum *work.*

However, there is a special significance to *circle*. This significance involves no Euclidean or quasi-Euclidean ideas, but only *topological* ones. The circle is the basis for introducing the notion of *quantity* to *physical space-time*. The *period* of a single cycle of the minimum action generating the work effected, is the first step toward the quantizing of physical space-time.

We must also reference these few several points, in connection with potential population-density.

If "circular action" is the only self-evidently elementary action in physical space-time, then *action* is of the form of circular action acting reciprocally upon circular action in every arbitrarily small interval of each action: *multiply-connected circular action.*

Therefore: A straight line's intelligible representation is the generation of its existence by multiply-connected circular action. Similarly, a point. Thus, since both points and straight lines have a constructed, intelligible representation of their existences, neither has a "self-evident," axiomatic existence.

All of the Euclidean axioms and postulates vanish so, and, with them, deductive synthesis evaporates, too. *Only construction according to the Maximum-Minimum principle survives.*

The construction of the physical universe *in entirety* can be accomplished *implicitly* from the starting-point of "nothing but" the proof of the Maximum-Minimum principle. This has never been and will never be completed by mortal mankind, of course. Furthermore, what the human constructs so, to the extent it approaches the unreachable goal, is an *intelligible representation* of that incompletely described universe. Such is implicitly the task and nature of true physical science; such are the limitations of what might be accomplished. (See Appendix XIV.)

For our present purposes, a mere outline of some salient features is sufficient. We present these as successive steps of refinement.

Let us imagine ourselves to begin from nothing but "circular action" as the Maximum-Minimum principle defines that action. No "self-evident" forms of existence are assumed, either in the guise of asserted axioms and postulates or in any other way. Similarly, the axiomatic existence of a system of counting numbers is prohibited; *the existence of numbers and of so-called number-theoretical arithmetics must be derived solely from geometrical construction,* by no other direct means, and by aid of no additional means.

From "multiply-connected circular action," we generate the existence of "points" and "straight lines." From that phase, onward, all of the forms associated with Euclidean geometry are constructed as intelligible representations, without tolerating axioms, postulates, deductive methods, or any means but the principle of construction according to the Maximum-Minimum principle of Nicolaus of Cusa.

We can also construct a second class of objective existences, the *cycloids,* negative curvature generated by circular action performed upon such objects as straight lines and conic sections. These cycloids and their relationship to the system of evolutes and involutes, define the so-called *non-algebraic functions,* of which the crucial principle of true physical laws, *isochronicity,* is an example. (See Appendix VIII.)

It is to be stressed, that we are confronted by a different sort of "hereditary principle" than confronts us as the characteristic feature of argument among deductive theorem-lattices. Here, *the "hereditary principle" is Cusa's Maximum-Minimum principle* expressed as a principle of purely physical-geometrical construction.

The next phase of elaboration of a physical geometry, is the elaboration of circular action as self-similar spiral action. (See **Figures 1, 2, 3,** and **4.**) This augmentation, *self-similar spiral action,* becomes interesting the moment we take into account the fact that we are dealing, in terms of physical space-time, with *multiply-connected, self-similar spiral action.* (See **Figure 5.**)

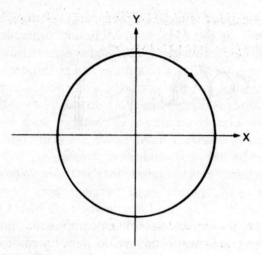

FIGURE 1. Circular Action
The smallest possible circumference (minimum) encloses the greatest
possible area (maximum). This "isoperimetric principle" is derived
from the Maximum-Minimum principle. It is a question here of the
characteristics of the *action,* and not of the circle as such, which is only
an image of the action.

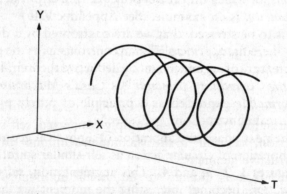

FIGURE 2.
Circular action along the time axis T.

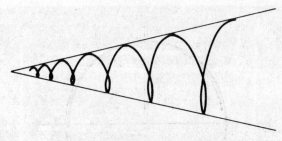

FIGURE 3.
Self-similar spirally formed action connects the circular action in time with growth.

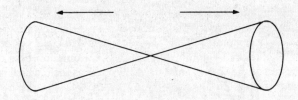

FIGURE 4.
The double cone, as it is defined by self-similar spiral action in opposite directions.

The relevant peculiarity of multiply-connected, self-similar spiral action, is that it generates *mathematical discontinuities (singularities)*. This is the characteristic of the Gauss-Riemann complex domain, whose intelligible representation is its construction, by means of multiply-connected, self-similar spiral action, according to the Maximum-Minimum principle of constructive geometry.

This takes us directly to the crucial ontological issue under consideration, the *physical efficiency* of the singularity as the characteristic of the transformation (A→B).

An entire class of *nonlinear continuous* functions, which can not be supplied an intelligible representation by lesser

FIGURE 5. Hyperboloids
Hyperboloids are the simplest expression of multiply-connected, self-similar spiral action. Shown is one of the pedagogical models used in the teaching of the LaRouche-Riemann method in *physical geometry*.

means, can be represented from the standpoint of the Gauss-Riemann complex domain understood as an expression of non-Euclidean, *synthetic geometry*.[1] This solution is typified, with a certain limitation, by the *Riemann surface function*. So, such continuous nonlinear functions are represented *intelligibly* as continuous functions (see **Figure 6**).

The picture we have sketched so briefly sums up a crucial point. The prevailing popular and classroom view, is the erroneous assumption, that modern science begins with either Francis Bacon, Galileo Galilei, Descartes, or Newton. To the degree that any contribution of Leonardo da Vinci, Kepler, Gilbert, Huygens, Leibniz, Carl Gauss, or Riemann is acknowledged, that contribution is stated within deductive terms of reference. So, the fallacy is popularized, that today's popularized classroom mathematics, the mathematics of de-

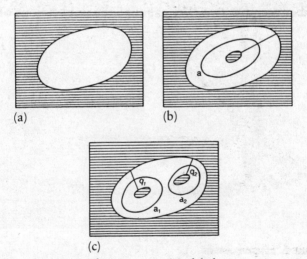

FIGURE 6. Riemann Surface Function Modeled
A simple geometrical model of a Riemann surface function. Here, we see (a) a simply-connected surface, (b) a doubly-connected surface, and (c) a triply-connected surface.

Source: David E. Smith, *A Source Book in Mathematics* (McGraw Hill Books Co., Inc., 1929), p. 409.

ductive formalism, is the *only* language of mathematical physics.

At this point, we should emphasize, once more, that the literature of modern physical science is divided into two camps. The first camp, which founded modern physical science in terms of reference to Plato and Archimedes (287–12 B.C.), is the school of Nicolaus of Cusa, Leonardo da Vinci, Johannes Kepler, G.F.W. Leibniz, Gaspard Monge (1746–1818), Carl Gauss, and Bernhard Riemann, based upon the "hereditary principle" of *synthetic physical geometry*. The second camp, which invaded the province of physical science from the outside, during the seventeenth century, about 200 years later, is based upon the "hereditary principle" of the deductive theorem-lattice.

Although the literature of the two camps often appears to coincide, on closer scrutiny of both, there is an unbridgeable gulf between the two. We examine the implications of this.

VI | Entropy versus Negentropy

Earlier, we examined some features of the series of respectively inconsistent, deductive theorem-lattices, denoted by association with the sequence of moments A, B, C, D, etc. We defined each of these deductive theorem-lattices as *discrete* (*finite*), by virtue of the fact that each is bounded, fore and aft, by a deductively "impassable," "unbridgeable" barrier. This *barrier* we identified as in the form of a *deductive* (mathematical) *discontinuity*.

Later, we supplied a more general term, *singularity*, which subsumes *formal mathematical discontinuity* as a special case under the general classification of *singularity*.

Here and now in this dialogue, we have introduced a new sub-class of *singularity*, the *impassable but not unbridgeable* gulf separating two "hereditary principles": The first of these latter is the principle of deduction-in-general; the second is the principle of (*true non-Euclidean*) constructive geometry, the latter according to Nicolaus of Cusa's Maximum-Minimum principle.[1] We must consider now the apparent similarities and the qualitative differences between the latter, higher order of constructive singularities, and the lower, deductive orders.

First, consider the differences in the manner the two different species of theorem-lattice are separated by singularities.

In the deductive cases, the discontinuities separating distinct theorem-lattices are *formally* "unbridgeable chasms."

Constructive physical geometry is a domain subsuming *relatively* distinct theorem-lattices. For example:

(a) Elementary *synthetic geometry*, as typified by Professor Jacob Steiner's (1796–1863) outline: essentially, the scope of deductive, Euclidean geometry "tiled-over" in a non-Euclidean, constructive way.

(b) An advanced elementary synthetic geometry, referencing the Maximum-Minimum principle in terms of multiply-connected isoperimetric principles.

(c) An advanced elementary synthetic geometry including the generation of the *cycloids,* and of the relationship of cycloids in terms of *evolutes, involutes,* and the Leibnizian and Mongeian *envelopes*.

(d) An advanced synthetic geometry premised upon multiply-connected, self-similar spiral action, the complex domain of Gauss, Dirichlet, Weierstrass, Riemann et al. This is the complex domain of the Riemann surface function. (See Appendix IX.)

(e) The more advanced synthetic physical geometry of the complex domain, which supersedes a Riemann surface function by a Riemann-Beltrami surface function. This takes into account the deeper implications of the non-algebraic curvature in physical space-time, as Riemann's collaborator, Eugenio Beltrami, referenced this needed, corrective amplification, negative curvature, for a more adequate version of a Riemann surface function.

Among the distinct deductive theorem-lattices, A, B, C, D, etc., no pair-wise consistency is possible. The singularities are "unbridgeable." This difficulty vanishes in the domain of construction according to the Maximum-Minimum principle.

First, the difference between any mode, immediately above, and its successor, is the act of physical-geometric integration: the addition of a "degree of freedom," "dimension" to the form in which elementary generative action is represented. So, the singularity is defined *perfectly,* but remains bridgeable.

Second, the standpoint of synthetic geometry permits it to bridge the formal discontinuities separating A, B, C, D, etc. in the manner implied by the Riemann surface function, whereas deductive method can not. This relationship is shown by aid of elementary (Riemannian) conformal, stereographic projection, between a plane in a Euclidean mode and a sphere in a constructive mode.

This permits us to proceed now as follows.

This brings us to a crucial intermediate objective, in our journey toward a true, modern definition of common sense. Our topic now is *entropy* versus *negative entropy* (*negentropy*). We begin this topic by reference to Isaac Newton's identification of today's popular university classroom mathematics as the culprit in the case.

Newton warned his readers, that he had embedded a wildly absurd picture of the universe into his fabrication of a mathematical physics. This absurdity is what today's university classroom terms "universal entropy," or a so-called *Second Law of Thermodynamics.* Newton warned his readers, that it is absurd to portray our physical universe as "running down," in the sense of a mechanical time-piece. He explained, that the source of this absurd idea in his physics did not come from experimental evidence; it came from the kind of mathematics he had chosen to employ.

The later defenders of "universal entropy," such as Kelvin, Maxwell, and Boltzmann, were admirers of Newton, but obviously either less intelligent or less honest.

The idea of universal entropy is inherent in the use of any choice of *consistent deductive theorem-lattice* as the basis for attempting to construct an intelligible representation of physical processes. Thus, all those, such as Galileo, Descartes, Newton, Laplace, Cauchy, Clausius, Kelvin, Maxwell, Boltzmann et al., who adopted deductive formalist mathematics as the basis for design and interpretation of experiments, seem to show a *universal entropy* in physical processes. Those, such as Nicolaus of Cusa, Leonardo, Kepler, Leibniz, Monge, Gauss, and Riemann, who employed the alternative, constructive method, have shown us a universe dominated and permeated by *universal negentropy*.

It were almost impossible to see such an ontological implication embedded in any one kind of mathematics as long as only one mathematics, the deductive, were imagined to exist. Only after some of the crucial distinctions between *deductive* and *constructive* methods are recognized, can we follow Leibniz's observations on the subject of Newton's cited warning.[2] So, this topic is situated at this place in our ongoing dialogue.

Consider two facets of each and every effort to construct a mathematical physics according to the principle of any deductive theorem-lattice.

Deduction begins with the axiomatic assumption that the existence of the *ideal point* is self-evident, *not subject to analysis*. This leads, however, to the absurdity of the *point-mass*, which, if it has any *mass* at all, represents *infinite mass-density* per cubic centimeter.

In constructive method, the point causes us no paradoxes. It is created, *not self-evident;* it is generated, as a singularity, by *multiply-connected least-action.*

In deductive physics, the least-action form of minimal assumption is point-mass, or other propagation along a self-evidently "straight" line. This assumption defines physics to be an ideological parody of deductive Euclidean geometry, as Descartes's analytical method illustrates that point. In constructive physical geometry of Cusa et al., we have non-Euclidean physical space-time from the outset.

For the same reason, all mathematical physics based upon the model of deductive theorem-lattices tends toward an ideal representation as solutions to systems of linear inequalities, as the economically cretinous John Von Neumann proposed for economics.[3] Hence, deductive method prohibits the construction of any intelligible representation of a nonlinear continuous process, of the sort represented more or less adequately by a Riemann surface function.

This is sufficient to ground a point: The deductive method imposes upon mathematical physics *a set of ontological assumptions*. If we subject the same experimental evidence to constructive method, instead, we have a different ontological picture. Thus, Newton's warning: The use of deductive method imposes certain linearized ontological assumptions upon mathematical physics. The idea of *universal entropy* is nothing more than another way of stating the ontological assumptions of the deductive method.

To define the kind of continuous function for *successful survival*, requires a deeper insight into functions based upon *universal negentropy*. We now concentrate upon *universal negentropy* as such. We begin with an outline of the modern history of that topic.

VII | The Golden Section

The appearance of *universal negentropy* as the characteristic feature of (constructive-geometrical) mathematical functions began with the focusing of modern physical science in Italy, during the first half of the fifteenth century. The required scientific method was elaborated by Nicolaus of Cusa, as we have noted above. Cusa's work led, both directly and by way of Leonardo da Vinci et al., to the first elaboration of a comprehensive mathematical physics, by Kepler. The topic of *universal negentropy* is inseparable from the topic of the *Golden Section,* as such a connection is reflected in Luca Pacioli's *Divina Proportione,* and in exposition of this by Kepler later. (See Appendix III.)

The connection between the idea of universal negentropy and that of the Golden Section, was made possible by Cusa's elaboration of a new basis for mathematical physics, as in, initially, his *De Docta Ignorantia:* the Maximum-Minimum principle as the basis for a true non-Euclidean, constructive physical geometry.

As we have indicated in the section immediately preceding this, it is knowledge that there exists a method completely alternative to the deductive, which encourages us to discover *the axiomatic quality of ontological assumptions embedded in the deductive method.* Similarly, we are enabled so to recognize also the elementary ontological implications of constructive physical geometry. So, we are enabled to think about science as a whole in a new, more profound, more comprehensive, and vastly more efficient way.

Therewith, to the modern implications of the Golden Section.

Cusa recognized already the central among the implied ontological features of his Maximum-Minimum principle. He associated his method with a directed, upward-evolutionary characteristic of *a noetic universe*. Otherwise, his work on the founding of modern physical science, made possible the work of his direct followers, such as Leonardo da Vinci, Kepler, and Leibniz.

As soon as we recognize that each kind of mutually inconsistent mathematical method has an embedded, exclusive set of axiomatic quality of ontological assumptions, we recognize that this exclusory feature is *universal* ("hereditary") to it. This exclusory aspect is otherwise what is called the specific kind of *self-bounded* character of the physical universe implicit in each specific kind of mathematical method chosen.

Once we begin to examine a mathematics as an implicit physics, with the foregoing in mind, we look at mathematics itself in a new way.

Consider briefly, as illustration, the subject of *prime numbers*. There are two ways of viewing this phenomenon, one deductive, the other based upon constructive geometry.

The popularized, and futile, classroom view is that deductive view consistent with Leopold Kronecker's cabalistic faith in the magical self-evidence of the counting numbers. This deductionist view leads inquiry into prime numbers into a hopeless quest, seeking the non-existent end of the Orphic number-theoretical labyrinth.

The opposing, constructive geometrical approach, is exemplified by the successive work of Leonhard Euler (1707–83), Lejeune Dirichlet (1805–59), and Riemann, culminating in the Euler-Dirichlet-Riemann function for determining the density of prime numbers within an interval. The deductionists, obliged to concede that Riemann's function succeeds, ignore the means by which this function was constructed and, instead, use digital computers to test Riemann's function by *deductive number-theoretical standards.* Those deductionists have learned successfully how to evade understanding anything of fundamental importance in respect of scientific method!

Riemann shows that the determination of the "primeness" of prime numbers belongs to that Gauss-Riemann complex domain which is constructed on the basis of *multiply-connected, self-similar spiral action.*

Similarly, we have other forms of crucial-experimental quality of phenomena which bear upon the axiomatic incapacities of the deductive method. For example, using Euclidean geometry, construct an intelligible representation of the *necessary existence* of a regular heptagon; it can not be done in that geometry, but is easily accomplished, in an elementary way, in a constructive geometry based upon multiply-connected, self-similar spiral action.

The most crucial phenomenon of the Euclidean domain in geometry, is the *Golden Section.* The significance of this is forced to our attention by the fact, that it is not possible to construct more than five species of regular polyhedra in Euclidean space, or in a constructive space based upon triply-connected circular action. Of these five regular (*Platonic*) solids, only one is truly unique, the 12-sided *dodecahedron;* the other four are simply derived from the constructed dodecahedron. (See **Figure 7a-b.**) The construction of both the solid dodecahedron and of the equal, regular pentagons which form its faces, is determined by the construction of *multiply-connected circular action's characteristic Golden Section.*

This shows us that both Euclidean geometry and the physical space-time of triply-connected circular action share a corresponding limitation. We say, thus, that such spaces are *finite, externally bounded* in reality, but *self-bounded,* too.

These evidences, such as prime numbers, the regular hep-

FIGURE 7A AND B.

(a) The five Platonic solids

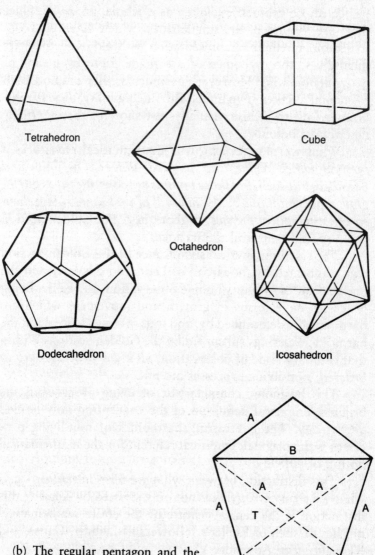

Tetrahedron

Cube

Octahedron

Dodecahedron

Icosahedron

(b) The regular pentagon and the Golden Section (A:B).

tagon, and the Platonic solids, express *discontinuities* setting off the internal world of these geometries from the larger, universal reality in which those geometries are situated. Hence, those geometries, taken as a whole, are *finite*. Since they are bounded by singularities which have no constructable, intelligible representation within those geometries themselves, those geometries are made *finite* by means of *external bounding*. However, inasmuch as they can not reach to encompass their own limits with intelligibility, they are *self-bounded* up to a limit virtually just short of the externally determined bounds.

What we require, as a true mathematical physics, is *a method which defines our physical universe as finite and bounded, but such that those bounds are constructed as intelligible representations by the method of that same comprehensive mathematical physics*. In that quest, the Golden Section has a special empirical significance.

The startling physical significance of the Golden Section was elaborated by the circles of Leonardo da Vinci: On the macro-scale of ordinary human observation, all healthy living processes have forms of growth and movement which are harmonically determined by, and thus in congruence with, the harmonic orderings subsumed by the Golden Section. Within that ordinary scale of observation, *all living processes are so ordered;* non-living processes are not.

This harmonic characteristic of living processes is the original, empirical definition of the conception now termed *negentropy*. The contrasting harmonics of non-living processes is the original, empirical referent for the mathematical notion of *entropy*.

This distinction between what we term *negentropy* and what we term *entropy,* was not only stressed by Kepler[1]; this distinction is the central feature of the entire mathematical physics of such of Kepler's followers as Leibniz, Gauss, and Riemann. (See Appendix VI.)

This view of *negentropy/entropy* assumes richer colors

as we extend our range of observation to the respective extremes of astrophysics and microphysics. The conclusive vindication of Kepler's astrophysics for the case of the asteroids, as shown by Gauss (1777–1855), shows also, and conclusively, that the universe as a whole is governed and permeated by a principle of *negentropy,* in exactly the sense that fanatical bunglers such as Clausius, Kelvin, Maxwell, and Boltzmann insisted upon the directly opposite view. So, Ludwig Boltzmann (1844–1906), faced with a choice between life (negentropy) and death (entropy), chose death.

The same result confronts us in microphysics: The geometrically characteristic orderings of the relatively elementary processes of the atomic and sub-atomic scale, are "Keplerian."

Thus, in three aspects the universe as a whole is *negentropic: life, astrophysics, microphysics. Entropy* exists in such a universe, but only as a necessary, subordinated feature of universal *negentropy.* That negentropy is the nature of *substance.*

This brings us back to functions expressing a rate of increase of *potential population-density.* This function, expressed as a *continuous function,* has the harmonic characteristics of a *negentropic*—e.g., living—*process.* The increase of mankind's power over nature, expressed as an increase of potential population-density, is *negentropy.*

To locate the source of mankind's gain of negentropy, we must investigate the source of scientific and technological progress. So, we pick up again our scrutiny of Kant.

VIII | Case Study: The U.S. Economy

To define true common sense, we must know what conditions of opinion-making render a society capable of successful survival. We must know how man—society—is capable of mastering the universe to the effect of accomplishing successful survival. We must locate man's capacity to know that which he must master to such effect; we must locate the conscious choice of means by which the individual in society might recognize such a capacity, and employ it effectively by choice.

To the purpose, we continue our exploration of some of the more crucial among the relevant features of mind and nature.

It is already indicated here, that creative mental activity is expressed by the transformation (A→B), on condition that this transformation generates, at moment B, a condition (potential) determining *future* successful survival in terms of the effects of transformation (B→C), (C→D), etc.

The awesome difficulty in defining an appropriate conception of *potential* at member B is no more than an apparent one. The concept of "hierarchical" *transfinite orderings* points toward the needed solution.[1] (See Appendix VII.)

Shift, now, the focus of our attention to this point, to the domain of statecraft.

Over the period 1963–89 to date, a dominant, pro-"New

Age" faction within the Anglo-American Liberal Establishment, has been pushing openly the imposition of a Nietzsche-Crowley "Age of Aquarius" *countercultural* revolution, upon the institutions of government and popular opinion-making otherwise. During this period, there have been frequent changes in policy, even marked ones, but virtually always in the same direction: away from the moral, cultural, and economic traditions of Western European Christian civilization, toward a gnostical or outrightly satanical, Malthusian, "post-industrial society" utopia.

Let us treat the political-philosophical standpoint of President Kennedy's investment tax-credit incentives and aerospace "crash program" as the characteristic of moment A. Let $1961 \equiv A$, and 1966–67 fiscal year represent B, for the downward, "entropic" transformation $T(A \rightarrow B)$.

President Lyndon Johnson's adoption of Malthusianism (and take-down of aerospace), brought the U.S. economy to a lowered rate of real growth than under the prior, Kennedy policies. More important than the rather disastrous changes Johnson made during 1966–67, was the accompanying *change in direction* of U.S. and Atlantic policy-shaping. *Civilized standards for policy-shaping were replaced by the revival of gnostic religious dogma in its guise as "environmentalism" and "consumerism."*

So, with rare exceptions (which proverbially prove the rule), human survival depends relatively less on society's policies of the moment, and absolutely upon the choice of *policy-shaping direction*.

View the economic process as such in the foregoing terms.

For purposes of discussion, let us identify two consequences of *moment* B, one the objective, physical changes in the economic process, during the course of moment B, and the other the synchronized ("subjective") state.

During some point within the continuation of moment B, the policy in operation since the beginning of B will have produced some effect. The mind must respond to that effect.

So, we must consider what that physical effect will be and also the variable way in which the mind might react to that physical effect. On the latter account, the *philosophical characteristics of that variable mind-set* show their decisive bearing in determining *successful survival*. We are near to defining true common sense.

For example, the following.

The U.S. policy-trends of 1963–69 have been a drift toward the anti-Western, radical counterculture's increasing influence; this has included a composition of anti-agroindustrial usury and Malthusian "post-industrial" utopianism. Every policy-response influenced by this combination of subjective trends and trends in practice, must be *toward* the effect of collapsing the society *physically*, entropy, and of fostering a deepening immorality and cultural-pessimism in the population.

How might the twofold "determinism" be expressed, to the purpose of constructing an intelligible representation of our continuous function for potential population-density? Consider next, the definition of "creative mental action," as seen from the vantage-point of the cited Kantian paradox.

Creative mental action is expressed most simply as the singularity which is characteristic of the transformation: $T(A \rightarrow B)$. Another definition of that same singularity, is that which Plato causes the Socrates (469–01 B.C.) of the dialogues to name "my dialectical method"—i.e., the method represented by each and all of Plato's *dialogues*, the *dialogical* method. Examine this dialectical method.

Instead of rebutting a proposition directly, the Socratic dialogue inquires: *What axiomatic assumptions necessarily, implicitly underlie the proposition, "hereditarily"?* In turn, it then examines those axiomatic assumptions from the same viewpoint expressed by constructive (physical) geometry. In general, all Plato's dialogues are, in this way, a repudiation of both the deductive method and of the ontological assumptions implicit in the latter method.

Plato prompts the Socrates of his dialogues to affirm that

any conception constructed by means of the dialogues were adequately represented and more, by the suitable (physical) geometric construction. So, in that same spirit, and from a modern occupation of that same viewpoint, we emphasize that *the singularity* of $T(A \rightarrow B)$ is not only the efficient course, but also the *substance* of creative mental action.

Refer to our earlier outline of the Kantian paradox. Plato's dialectical method and those creative processes of the mind which generate scientific progress are congruent. At least, this is true in the manner one speaks of a sense of direction; the Platonic method is, in intent and in approximation, an intelligible representation of the creative process. It is intelligible in the sense that Kant insisted that such a representation were impossible.

As we have indicated already, this dialectical method addresses two "levels" of "axiomatic thinking."

First, it addresses the relatively more superficial of the two levels. In respect to deductive modes, this signifies the necessarily implicit set of axioms and postulates underlying the proposition. Respecting constructive method, this signifies the choice of constructive principle. On this "level," the mind is, at best, designing the future crucial experiments which might overturn theorem-lattice "A."

On the second, relatively deeper "level," this same dialectical method addresses the assumptions necessarily subsuming both deductive and constructive views of physical geometry, and also governing the choice of one over the other. It is the class of assumptions which are the subjects of this level, which correspond to policy-shaping assumptions.

The first "level" is the generation of a set of transfinite orderings: a "hereditary principle" subsuming, by "ordering," a set of "hereditary principles."

From the "deeper" (higher ordered) of these two levels, the following are to be emphasized at this juncture:

1) All deductive systems are *entropic,* as we have noted already;

2) All constructive systems satisfying the requirements of

Nicolaus of Cusa's Maximum-Minimum principle, are implicitly *negentropic,* but also subsume the generation of *entropic sub-phases* of such negentropic processes.

Consider a simple illustration from the pages of Physical Economy, as follows:

At first impression, an increase of potential population-density signifies merely an increase of the average physical productivity of labor, per capita and per hectare. Within limits, it is feasible to increase average productivity so, by increasing the physical intensity of labor, while maintaining constant per capita and per hectare capital intensities, and without accompanying technological progress.

However, *over a longer term, the result of such an "improvement" in average productivity is a collapse of productivity!* Relative to the relationship among the "two factors," here, a philosophy which ensures a medium-term decline in productivity has caused a short-term rise in apparent productivity.

The use of physical intensification of labor, as a substitute for satisfying those constraints of increased potential population-density which we listed earlier, is a retrogressive policy of practice over the medium to longer term. It is, in fact, an *entropic policy-shaping* practice.

Thus, if T(A→B) is of this entropic form, then the desire for improved nominal productivity at moment B, will prompt an analogously and entropic T(B→C), and so on. The austerity policies of Hjalmar Schacht under Weimar Germany created Hitler's Nazi regime in just this way: The continuation of Schacht's entropic "conditionalities" policies under the 1930s Hitler regime, produced the mass-murderous Nazi slave-labor system. So, U.S. support for IMF "conditionalities" policies today projects mass-murder rates which are approaching rates two decimal orders of magnitude greater than anything which the Nazis either did or of which they were accused, even by the most perfervid Germany-haters.

If only in abstraction, every transformation of the form T(A→B) has two moments: the ostensible, immediate change in rate of increase (or, decrease) of potential population-density, and the character of the implied policy-shaping trend under which the momentary development has occurred.

That policy-shaping trend will determine how the society will react, with policy innovation, to the practical effects of moment B.

Consider the following, amplified recapitulation of the foregoing illustrative case. Consider first several aspects separately, and then in terms of their combined impact.

The popular, official method of national income accounting today is the *Gross Domestic Product* (GDP) or *Gross National Product* (GNP) system. This yardstick is based upon the popular definition of *Value Added:* the price of that marginal ration of accrued sales income, after deducting "exports" (accrued earnings) from "imports" (costs of materials, supplies, machinery, equipment, tools, maintenance, semi-finished product, and services) purchased from either other nations or simply other enterprises.

Under certain conditions, U.S.A. GNP data, although corrupted, was useful. The data was made available by the U.S. federal government; and other data available, from different sources, was designed to mesh with the national income accounts definitions used by the U.S. government. The United Nations Organization-sponsored GDP system used by other nations, was more or less compatible. Provided one understood the leading built-in fallacies of the Value Added system, one could use the data with the proper caution.

The reliability of GNP and related data began to erode approximately fiscal year 1966–67, under President Lyndon Johnson. The situation grew worse over the following decades. The federal government and Federal Reserve System resorted more and more to the sort of political "cosmetics" which ordinary mortals would prefer to term "faking the statistics." During 1983, and especially since the beginning of 1984, the report of a U.S. "economic recovery" has been based upon wildly faked, virtually useless data.

Despite the growing margin of fakery in GNP statistics prior to 1983, the federal agencies faked their data in a fairly consistent way. There was sufficient consistency that one could adduce meaningful quarterly changes in annual trends. Although the absolute values of government statistics were false, the margin of arbitrary error was increasing gradually, from year to year, and therefore, meaningful *trends* could be detected and forecast.[2] After 1983, the fakery of data, by the Federal Reserve and other agencies, became so wildly arbitrary, that no meaningful trends could be adduced from their data, for the real, physical economy hidden behind the fraudulent statistics published.

Since 1970, especially since the inauguration of the Trilateral Commission's Jimmy Carter as President, the U.S. real economy has been collapsing without interruption. There has been some variation in the rate of collapse, but there has never been a year during which the collapse did not continue. Since 1966–67, in fact, no economic recovery has occurred within the U.S.A.

The net capital value of U.S. basic economic infrastructure reached its high point, in rate of growth, prior to fiscal year 1966–67. This rate sank to zero during 1970, and has been in continued decline since 1970. To achieve that per capita and per hectare quality of basic economic infrastructure today, would probably represent a repair bill of $4 trillion.

Since the relatively happier days before Jimmy Carter, U.S. agriculture has been drawn into nationwide, spiraling bankruptcy. Our once powerful manufacturing belts have been transformed into "rust belts." The physical market-basket value of per capita households', farmers', and manufacturers' consumption has collapsed.

What has grown, apparently, is per capita financial income, largely as growth of indebtedness and looting of all combined categories of income by debt-service charges. It is drug-trafficking revenues combined with cancerous forms of parasitical expansion of financial "industry" income, which is the chief source of *nominal* growth.

So, we have the case that, in physical terms, the U.S.A. has been collapsing since 1970, in per capita and per hectare potential and output, while the GNP has grown greatly on balance. The difference between collapse of per capita *physical* productivity and growth in per capita, nominal GNP, is a margin of *fictitious* growth, *fictitious national income*.

This brings us to this crucial point: Those who are managing U.S. economic, fiscal, and monetary policies, are seeking to "optimize" GNP, by emphasizing the growth of margins of fictitious income, as is consistent with dedication to a "post-industrial" utopia. Thus, that which "optimizes" fictitious income is deemed success; to each relevant problem, as challenge, the policy-shaping response is to modify policy to the effect of increasing the growth of *fictitious income*.

The emphasis on fictitious income is, rather obviously, an *entropic* policy. Since GNP counting is *linear,* all policy-shaping under the influence of GNP-style national income accounting is also *entropic* to that degree. It might be the case, that the promotion of some military or other, relative broad-based investment in technological progress might offset the entropic influence of GNP-style policy-inpulses. *Monetarism,* and all consistent expressions of Adam Smith's (1723–90) dogma of "free trade" are inherently, axiomatically, "heredi-tarily" *entropic* policy-impulses. Their predominance ensures economic ruin if tolerated sufficiently long.

So, any economic policy emerging from T(A→B) has a two-faceted effect. On the lower level, more immediately, more obviously, it determines *a momentary, potential* population-density, at least apparently, with respect to the *actual* population-density. On the higher level, at the same time, it effects the philosophy of the policy-shapers in ways which must tend to govern choice of policy during T(B→C), and so on.

In other words, in respect to T(A→B), we must consider not only the changes in *what we think,* appearing between moments A and B. We must consider also the changes in *how we think*—the efficient, present, and future effects of changes in how we think.

IX | Curvature of Physical Space-Time

It were essential that the following material be summarized now, although most readers may be able to follow much of this only superficially.

There is another, implicitly quantifiable way of representing both distinction and interrelationship between changes in *how we think* and changes in *what we think*. Look back to Kepler.

Nicolaus of Cusa, and also such followers of Cusa as Leonardo da Vinci and Luca Pacioli, showed Kepler a notion to which we refer today by the term *curvature of physical space-time*. We refer to that connection here again, to show the intelligibility of a universal principle of negentropy, and thus to show also the meaning of *curvature* as applied to policy-shaping conceptions, in Physical Economy, for example.

Most readers who have encountered the term *curvature of physical space-time* have learned it as a spillover of that discussion of special relativity theory, which erupted near the onset of our century. Unfortunately, few who have used that term, including one so gifted as Herman Minkowski, understood the proper meaning of the term. That latter observation is relevant to our defining true common sense; our representation of the failure of Minkowski et al. is in no way exaggerated.

57

Take "curvature of physical space-time" apart; there are two ideas combined as one. *Physical space-time* is one idea; the idea that physical space-time has *a curvature,* is a second idea. Consider *physical space-time* first.

Minkowski's error, in his celebrated presentation, was that he failed to grasp the essential distinction between a truly *non-Euclidean* geometry and a *neo-Euclidean geometry.* His choice was the *deductive,* neo-Euclidean representation of a Lobachevsky, hyperbolic geometry. Minkowski's exemplary error and failure, was that it is impossible to construct an intelligible representation of physical space-time on the basis of deductive method.

Minkowski's case is more useful because it shows a first-rate scientific thinker of his time, crushed by the follies of deductive method, even in his moment of what is otherwise exceptional vigor of achievement. The Minkowski case has the added value of focusing our attention on the respective problematics of both form (*curvature*) and substance (*physical space-time*).

"Hereditarily," all deductive method locates "matter" (*substance*) in the axiomatically self-evident existence of the point-mass, of a point of infinitesimal smallness. All efficient action in the physical domain is, similarly, associated with pair-wise interaction upon, by, and among point-masses.

Point-mass is a self-bounded singularity—a Kantian *Ding an sich*—and a *self-finitized* quantity, in this manner and sense. The finiteness so defined, ends the quality of "matter" (*substance*) within the self-finitization's bounds, such that, within the universe as a whole, the quality of "matter" does not exist beyond the self-bounds of axiomatic point-masses.

So, *space* and *matter* are made to appear as mutually exclusive qualities. Similarly, the contrast among *matter, space,* and *time.*

For analogous reasons, if we construct a deductive universe in which the quality of *matter* is applied to *space,* rather than to points, we can not determine the existence of singularities, and our mathematics becomes an endless, silent labyrinth, a tomb in which nightmare we may wander forever without consequence.

Contrast Minkowski's neo-Euclidean, deductive viewpoint with that of non-Euclidean physical geometry. *Substance* is that form of existence whose generation, persistence, and dissolution each and all are *causally efficient* respecting effects of the continuing process with which their existence is associated.

The first published mathematical treatment of the notion of special relativity is supplied by Bernhard Riemann's inaugural dissertation, published in 1854.[1] In the concluding portion of that paper, Riemann indicates the nature of an experimentally demonstrable, measurable effect of a relativistic change.

This measurement has been understood in two ways. Riemann understood all his contributions to mathematical physics from the standpoint of a constructive, physical-geometric generation of the Gauss-Riemann, non-deductive domain. In this domain, from the beginning of his work at Göttingen, Riemann understood the universe to be driven by a universally subsuming principle of *negentropy,* as we define "negentropy" in this paper before the reader now.[2] Minkowski read it in a different way, a deductive way: that is the cause of Minkowski's blunder.[3]

Riemann's collaborator, Eugenio Beltrami, to whom Riemann assigned the task of successfully exposing the devastating, paradoxical fallacy of Maxwell's (1831–79) electrodynamics, supplies us the neatest portrayal of the construction of a Lobachevsky geometry of negative space-time curvature; Minkowski should have studied Beltrami's work more thoughtfully.[4] Instead, Minkowski read "negative curvature," itself a very useful insight, otherwise, as a mandate to do no more than modify some postulates of that swamp of Kantian deductive formalism so much admired in today's science classroom: that of Descartes, Newton, Laplace, and Cauchy.

Minkowski's case is notable, because he was so vastly

superior to our century's ordinary, half-brain-dead formalists of "pure mathematics." Like most among the dwindling ration of today's better scientists, he was crippled by his addiction to the popularized mathematics of his day, but he was also capable of actually scientific, geometrical thinking. There is no evidence which permits us to doubt, that when he said "physical space-time," he was pointing in the direction we are pointing. The reason he failed, was that he erred in attempting to describe physical space-time in *deductive* terms of reference.

Instead of a *non*-Euclidean geometry, Minkowski used a *neo*-Euclidean one: a Euclidean deductive method, but with some modification of the postulates—a wrong choice of $T(A \rightarrow B)$. Thus, his choice of deductive mathematics led him back to the same failed ontologies from which he desired to escape.

The last ditch objection for those who would cling to deductive method, assumes such forms of expression as, "How shall we measure?" *If we permit nothing but geometrical continuity to be employed anywhere in the construction of a physical geometry, whence do we secure the yardsticks a physics requires?* In this connection, the Golden Section is crucial.

We have already noted that, beginning with the Maximum-Minimum principle, the entirety of constructive physical geometry is elaborated by the route of multiply-connected circular action. That curve, defined topologically, is the only elementary source of the notion of *quantity* in the entirety of a physics so constructed. It begins with the notion of the circular cycle and is elaborated in terms of construction of divisions and multiples of that circular cycle.

Thus, multiply-connected circular action generates the forms of existence known as the "point" and "line." Thereafter, beginning with the creation of such primitive singularities in such a fashion, the generation of all singularities is measured in terms of ratios of a single, or multiple cycle, or cycle of circular action. As early as Plato, it was known, that the most crucial of all among these *rations* or *sections* of circular cycles, is the so-called *Golden Section*.

Plato referenced, relatively often and crucially, the discovery of a proof, that only five species of regular polyhedra are constructable in the space defined by human vision. The notion of *Platonic Realism* is centrally referenced to this proof; the development of modern physics, from Nicolaus of Cusa onward, is referenced to the implications of this. The fact, that only five species of regular polyhedra are constructed so, shows that physical space-time so conceived is finite, bounded, and that by virtue of this manifest delimitation.

To the best of our knowledge thus far, Plato was the first to prescribe the construction of a comprehensive mathematical physics upon the implications of the uniqueness of the five Platonic solids. Plato emphasized musical harmony, in prescribing the laws of action in the universe to be governed by those harmonic orderings which are implicit in the Golden Section. This is the germ of the first successful modern form of a comprehensive mathematical physics.[5] (See Appendix VI.)

We have noted and emphasized, that Leonardo's circles had demonstrated that the Platonic harmonic orderings were the characteristic harmonic orderings which set healthy living processes apart from dead ones. We have noted and emphasized, that the five Platonic solids demonstrated the *finiteness*, the *boundedness*, and the *self-boundedness* of the physical space-time seen as visual space-time. We have emphasized, that any process which is harmonically ordered in congruence with the Golden Sections of circular action is *negentropic*.

Kepler's founding of a comprehensive mathematical physics, based the lawful ordering dominating and pervading that universe upon the distinction between negentropic and entropic harmonic ordering of processes, and *defined the universe as a whole as negentropic*. (See Appendix V.)

The *algebraic form* in which Kepler's three laws of universal "gravitation" are represented, is an algebraic description of constructive-geometric *locus-functions*. First, the planetary bodies *generated by the formation of the solar system* are situated in orbits, relative to one another, at distances determined (to put the matter most simply) by the ordering of the five Platonic solids. Second, the distribution of relative angular velocities among these elliptic orbits subsumes a *musical* species of harmonic intervals, as ratios. These ratios are ordered, to a significant degree of good approximation, in conformity with the characteristic series of intervals of the well-tempered polyphony of the classical composers.[6] It is from this standpoint, that the notion of *the characteristic curvatures of physical space-time* is already central to Kepler's physics.

The work of Leonardo, Pacioli, et al. is an immense value in this connection. To conceptualize the ontological implications of a deductive method, it is indispensable to be able to see the contrasting ontological implications of the constructive method. Only by aid of contrasts can we teach the principles of physics and metaphysics with confidence that we are communicating to the students. Only by contrasting entropic and negentropic harmonics, as Kepler makes use of the work of Pacioli et al., can we teach confidently the elementary notions of *curvature of physical space-time*.

Once that primary distinction is clear, we may next recognize, without confused minds, the meaning of an effort to measure the relative degree of negentropic or entropic curvature. The bearing of this upon the fallacies of "GNP" are implicit; we must render the functional connection explicit.

X | Isochronicity, Least-Action, and the Parmenides Paradox

We continue the pathway we were traveling in the preceding moments of our dialogue. It is time to bring the intermediate, mathematical portions of exploration toward their conclusion. Our proximate objectives at this moment are three types: *isochronicity, least-action,* and *the Parmenides Paradox.*

Leibniz emphasized the crucial implications of so-called *non-algebraic curvature,* the phase of physical-space-time curvature associated with such constructions as *cycloids, evolutes,* and *envelopes.* (See Appendices XIII and IV.) Non-algebraic curvature became promptly one of the crucial points of irreconcilable difference between Leibniz's circles and the opposing deductionists. The two aspects of this topic most directly relevant in this dialogue on successful survival, are the sub-topics named, respectively, *isochronicity* and *negative curvature* of physical space-time.[1]

Consider **Figure 8.**

$\overline{L_1L_2}$ is a "virtually frictionless" rolling track with the curvature of a simple cycloid. O is the lowest point of that track. P is any point on that track other than O, and P′ is any point on the track neither O nor P. $\overline{P'O}$ is an alternate, "virtually frictionless," straight-line, inclined-plane rolling track.

Release two balls, to roll freely, simultaneously, from point P′. One ball, rolling along the arc of the curved track

63

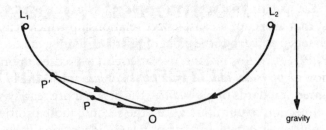

FIGURE 8. Least and Equal Time
This figure illustrates the principle of least-action and isochronicity. A sphere rolling down this cycloid from P′, reaches the middle point O *at the same time* as one which is released from P. Even though the curve P′O is longer than the straight line $\overline{\text{P′O}}$, the sphere arrives more quickly at O if it rolls along the curved path, because it is a path of *least-action*.

$\overset{\frown}{\text{P′O}}$, reaches O before the ball rolling along inclined plane $\overline{\text{P′O}}$. Arc $\overset{\frown}{\text{P′O}}$ is a *least-time* track. Release two balls simultaneously, this time at P′ and P, respectively. Both will roll freely to O in the same lapsed time. The curvature is, we say, *isochronous*.

In the more superficial treatment of that or analogous examples, it is stressed that the equal times of free rolling between arcs $\overset{\frown}{\text{P′O}}$ and $\overset{\frown}{\text{PO}}$ is a reflection of the factor of *acceleration*. The truth is directly opposite. It is not the principle of acceleration which determines such a curvature; it is the *isochronic curvature* characteristic of physical space-time, which is reflected as the phenomenon of acceleration.

Imagine a universe defined according to the curvature of the rolling track in this figure. Imagine all possible functions in a universe of such a physical space-time curvature. Express such functions' effects in respect to point 0, and all proximate causes for these effects in terms of points $P_0, P_1, P_2, \ldots$ etc., alternately. That aspect of such a cause-effect relationship which does not change as we substitute, successively, points $P_1, P_2, P_3, \ldots$ for P_0, is the aspect of the cause-effect relationship which interests us now.

Let us term that feature the *invariant* feature of the process, that aspect of a cause-effect relationship which remains unchanged as we shift the point of reference among the choices $P_0, P_1, P_2, \ldots$. In this instant, our attention is focused upon the notion of *physical law*. A *true physical law* must satisfy two required standards of *invariance:* the first is more readily seen in connection to our illustration; the second, more profound, we shall explore later, after but merely identifying it here:

1) A *physical law* must be invariant with respect to the processes and positions in which its action is observed.

2) A *physical law,* properly defined in respect to its bearing upon processes in some part of the universe, must be true isochronically at that instant for the altered state of the universe as a whole.

As we shall note, later on, the latter bears upon the Parmenides Paradox, as the treatment of that paradox appears, for example, within Nicolaus of Cusa's *De Docta Ignorantia.* That said, turn now to some other fish to fry and *negative curvature.*

Earlier, we considered one implication of Cusa's Maximum-Minimum principle: *the minimal action required to generate the relatively maximum work (e.g., "volume") accomplished.* Now, consider the complementary notion: *the minimum work required to generate the relatively maximum action.* Let us associate the first with the obvious choice of term, *positive curvature.* Let us associate positive curvature with the term *weak forces,* and negative curvature with *strong forces.* Let us examine this array, first, in light of the Riemann surface function, and then, the prospect for constructing the more adequate *Riemann-Beltrami* surface function. Precede that with a few relevant historical observations.

Appreciation of (and controversy respecting) *non-algebraic curvature* is usually dated to the period of Christiaan Huygens's (1629–95) and Gottfried Leibniz's Paris discoveries of the 1670s.[2] (See Appendix VIII.) Not only were both indebted to B. Pascal for preliminary work; the physical significance of negative curvature was not only recognized, but employed successfully by Filippo Brunelleschi and Leonardo da Vinci, two centuries and more earlier.

Formally, the study of non-algebraic curvatures begins, in constructive (e.g., projective) geometry, with the most elementary of the cycloids. (See **Figure 9**). This proceeds to the cycloids generated by regular curves (conic sections) other than straight lines. This proceeds through *evolutes,* and into the domain of Leibniz-Monge *envelopes.*

Construct a figure, the surface of which has everywhere a continuous positive curvature, a figure which represents the largest volume for a curved surface of that area. Put the same task more simply: Rotate a circle; a *sphere.* Compare this with a *pseudosphere* (**Figure 10**). With positive curvature, we associate the generation of the greatest *work* (volume) by the minimal *action* (surface); with negative curvature, we associate the generation of the relatively largest *action* (surface) by the minimal *work* (volume).

Together, they generate *negentropy.* Generate the relatively greatest *work* with the minimal *action;* with that *work,* generate the greatest surface (*action*). Look at a Riemann surface function and its problems so.

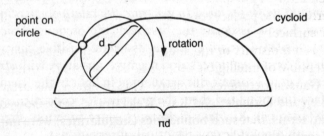

FIGURE 9. The Cycloid
Lay a rolling circle along a straight line. Where the circle is, initially, tangent to the line, mark a point on the circumference of the circle. Now roll the circle along the line; the cycloid is the locus of the point, generated as the circle moves through one rotation. Amateurs: Test for yourselves: Is this locus, turned upside-down, isochronic?

The *negentropy* of Kepler's universe is defined by the characteristic harmonic ordering of the laws of that universe, a harmonic ordering congruent with the *Golden Section* of circular action. By replacing circular action by self-similar spiral action, we generate the Gaussian complex domain as the higher domain, of which Kepler's is a projected image.

The formal problem is twofold. Any domain constructed by means of continuous, multiply-connected, self-similar spiral action, is richly populated with singularities. Indeed, all elementary physical processes can be represented adequately only by functions which are *continuously nonlinear* functions on this account. This leads us to what is known as (Lejeune) Dirichlet's Principle.

Refer in an earlier portion of this dialogue to our reflections upon the subjects of indistinct and distinct forms of singularities. The formal characteristic of continuous impulse of multiply-connected, self-similar spiral action, is to "interrupt" the appearance of continuousness of that action by means of a harmonically ordered series of singularities. These singularities are in the form of hyperbolic lines or sheets. On a sphere of conformal projection, the discontinuity itself takes the form of what we might regard as a point or perturbing indentation of the sphere's smooth surface. That latter locality interrupts the perfect continuity of movement (action) along that surface.

Is it a distinct or indistinct singularity? Is the singularity susceptible of intelligible representation within the bounds of the Gaussian complex domain? This question was attacked by Lejeune Dirichlet and Karl Weierstrass; both showed, in broad terms, that such boundaries (singularities) were implicitly comprehensible from the vantage-point of the Gaussian domain itself. By accepting the existence of a relatively higher order of physical space-time at that point, the singularity persists, but *not* its "property" of discontinuity; we "bridge" the discontinuity, in the manner a Weierstrass function suggests. This is the kernel of what Riemann identified as *Dirich-*

FIGURE 10A, B AND C.

Three surfaces of rotation with constant negative curvature, which are all formed by the rotation of different cross-sections of the tractrix:

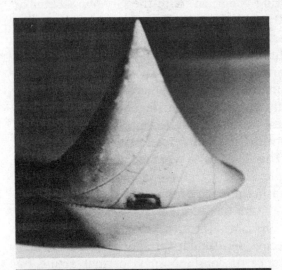

(a) the elliptical-pseudospherical surface of rotation, which corresponds also to the surface of rotation of the caustic;

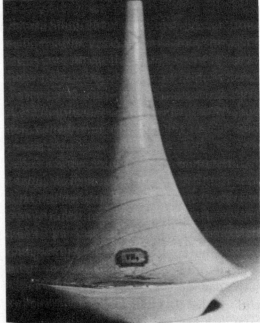

(b) the parabolic pseudospherical surface or "pseudoshere";

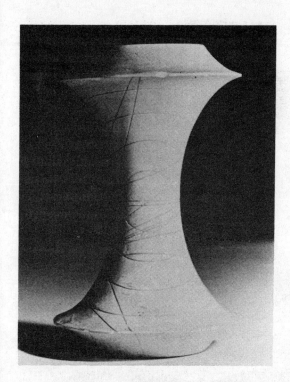

(c) the hyperbolic pseudospherical surface of rotation. The photos, by Dino de Paoli, show Beltrami's original models, conserved at the University of Pavia in Italy.

let's Principle. Riemann employed that principle to unleash the fuller potential of the Gaussian domain; hence, we have the Gauss-Riemann domain.

The typification of this employment of the Dirichlet Principle of topology, is the Riemann surface function. (See Appendix IX.) (**Figure 11.**).

It is the method of Riemann, that whenever the representation of a physical process encounters such a singularity, he says, in effect, "If the process continues, through and beyond this point of apparent interruption, it will do so in the form the Riemann surface function suggests." A useful illustration of such a tactic, is the case of the 1859 dissertation, "On the Properties of Plane Air Waves of Finite Magnitude." This is already clear implicitly in his 1854 inaugural dissertation, "On the Hypotheses Which Underlie Geometry."

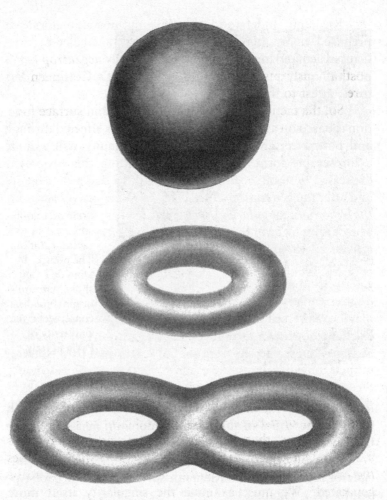

FIGURE 11. The Riemann Surface Function Applied to Structures in Space

The topological projection of a circle (constant positive curvature) is simply connected; there are no singularities (holes), nor poles. The projection of a torus with its hole in the middle is triply connected, and the projection of the third form (pretzel shape) with two holes is fivefold connected.

LaRouche cites Beltrami's work to offer the hypothesis that these singularities are not simply points or holes—empty space—in an otherwise continuous positive curvature, but rather regions whose physical geometry is characterized by negative curvature.

Riemann understood this, his optimistic method, as premised upon the demonstration that a Kepler-Leibniz-Gauss-Riemann universe is characteristically *negentropic*. His posthumously published notes on Herbart's Göttingen lectures, attest to that.

So, the tactic associated with the Riemann surface function is based on asserting, that if continuity is effected through and past a certain singularity, this continuity will assume a foreseeable form. With this qualification, this process is expressed in terms of functions of positive curvature, before and after the generation of the relevant singularity. However, *the hyperbolic singularity itself expresses a negative curvature.* The development of the transsonic shock-front, in Riemann's referenced 1859 paper, illustrates the appearance of the negative curvature.

In short, *in the case that the singularity ceases to be a discontinuity in the appropriate function,* the following result may be expressed in terms of reference to positive curvature. Riemann does not address the negative curvature *functionally;* he limits himself to showing the form accepted by a solution to the problem, but not the explicit, functional solution itself.

A truly *adequate* solution is sought. When, for example, is the solution proffered by Riemann's method *the result which must occur, of necessity?* It is not sufficient to note and to react to the fact, that the generation of a singularity requires a solution of that form. Possibility is not sufficient. Does *the existence of the singularity* cause such a solution to be generated? We must examine the singularity itself more closely.

That singularity is no indistinct form of boundary, no mere point, no mere logical abyss in the form of a hole. The way by route of which this singularity is generated by the continuing impulse of multiply-connected, self-similar spiral action, causes the singularity to be *hyperbolic* (or, *hyperboloid*), and to have a characteristic *negative curvature* of this kind. Eugenio Beltrami showed how this must appear as a region of singularity in a Riemann surface (**Figure 12a-c**).

FIGURE 12. The generation of a singularity in a Riemann surface.

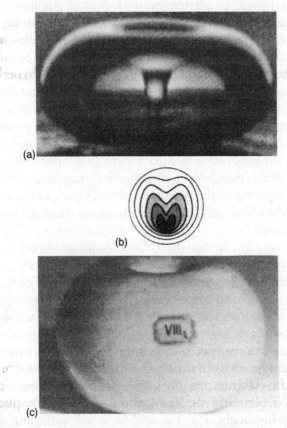

(a) High-speed photograph of the implosion of a liquid irradiated with ultra-sound. (b) The drawing shows the process of change of form and the shrinking during implosion. (c) Beltrami's model resembles the photo (a). Characteristic is the indented, negatively curved surface.

These ultrasonic experiments may represent more than a simple analogy to Beltrami negative curvature space. Recent observations have shown that such ultrasonic experiments lead to the generation of light (sonoluminescence) with the greatest amplification process seen in nature, and with extremely coherent outputs (*Nature*, Vol. 352, pp. 318–320, July 25, 1991). Others have indicated that the process leading to the light emission may involve spherical Beltrami vortices. See *21st Century Science & Technology*, Winter 1990, Vol. 3, No. 4, pp. 58–78; and *Proceedings of Plasma Focus Workshop*, May 1990, Stevens Institute of Technology, Hoboken, N.J., "A Note on Beltrami Fields Applied to Confined Currents and Charges," by Jack Nachamkin.

Consider the *negentropy* of Kepler's universe. Note that this negentropy is characteristic of the Gauss-Riemann complex domain as a whole. This invariance of the two regimes is associated with the Golden Section and its harmonies, as the metrical characteristic of self-similar spiral action. Similarly, it is the harmonic ordering of the generation of hyperbolic singularities, by continuous impulse of multiply-connected, self-similar spiral action, which expresses the characteristic negentropy of the Gauss-Riemann and Gauss-Riemann-Beltrami complex domains. That latter characteristic, is *the harmonic ordering of increase of (accumulated) density of singularities per interval of (cycloid) action.*

That characteristic is implicitly the measure of *increase of technology,* and of *potential population-density.*

Yet, those various measures of *negentropy* define processes which are *bounded by* negentropy, without representing the negentropy itself. Once we shift our focus to the causal sequence of alternating *weak* and *strong* "forces," the intelligibility of negentropy becomes a distinct geometrical idea; the negentropic process is then represented intelligibly as *a self-bounded process.*

This, all taken together, defines a different set of ideas about *measurement* and *substance* than those in common use in either the classroom or ordinary conversation. With these changes, Nicolaus of Cusa and Leibniz would and, implicitly, did agree. Implicitly, Plato would have followed our argument, would have understood and agreed. If we do not recognize such agreement, we understand nothing essential respecting the purpose and content of Plato's dialogues.

A few reminders:

Metrical ideas feature inclusively:
1) *circular action* (isoperimetric action)
   (a) constant rate of continuous circular action
   (b) counting in cycles

    2) *sections of circular action*
- (a) Golden Section (negentropic) harmonics (curvature)
- (b) entropic harmonics
- (c) invariance of negentropy in projective correspondence of Golden Section and self-similar spiral action

    3) *negentropy*
- (a) increase of density of singularities per interval of action
- (b) alternating *weak* and *strong* "forces"

    4) *isochronicity*

Substance: physical space-time, as, essentially, universally negentropic, but subsuming entropy in the guise of "negative negentropy." *Efficient cause* is substance; *substance* is efficient cause.

XI | On Human Mortality

A famous fool said, "I think, therefore I am." An utterly degraded wretch insisted that self-interest is defined by the seeking of pleasure and the avoidance of pain.

An individual mortal life is a moment between conception and death. All sensual pleasures' and pains' memories are interred with the deceased. What is there of value in such mortal human individuals, that society should consider the existence of any among them worth tolerating? The definitions of self by Descartes or Adam Smith put mortal individuals on no higher level than a talking swine.

The essence of the mortality of the person is that that life is brief. Once that life had ended, what value persists for the society? What such value to society inheres potentially in an individual life, that society might desire the conception, birth, maturation, and adulthood of yet another such individual?

Do we not touch here, again, the complementarity, the sameness of *isochronicity* and *natural law*?

The self-interest and the proper definition of individual, personal identity, is rooted in *the quality of being human*, that which distinguishes *being human* from being that which is most nearly human, but is not human. It is sufficient that this distinction be that which separates mankind from all the species and varieties of beasts. That distinction is defined as mankind's capacity for generation, transmission, and efficient assimilation of scientific and technological progress. By *efficient assimilation*, we signify the effect of increasing *potential population-density*, as we have defined *successful survival*.

In first approximation, that distinction is *a singularity*, as illustrated by the singularity of the transformation T(A→B). It is *that singularity which reflects the essential, functional difference between mankind and each and all species of beasts*. This singularity sets mankind *absolutely* apart from, and *absolutely* above, all of the species of beasts.

By its nature, this transformation and its associated singularity, are the activity of the mental life of a human individual, and thus set each individual apart from, and finite in respect to, each and all other persons in all history and for all future time.

Here lies the key to the nature of individual personal identity. As the definition of the identity of the individual personal *self* delimits the definition of *self-interest*, this singularity provides the delimiting definition of the distinctions, and the finiteness of individual self, and of individual self-interest.

This singularity could not exist, but to the degree that it is brought into existence and also nourished into persisting existence by, and as the "nonlinear" form of activity typified by, the transformation T(A→B). The existence of self and the existence and fulfillment of individual, personal self-interest each lies within the process of developing and expressing the capacity expressed as the activity of generation, transmission, and efficient assimilation of scientific and technological progress.

If and when this activity is viewed in the narrowest sense of "scientific and technological progress," we see but an aspect of a larger process. It were better said, that we are viewing then but one facet of this precious gem. It were permissible to announce now, that all other facets of the gem cohere, as if by constructive geometric "hereditary principle," with the qualities of the facet upon which our attention is now, more narrowly focused.

The generation, transmission, and assimilation of scientific progress occurs by and among individual human minds. The occurrence of the transformation function, typified by $T(A \rightarrow B)$, must be, initially, as generation, as the internal activity of an individual human mind; it can occur in no other, no different way. The receipt of the transmission is by the individual mind, and the efficient assimilation, as transformed human behavioral disposition, is also by the individual human mind.

In short, the occurrence of the singularity $T(A \rightarrow B)$ typifies that within the individual's mental life which sets the human species apart from and above each and all of the species of beasts.

At this juncture, let us detour but slightly and momentarily, to register an important point respecting the definition of true common sense. Common sense does not lie within the domain of deductive logic; that on which the species-nature of persons and society depends essentially, absolutely, is the proverbial "essence" of the sort of "nonlinearity" which defies comprehension by deductive methods.

So, we must contrast true *reason* to mere *logic*.

That which exists above and beyond the grasp of mere logic and yet governs a sound logic, is *reason*. This metalogical, *metaphysical* domain of *reason,* is typified by that activity of efficient scientific progress, which causes a general increase of society's potential population-density, and accomplishes this as formal (deductive) logic does not permit such a beneficial result.

So, since successful survival of society depends upon this higher, "nonlinear" agency, the quality of reason, true common sense must be nothing other than a common disposition for the rule of individual social behavior by reason, reason as opposed to Adam Smith's empiricist, hedonistic irrationalism and as superimposed, as higher authority, upon mere logic.

That definition, although not adequate, is useful in a preliminary way. Focus now upon the paradoxical relationship between the person and society. Consider, first, the simpler aspect and, later, the *isochronicity* which binds the person of the present to the society of past and future, in addition to the present.

Potential population-density is a characteristic of the society as a whole. It determines how many persons may exist and at what level of quality of existence. It determines the relative power of society to resist and repair the damage done by natural and other catastrophes. It is also a power of the individual, as *level of the individual scientific knowledge and technology of practice* illustrates this point.

Within that arrangement of individual and society, the following fact is singular. The contribution of a valid, fundamental scientific discovery of an individual mind benefits not only all of present and future generations of mankind, but the dead of past generations, too. So, in this way, too, the dead need never die.

The contribution of such a single, valid, fundamental discovery by the individual, is *a universal act:* By means of such an act, the single individual is acting *directly* upon the society as a whole. This act transforms the character of the entire society in a "nonlinear" way. The person, as individual, is acting upon the society also as a conceptually indivisible individuality, in a direct way.

This elementary conception requires careful elaboration.

XII | The Individual and Society

Society can and must be conceived as an indivisible individuality. Most simply, the addition or subtraction of any ration of individual persons does not alter the most essential feature, quality of a society: that it exists as a society, that it exists only as it is brought into existence as a society, and is perpetuated as a society. With this we associate certain functional ideas which are unique to the notion of society as indivisible in this sense.

So, the indivisible individual person is to be seen functionally as interacting with nothing other than that notion of the whole society as a functional indivisibility. It is the leading implications of that functional relationship, which now occupy the center of our attention.

Although only in first approximation, we have now identified, in principle, the solution to the Parmenides Paradox. What we have considered and are about to consider, should be read against the background of Nicolaus of Cusa's *De Docta Ignorantia,* where the same principled solution is employed extensively and with an essential relevance. This should be considered also against the background of our review of the *Riemann,* and proposed *Riemann-Beltrami* forms of continuous representation of a "nonlinear," negentropic *surface function.*

With respect to that concept of society as an indivisible,

continuous process, the individual person within the society
is a true singularity, topologically. This singularity's existence
destroys the continuity (connectivity) of the process, at least
in the formalist's view of topology. This damage is repaired
by human creative activity, as we have defined that in broad
terms.

For example, the act of effecting and imparting a valid,
fundamental scientific discovery, establishes a perfect, pervasive, efficient connection to the society as an indivisible process. This "repairs the damage," and accomplishes that result
in a manner depicted by the Riemann surface function. (Or,
more adequately, by the required Riemann-Beltrami surface
function.)

Those two preceding paragraphs may be read, also, as a
statement of the Parmenides Paradox and its solution.

This action we have described, has the effect of increasing the potential population-density of the society as an indivisible whole. Take not only $T(A{\to}B)$, but also,
$T[(B{\to}C), (C{\to}D), (D{\to}E)]$ as corresponding to $T(A{\to}B)$, in
the sense we have described *successful survival*. In that sense,
the function for the individual discoverer, $T_0(A_0{\to}B_1)$, inter-
sects the transformation $T_0(A_0{\to}B_2)$, where B_1 is the new momentary state of the individual, and B_2 the effect of the same
act of transformation, by that individual upon the society.

So, the creative mental processes of the individual are
acting directly upon the functional state of the society as a
whole, as $T(A{\to}B)$.

Is this the individual particular, the singularity, acting
directly, efficiently upon the whole, to the effect of changing
that whole? Is it, rather, as some formalists would argue, a
mediated connection? If I project a laser's pulse through a
medium, is the impact upon the selected target then a relationship *mediated* by the medium?

The guidelines for use of lasers in the U.S. Strategic Defense Initiative (SDI), for example, prescribe that the laser
pulse shot at the incoming Soviet weapon from the Earth's

surface, must have the minimal interaction with the medium—
e.g., the atmosphere. This policy may be served in either or
both of two most notable ways: One may decrease the amount
of atmosphere between the laser and the target; or one may
tune the laser pulse to bypass the frequency-channels of interaction of the atmosphere. There are other notable tactics, but
these indicated suffice to illustrate the point at hand.

Another useful analogy is the method for effecting thermonuclear fusion of, for example, small "cannonball" pellets
of *tritium* and *deuteride* to maximize *isentropic compression*
in the manner implicit for plasma physics in B. Riemann's
1859 "On the Propagation of Plane Air Waves of Finite Magnitude."

To cause the *singularity* (the relatively small particular)
to exert the maximal impact upon the process as a unity, *tune
the action harmonically to follow the isochronic pathway of
physical least-action.*

Merely note what is said explicitly; catch another implication out of the corner of your eye and then bring your
concentration fully to bear upon the latter implication. Is it
necessarily the case, that action transmitted by way of an
interposing medium is a *mediated* relationship, *merely* a mediated relationship? That is the point immediately at issue.

The "idea" associated with individual's $T_0(A_0 \rightarrow B_1)$ has
the relatively universal effect $T_0(A_0 \rightarrow B_1)$. Is the transmission
of that "idea," to become the efficiently increased potential
population-density of the society, a *mediation* in the sense
statistical gas theories, for example, might desire?

Think of a great tidal wave, borne across most of the
expanse of the Pacific. Assume a seismic disturbance near the
Andes to send such an effect to some island of the Asiatic Rim.
Is the latter effect to be described meaningfully as a "mediated
effect"? In a hydrodynamics conceived from the viewpoint of
a constructive "non-Euclidean" geometry, it is not a mediated
effect, but a direct one, (e.g., retarded potential of propagation
of effect). More emphatically, once we focus directly upon the

role of *isochronicity* in defining the principle of (physical) least-action and the nature of all valid representation of elementary physical law, the relationship is not a mediated one.

Unmediated, negentropic transformation of society by the singular individual person's sovereign, creative mental powers, is possible in terms of the functions implicit in $T(A \rightarrow B)$. This is the only means by which that negentropic transformation may be effected. The existence of the sovereign, singular, creative mental processes of the individual person, is not merely a source of such advantage to society; it is the only source of that advantage, an advantage upon which the continued successful survival of society depends absolutely.

Any society, in history, which has abandoned the commitment to fostering scientific and technological progress, has suffered some or all of these consequences:

1) The society has been taken over by a native or foreign usurping power.

2) The society has collapsed into a dark age, as early fourteenth-century Europe did.

3) The society has become extinct, or virtually so.

The suppression of scientific and technological progress, by suppressing the singular, natural distinction of human nature, eradicates from practice that which sets society above the troops of baboons, and tends so to transform the society's population to become that of a barbarous, brutish collection of "yahoos."

It is shown in the science of Physical Economy, that any fixed mode of production depletes marginally the improved natural resources upon which that mode depends.

This marginal depletion is overcome solely by means of technological progress. Those means center, most visibly, around the density of consumption of usable energy, as this is measurable three-foldly: (a) per capita, (b) per hectare, (c) per Angstrom unit (or micron) of application to working area of production. The possibility of generating and assimilating efficiently such increases of "energy-flux density," is constrained by the rate of advance of level of technological progress.

So, in the latter account, a "zero technological growth" society is an *entropic* economic process, a dying culture, a people which has lost its moral fitness to survive.

Hence, the value of the singular individual person is located uniquely in that which sets that person and all mankind apart from and above each and all species of beast. The existence of society depends uniquely upon this value. That is the *value of self;* that defines true self-interest, to preserve and enhance one's true value; that defines what one requires be fulfilled as needs of personal self-interest.

"I transform and do this singularly; therefore, I exist," were a rebuke to Descartes. "My value were measured by the practical consequences of my present existence for the present, future, and past generations, each and all considered as an indivisible whole," were the proper rebuke to the immoral irrationalist Adam Smith.

It is now time to turn to the matter of present, future, and past, considered as an indivisible unit.

As to the present society, it depends upon our singular function for its quality, its potential for successful survival. As to the future society, however far it advances above the level of the present, our singular contribution remains actively embedded in the foundations of that future, more advanced state. As to the past generations, by acting upon the singular contributions of each, directly, or even only indirectly, we enrich those with our own singular contribution.

So do we bridge the gulfs of time, to establish so a direct form of transaction with persons from the most distant places in the remotest corners of mankind's past and future time.

As the relationship exists, so is all humanity presented to us as an indivisible singularity of our universe. The expanse of past through future time, which is the span of mankind's existence, does not render mankind divisible into periods of time. As I am in direct relationship to my remotest ancestor and your remotest posterity at once, the image of absolute time dissolves, replaced by the appropriate notion of relativity in physical space-time.

Similarly, all mankind's relationship to the universe as a whole, mankind's efficient, negentropic existence as a singularity of that negentropic universe. So, in our mind, can we approximate the Creator's view of His universe and so inform our minds better to His work in that universe.

In our singular, almost timeless relationship to the entirety of our universe, through the means of our creative mental power's singular, negentropic relationship to all of past, present, and future mankind, we should see the meaning of universal physical law and the characteristic isochronicity of physical law properly represented.

In the last, highest—or next-to-highest—view attainable by mortal minds, the extent of physical space-time becomes as if a timeless, indivisible single entity, with which all of mankind's existence is a singularity and each of us a singularity in indivisible physical space-time, because each of us is a singularity within the existence of mankind.

"Hello, Socrates! Let us now act in concert to save mankind from the peril of A.D. 1992." That is a practical proposition, if we but understand the means by which that is rendered efficiently true.

XIII | Agapē

Those among us who are habituated to creative mental life, are situated to observe some curious, provocative, and most pleasing aesthetical congruences among scientific discovery, classical music and other matters. Love of God, love of mankind, love of truth, and love of classical beauty, partake of a single, empyreal quality of joyfulness, called *agapē* in the New Testament's original Greek, *caritas* in the Latin translation, and therefore *charity* in the King James's I Corinthians 13.

This emotional quality, *agapē,* may be addressed as emotion by aid of reference to several types of relevant experiences. It is known as the "tears of joy," evoked by witnessing a child's happy discovery of the rational solution to a problem of construction. It is the warm glow experienced, the sense of a light turning on in one's head, as a valid discovery is made in science, in music, in classical plastic art-forms, and so on. It is the joy of love toward the deceased, as this is captured in a valid performance of Mozart's (1756–91) *Requiem.* It is the emotion associated with a conscious practice, congruent in guiding conception with the view of self, mankind, and universe outlined in summary just above.

For the classical culture of Plato's Athens, aesthetical beauty was inseparable from congruence with the harmonic orderings subtended by the Golden Section. This was the rule

85

of aesthetics applied explicitly to the plastic fine arts and to well-tempered musical polyphony as it was known in Plato's lifetime. Today, we know much more than did Plato, but that enables us to admire Plato's work the more adequately.

Simply, the harmonic orderings congruent with the Golden Section are beautiful, because they express life, in opposition to death. Any artistic composition not congruent with that harmony is ugliness and thus, no true art. Life expresses and portrays *negentropy*. Life, as negentropy, expresses the referent for the verb, "to create."

Kant, in his last *Critique,* the *Critique of Judgment,* restated his argument, that creative processes are not comprehensible. In the same location, Kant argues, from the same premises, that there exists no rational yardstick of artistic beauty and that, therefore, no yardstick exists excepting an irrationalist's pragmatic one.

Although Franco-Swiss varieties of eighteenth-century Romanticism were premised upon the "oedipal" irrationalism of the correspondents of Voltaire (1694–1778), the spread of Rousseau's (1712–78) moral degeneracy into Germany as nineteenth-century Romanticism, was made possible by the influence of Kant's axiomatically irrationalist dogma of aesthetics. On this account, the anti-Romantic poet Heinrich Heine (1799–1856) denounced the evil Madame de Staël and Kant with surgical precision of insight. The so-called "neo-Kantians," unleashed in Germany both to defend the radical irrationalism of positivism and to corrupt and destroy the morals of Bismarck's Germany, are of special historical relevance to our practical concerns here.

Today, wherever the dogmatic irrationalism of Francis Bacon, David Hume (1711–76), Voltaire, Rousseau, Kant and the positivists prevails, the absurd aesthetical dogma of "art, by the artists, for art's sake," prevails. The fanatical irrationalists of art insist, because they have been brainwashed to do so, that Kepler is irrelevant to musical tuning, because, as Savigny decreed, there is an unbridgeable gulf between

Naturwissenschaft and *Geisteswissenschaft,* the dogma that scientific reason must be barred from the axiomatically irrationalist domain of professional artistic opinion. (See Appendix X.) That Romantic psychosis has reigned, side by side with Helmholtz and Ellis, in the musical and plastic-arts curricula of most of this century. So, in music, cacophony and ugliness for the sake of evil ugliness prevail, driving *bel canto* and the classical heritage into shrinking exile places of refuge.

The relevant point being stressed here in this way, is that the appreciation of the quality of *agapē* in great classical art is dependent upon the artistic expression of two interdependent qualities of composition.

First, all classical artistic compositions are demonstrated throughout by physical space-time congruence with Golden Section harmonies; any work which departs from that rule is not classical, is arbitrary, irrationalist, and ugly, to exactly the degree such violation occurs.

Second, mere congruence with Golden Section harmonies is not sufficient. That congruence defines only the rules which must be satisfied by draughtsmanship. It is not art unless the composition is subsumed by an artistic idea of the same quality as a mentally creative, valid, fundamental scientific discovery.

A dead human form, at the moment of death, has, like the human skeleton, the harmonic organization supplied to it by life. It is not alive, as a harmonically sound plastic or musical composition is not art, but only student's exercise in rules of draughtsmanship, unless the spark of creative genius is embedded in the harmonies.

Thus, in classical art, the composer is prohibited from exercising what anarchists generally, including such irrationalists as the Romantics and overtly satanic modernists, regard as "individual artistic freedom." There is no "freedom" permitted in the composition of classical art, but *true human freedom,* valid creative discovery, as the highest practice of physical science, sets a standard for this.

Thus, as to artistic "emotion."

The true (classical) "artistic emotion" is that state which is evoked by excitation of the faculty associated with T(A→B). This is evoked by moments of generating, transmitting, and efficiently assimilating valid fundamental scientific discoveries. This evoked emotion is the power of sustained, protracted concentration we require for such mental activity. It is the emotion, such as "tears of joy," evoked by fixing upon that singular human quality we should love in all other persons. It is, therefore, the emotion of love of truth and of classical beauty.

We humans require scientific and technological progress as the continuing, ruling impulse in the policy-shaping of society. Without it, successful survival is impossible to attain. We require it to sustain human life; but the question arises, "Why should human life be sustained? Why should the Creator assure that it exist, and be sustained?"

By sustaining life through scientific and technological progress, we require society and its members to cultivate those mental, creative potentialities by means of which society sustained a stream of generation, transmission, and efficient assimilation of successive waves of such revolutionary progress. To foster that activity, we must motivate that activity as the apparently spontaneous, characteristic response-disposition of many individual members of the society.

In other words, the society requires an all-sided way of life congruent with this task: *culture* in the best definition of the term. That culture has the effect of cultivating the individual and society in the mode most consistent with the distinction of being human and in the exercise and development of that quality, to create, which is in the image of the Creator.

The universe has no need of a human species whose practice does not express, foremost, that creative mental activity which sets man absolutely apart from and absolutely above each and all species and varieties of beasts. Should society's practice disregard that conditionality of our species's existence, we risk producing the circumstances, as we are tending to do today, under which the laws of the universe will act, lawfully, virtually to obliterate a human species which has rendered itself as immoral and useless to Creation as "Malthusianism" has caused the United States of America to become during the recent 20-odd years.

This choice of disposition, between commitment to "Malthusianism" and to promotion of scientific and technological progress, corresponds to a policy-shaping disposition, the quality of impulse which regulates the way in which we formulate policy-initiatives successively, in response to changing circumstances. Thus, such a policy-shaping disposition corresponds to the determination of *successful survival*. In contrast, a successful choice of particular policy of the moment, represents *momentary survival*.

Obviously, a policy-disposition corresponding to successful survival, is only a general way of expressing an efficient commitment to fostering a *negentropic* increase of potential population-density. This generality subsumes varying degrees of that general quality. Conversely, *entropic* policy dispositions vary in degree.

Although we have put the emphasis upon physical economy as such, we have indicated, perhaps adequate for our purpose, that by a commitment to a negentropic policy-shaping disposition, we mean to imply the cultural correlatives we have indicated here as cohering with the singular, creative mental act of "nonlinear" transformation, that latter as the means of increasing potential population-density through the generation, the transmission, and the efficient assimilation of scientific and technological progress.

Once we have begun the permanent colonization of Mars on a sound basis, as we might approximately 40 years from now, the philosophical standpoint in statecraft, which has been reflected here, would be hegemonic for humanity. Once we commit ourselves to becoming man in the universe, there is a great and good change in our way of viewing ourselves. Today, the problem is to reach the point, when the policy disposition I have described here will be recognized as *the only true common sense*.

The Christian Philosophers

St. Augustine
(354–430) Bishop of
Hippo

St. Thomas Aquinas
(1225 (ca.)–1274)
denounced usury as
anti-Christian

The Ancients

Aeschylos (525–456 B.C.) Greek
dramatist

Socrates (469–401 B.C.) Greek
philosopher

Plato (427–347 B.C.) Greek
philosopher

Aristotle (384–322 B.C.) Greek
philosopher

Archimedes (287–212 B.C.) Greek mathematician, shown here at the moment before his murder by Roman soldiers invading Syracuse.

Fathers of Western Civilization

Charlemagne (742–814)
Holy Roman Emperor

Dante Aligheri (1265–1321) author of *The Commedia*

Great Minds of the Golden Renaissance

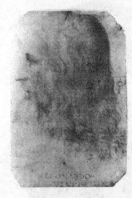

Leonardo da Vinci
(1452–1519)
universal genius

Pietro Francesco Alberti (1584–1638)
Italian architect

Pietro Toscanelli
(1397–1482)
Botticelli's famous
portrait of St.
Augustine is believed
to have been
modeled on this
Italian astronomer
and cartographer.

Giants of Physical Science

Johannes Kepler (1571–1630)
astronomer

Leonhard Euler (1707–1783)

Carl Gauss (1777–1855) German mathematician

Bernhard Riemann (1826–1866) German
mathematician

Georg Cantor (1845–1918) German
mathematician

Great Thinkers of Modern History

Nicolaus of Cusa (1401–1464) father of modern science

G.F.W. Leibniz (1646–1716) founder of Physical Economy

Friedrich Schiller (1759–1805) German poet and dramatist

Project A

LaRouche Discusses His Solution
To the 'Riddle of the Ages':
The Parmenides Paradox of Plato

Foreword

These essays were composed at a time when the United States of America appeared to be plunging toward its self-destruction. The Bush administration appeared to be as mad as the biblical King Nebuchadnezzar, and this for similar causes. Such madness is the characteristic feature of a "Thornburgh Doctrine," which elevates the mere whim of a U.S. President above all international law, even higher than the natural law of Almighty God.

Since the spring of 1989, it has become increasingly obvious that, using the imagery of the ancient Chinese philosopher Mencius (314 B.C.-?), "the Mandate of Heaven" has fallen away from each and all of the three empires lately dominating our planet: the Anglo-American ("Animal-Saxon"), Moscow's, and the Communist Chinese dynasty. As all three are visibly doomed, so, "whom the gods would destroy, they first make mad."

The ongoing economic and moral breakdown of those three empires may suggest, that the dreary object-lesson of this waning century is the common worthlessness, and consequential folly, of those ideas associated, respectively, with the names of Adam Smith (1723-90) and Karl Marx. If we examine the same contemporary facts from a more appropriate standpoint, the preceding 25 centuries of European history as a whole, we are led to those deeper truths which are the subject of the essays in view here.

All European history, including European civilization's unfolding in the Americas, is characterized by a single principle of conflict, a conflict between *republicanism,* on the one side, and *oligarchism,* on the other. Such was the conflict between the young United States of America and the oligarchical regimes of King George III and the Holy Alliance powers. Since the Achaemenid empire of oligarchical aggression against the Ionian Greek city-state republics, the only real issue within European history as a whole, has been the conflict between the *republican* followers of Solon (638-558 B.C.), Socrates (469-01 B.C.), and Christ, on the one side, and the opposing, usury-ridden heritage of Babylon, Canaan, and pagan Rome.[1]

This pertains not merely to political history, but to every important development in the arts and sciences for as far back in the existence of mankind as our knowledge can reach.

Most simply, *oligarchism* signifies a division of the families of which every society is composed, a division between a relative few, powerful, ruling families and a relatively great mass of the oppressed families which are the mere objects of rule by the ruling families. The apotheosis of oligarchism is the Greek pagan, Olympian pantheon of Zeus and other immortals, playing with merely "mortal" men and women in the fashion a cruel, bullying, capricious child plays with and breaks his dolls.

The distinction between *oligarchism* and *republicanism* arose in literary history with the defense of the Greek city-state republics against the oligarchical enemies from Babylon and Canaan. The idea of *republicanism* grew up and evolved during many successive battles for freedom. Thus, when a truthful historian speaks of the history of republicanism, he offers two primary sets of distinctions. He refers to the succession of struggles, beginning with the constitutions of the ancient Ionian city-state republics, continuing through the work of Solon of Athens, Aeschylos's (525-456 B.C.) *Prometheus,* Socrates, and Plato (427-347 B.C.). The historian concurs with St. Augustine's (354-430) relevant letter, on the point that Christianity adopts *a Christian Socrates* as to scientific method in arts and science, but sees a crucial single flaw in a merely pagan Socrates. So, we have the history of republicanism and the crucial distinctions emerging in the course of that history.

The essays before us peer into the deepest features of the historical conflict. The mind of the oligarch sees "God," "man," and "nature," in an entirely different way than does the mind of the republican. It is the axiomatic quality of those deep epistemological differences which the essays address, thus continuing the work of the 1989 book *In Defense of Common Sense.* The object of the present essays and the indicated predecessor, is to demonstrate the possibility of *intelligible representation* of an entire class of conceptions. These conceptions share the common quality of showing that the choice between an oligarchic or republican political-philo-sophical world-outlook leads, as a consequence, toward a congruent set of ideas in every field of rational thinking and discourse, including both art and physical science.

What the author has done, in connection with the two sets of philosophical essays referenced, is to revive the Socratic method by recasting it, as it were, *de novo,* and doing this from the standpoint of the best knowledge available in the present century. Thus, *In Defense of Common Sense* was written in the form of such a commentary upon the topics of Plato's *Theaetetus,* and also, implicitly, the *Sophist* and the *Parmenides.* The purpose was to illuminate the potential intelligibility of Plato's method and conceptions, by presenting a more advanced, twentieth-century vision of the same topical areas.

Relative to *In Defense of Common Sense,* the objectives of Project A are more specialized ones. In the latter, we address directly, chiefly, certain crucial problems of modern mathematical physics and also the underlying principles to be employed for effective conduct of winning republican "cultural warfare." Different as those two topics might appear to be, the text of the essays shows that they are, in reality, the same topic.

The crucial *formal* issue addressed in the essays, is the definition of the *ontologically elementary* in physics. The following comments conclude these preliminary, summary observations as a whole.

Modern classroom physics begins only after it has successfully ignored those topics upon which the very idea of a rigorous physical science might be premised. That is, mathematical physics begins from the starting-point of certain naively conceived and provably false ontological assumptions taken as axiomatic.

At the center of those such popular, ignorant follies upon which so much of modern classroom physics is premised *mathematically,* is the popular delusion, the axiomatic assumption, that the elementary form of "matter" must be *simple substance*. The essays identify the readily accessible, conclusive proof that such a popular assumption is false. The nearer to the very small we reach, the more that substance in the very small partakes of all of the complexity inherent in a *negentropic* form of the *universe as a whole*.

This view, just expressed here, was already implicit in the Socratic work of Plato and in the work of Nicolaus of Cusa (1401-64) and Gottfried Wilhelm Leibniz (1646-1716)—among others—in the founding and elementary elaboration of modern physical science. For them, as for Professor Bernhard Riemann (1826-66), the universe as a whole is "axiomatically" *negentropic* (mathematically), and substance in the very small reflects this negentropic quality, this "nonlinearity" of the universe as a whole. (See Appendix IX.)

These essays' approach to the most crucial among the problems of present-day physics, brings us back, directly, to the political issues as such, and does this in a most interesting and profitable way.

The proper basis for a physical science is found by means of an adequately rigorous reflection upon the question, "What is it possible for the mind of the human individual to know, and that by means solely of the individual's sovereign potential for creative reason, the sovereign potential which sets mankind apart from, and above, the beasts?"

This required demonstration is immediately at hand, as *In Defense of Common Sense* and these essays combine to show. The showing of the central role between, on the one side, a sovereign individual potential for (anti-Kantian) creative reason, and, on the other side, a *negentropic* form of existence of the universe taken immediately in its indivisible entirety of unitary existence, is the key.

By means of developed (individual) creative reason, we are each capable of making our own conscious thought a process rendered an intelligible subject of the same quality of conscious thought. In the language of the mathematician Georg Cantor (1845-1918), we are able, on a higher level of consciousness, to adduce the ordering-principle characteristic of a relatively inferior, observed aspect of our same conscious process. In mathematical physics, this is the "hierarchical ordering" of *transfinite* orderings. (See Appendix VII.) In this same way, we are enabled to become efficiently conscious of a *transfinite ordering* of a direct relationship between our conscious, sovereign powers of creative reason and an undivided universe as a *negentropic* form of *elementary existence*.

The exploration of that conscious appreciation of that transfinite connection between "monad" and universality, shows us that this transfinite process is the only form in which a true physical science is possible.

Then, by exploring the higher, "nonlinear" forms of transfinite ordering associated with this "maximum-minimum" connection, we are enabled to find in this transfinite realm the higher correspondent to the formal "hereditary principle" in the deductive modes. On that basis, we have begun to practice a truer physical science; on the same basis, we have established, at last, a true *political science*.

Finally, now, the following observations.

The map of the universe just identified, is peculiar to the deepest epistemological implications of the Christian form of Socratic thinking, of the Christian form of republican world-outlook. It is the physical science of a Cardinal Nicolaus of Cusa, a Gottfried Wilhelm Leibniz.

This fact is key to understanding modern physical science properly, as the complicated reflection of a 400 years' war within the ranks of science, between the opposing republican and oligarchical factions within science: the republicans Brunelleschi, Cusa, Leonardo, Kepler, Pascal, Leibniz, Carnot, Monge, Gauss, and Riemann, against the oligarchists Descartes, Locke, Newton, Cauchy, Kelvin, Clausius, Maxwell, Rayleigh, Boltzmann, and so on.

First, the oligarchical world-outlook is incapable of understanding the nature of creative reason and could never understand the most crucial conceptions of a Plato, Augustine, Cusa, Leonardo, Kepler, or Leibniz. The closest approximation to a science of which the oligarchical mind is capable, is the pseudo-rational, deductive formalism of an Aristotle (384-22 B.C.), Descartes (1596-1650), or Kant. (See Appendix IV.) (Otherwise, oligarchism is mere, arbitrary irrationalism, akin to that of a David Hume [1711-76] or a Friedrich Nietzsche [1844-1900].)

Second, the present form of mathematical physics is chiefly the result of the political power of the oligarchical faction over the monied institutions of science and education. It is the past hundred-odd years' rise to superior political power by the usury-practicing, "New Age," oligarchical faction, which has caused the classroom triumph of arithmetic-algebraic formalism over the more natural mathematics of non-Euclidean constructive geometries.

Third, the scientific inferiority of the oligarchical world-map, is a crucial, potentially fatal tactical vulnerability of the oligarchical political-philosophical faction as a whole. The included purpose of Project A, is to foster among republicans the knowledge needed to exploit that feature of the oligarchists' "genetically"-determined tactical inferiority.

Finally, the time has come, when the oligarchical faction's corrupting influence can be tolerated not much longer.

I | The Topics

This is a project which pertains to all kinds of tactical and strategic, educational, and scientific matters. It is stimulated partly by the work we are doing in the strategic area, as defined by a few electronic memoranda I have made on this recently. It reflects my struggling with some of the lingering problems in the scientific area of my work, which I have been reviewing afresh recently. For example: Winston Bostick's sequel to his "Plasmoid Construction of the Superstring" prize essay, which I have been looking at; Daniel Wells's paper, "On Quantization Effects in the Plasma Universe"; and various other things.[1] Similarly, an item recently came to my attention pertaining to the subject of noisy foam in astrophysical space; these matters are very messy and the reason they are messy is obvious to me.

Then we have the problem, that people sometimes are a bit wild on geometry, ignoring what I had insisted earlier following upon as the ordering principle, a *hereditary ordering principle* in constructive physical geometry. Let me just emphasize that for a moment.

People say, "Okay, we are going to find a geometrical construction which conforms to a physical design, a physical concept." That seems to satisfy the requirement which Prof. Felix Klein (1849-1925) had for his graduates back at Göttingen, but not quite. It certainly does not satisfy *my* design; and, it does not satisfy the requirements of physical science.

My essential discovery in physical geometry, and therefore, bearing upon physics generally, is that scientific conceptions are geometrically ordered in a *transfinite way*. That is, if you construct things, such that the same construction with one action added (one ply of action, to put it properly), is a multiply-connected manifold, this ply has the requirement not only of generating an additional singularity, but actually or implicitly redefining the entire process, the entire nonlinear function. So, the rate of generation of singularities is increased.

That is, if we set up any kind of a notion of a constant, arbitrarily small length, we might take Prof. Winston Bostick's reference to the Planck length, 1.6×10^{-33} centimeters. Take a length like that and say, *the increase of the density of singularities* within that arbitrarily chosen interval of action, is a measure of negentropy.

That obviously applies to Bostick's construction in all kinds of ways, provided you assume this is not a black hole. A Planck length is not really a very arbitrary choice of length, but will serve the same purpose as an arbitrary one. For that process, an increase of the density of singularities per unit of action, as referenced to that length, would do all kinds of wonders for that kind of representation; and it obviously is relevant.

The *hereditary principle* means that we start from the simplest notion of a multiply-connected circular action, with reference to the simple isoperimetric concept, but not limited to it.

I have emphasized earlier that that little critter is actually an envelope. We are obviously following the pathway of non-algebraic curvatures; and, continuing with all the things we have discussed, we are finding a pathway which is, for any point of reference, a consistent pathway of growth of density of singularities per interval of action. That would be a demonstration of, a specification of, a rigor for, a hereditary principle.

You don't have to be perfect; but you always have to be moving in the right direction, and whatever you do has to be based on what has gone before. You can do nothing which is not based on what has gone before. Otherwise, you are cheating: When you jump up out of the clear blue sky and say, "Ah, we can do this," you are cheating. If you do that often enough, you'll go absolutely mad, because cheating means thinking without a mooring; and the more attractive it is, the more dangerous it is. So, don't cheat; everything must go in a succession.

To that end and to other ends, it is obviously my responsibility, my pleasure, my duty, to address what I said before, on various aspects of this, once again, perhaps in a clearer way than before.

We have two things to consider, primarily.

First of all, as we did in connection with the recent Martin Luther King Human Rights Conference,[2] the discussion of principles: the sovereign, creative-reason potential of the individual, and the relationship of that creative potential, as a sovereign capability, to the totality of existence. The immediate, non-mediated relationship of that individuality to the totality, is the primary distinction which sets man apart from and above all other species.

All law, all natural law and all proper law otherwise, is derived from that consideration, *never goes away from it,* is always subject to it. So, any law which does not meet that requirement or is inconsistent with that requirement, is to be nullified as unlawful; that is its character.

The second aspect to consider is: How do we describe the mapping of that mind which knows, or the activity of the mind which is in the process of knowing, the universe, of knowing the law, of knowing art, and so forth?

I would emphasize again in this same setting the human factor, the individual human potential, the divine spark of reason.

Look at the physics of this. There can be no true law of

physics which is not in conformity with what I have indicated. That is, the ability of the mind to understand the universe, from the standpoint of not merely abstraction, but for practice, depends entirely upon this capacity of the human being: this sovereign potential of creative reason and, primarily, the unmediated or direct relationship between that potential and the whole, as well as all other kinds of relations. Therefore, there can be no law of the universe which is correctly presented in any contrary terms, which can be represented intelligibly in any contrary terms.

So, those are the two facets:

The one is the principle of law of the individual, the strategic implications of that as I have addressed that; and, second, as I have addressed earlier, the map of the mind in the act of knowing the lawfulness of the universe, including the lawfulness of physics, is of that form. There is no other way in which the mind could know the universe; therefore, all knowledge of the universe is expressed in that form.

This brings us back again to the sticking point: People say, "We must use accepted classroom mathematics to explain mathematical physics." Among relatively responsible, thinking, serious, rigorous people, most of the problems that arise, which are vicious (that is, of a persistent nature), are all of the form of trying to adapt what we know, empirically, experimentally, to the limitations of what is commonly described as "commonly accepted, classroom, deductive mathematics."

In no way can deductive method represent the process, except, as I show in *In Defense of Common Sense*, we can use the deductive method negatively. We can use the inductive method less reliably, but the deductive method negatively to show that what we are looking at is what it is not, what the deductive method is not.

That is the definition of the project; so, when I refer to "the project," please note that that scope of this introductory outline is that to which I refer.

II | The Crucial Fact

The general scope of our inquiry is the following:

1) The demonstration that creative reason sets mankind apart from and above all other species.

2) That this creative reason, this potential, when developed, or as developed, is *sovereign*. That is, that all creative acts of discovery occur within the sovereign domain of an individual intellect. This includes:

a) The generation of a discovery.

b) The transmission of a discovery. (In order to transmit a discovery you must in some degree assimilate it—a generated discovery—and you must, in a sense, regenerate it.) Though the requirements are not as rigorous as they are for generation, the effected transmission requires a very significant amount of use of the creative powers of reason of the person engaged in nothing more than even apparently mere transmission.

c) The assimilation of this discovery or this principle by the human mind for employment in practice or in reshaping, as if axiomatically, the practice of the individual who ingests this.

3) Although the sovereign, individual act of creative reason is conditioned by a social context, ultimately, the primary relationship within society is a direct, as if to say *unmediated*, relationship between the individual person and all of past, present, and future historical existence of mankind.

The social aspect is twofold. The social aspect, in terms of relations to other people as if they were particles in a Cartesian space, is: This image enables us only to show that others are acting upon the individual to develop individual potential; and it also shows that the individual is acting on others around him, so that there is a kind of radiation of an idea through transmission.

Nonetheless, the essential, efficient aspect of any discovery is its effect upon the potential rate of progress of the power to exist of the human species as a whole. This includes, as we have said before, the fulfillment of contributions from the past which are incorporated implicitly or directly in the discovery. Thus, every action in the present which is creative, as opposed to non-creative or deductive, acts upon the entirety of mankind's past. In the same way, but also in a different way, every truly creative act in the present, or the omission of such an act in the present when required, acts upon all future human existence, directly.

That is, it is not the communication that is mediated; it is what the communication transmits which affects the whole. That which is transmitted is what affects the whole; it is not the transmission as such which does so. That confusion, implied there, must be avoided: Communication is not knowledge. Communication is the transmission of a stimulation of a capacity within the recipient. It is that which is conveyed to creative reason, implicitly in this manner, which acts directly upon the universe; it is not the literal so-called message. *The medium is not the message and the message is not the medium,* even though communication is essential. Those are preliminary conditions.

Since the existence of the species depends upon its creative capacity, two things follow. Since, if only for purposes of illustration, animals could be substituted hypothetically for human beings, it is thus shown more clearly, that the propagation of human beings with this power of reason, is the most essential self-interest of the human species. The activation and development of this creative potential determines the potential, the relative potential of the human species to continue to exist.

This has many implications.

The human species is the only species which is in the living image of God, in the image of the living God, by virtue of being the only species which, as a species, as well as individual members of the species, is creative and verifiably so. Thus, in that sense, the human species is its own reason for existence. That is not true of any other species.

That does not mean that man sets himself against God; that means, in God's creation, man is the only species which in the eyes of God has a self-subsisting reason, sufficient reason for existence. And that is God's love of mankind.

From this flows natural law in its entirety. Natural law consists of propositions which satisfy the so-called equation which we have just set up between the totality of human existence and the creative individual, the sacredness of the sovereign, creative potential for reason of the human individual.

This applies to strategy, in the sense that all proper conflicts in society involve issues directly pertaining to natural law as I have just defined it—that is the second general topic. Therefore, strategy and strategic issues represent conflicts between cultural impulses, which are effectively institutionalized cultural impulses, which are either more or less negentropic, or more or less entropic, as opposed to negentropic.

Thus, oligarchism, which is inherently entropic for the existence of mankind as a whole, as demonstrated, is the enemy of mankind; and, the weakness of mankind is the tendency to have less negentropy than mankind requires in its battle against oligarchism. Hence, republicanism versus oligarchism is the basis for strategy. Propositions which are not stated in those terms of reference are not legitimate propositions.

And thirdly, which we will concentrate upon in my next note in this Project A, is the map of the human mind as it pertains to knowing and transmitting scientific knowledge.

III | Leibniz's Mind

Now, we turn to the question of scientific thought.

Some time ago, I made a reply to a paper in which Leonhard Euler (1707-83) had attacked Leibniz's *Monadology*. (See Appendix XI.) In that connection, I emphasized two things about Euler's attack on Leibniz, beyond the bare fact that it is simply incompetent. I emphasized the fact that Euler's argument was not physics, in the first place, and showed what its fallacies were geometrically, the nature of its geometrical fallacies. I also emphasized that the empirical basis for knowing the *Monadology* does not lie in some abstract, arbitrary, geometric construction, but rather, lies in a very simple demonstration of physics.

For example, it is shown that all creative reason, and therefore all knowledge of the lawful ordering of our universe, is associated with a sovereign power of creative individual reason in the individual personality. Hence, that individual is, as Leibniz emphasized, a *monad*. Hence, the organization of the universe is based on the action corresponding to creative reason by monads. That is physics. It can be demonstrated, that in no other way can we possibly achieve science.

The notion that a science, an empirical science, leads us to a different kind of view, i.e., the Euler view, is absurd.

For example: In a universe which undergoes change, we can demonstrate creative reason in the case of human behav-

107

ior, that is, historical behavior, as the creative lawful ordering of change. In such a universe, one can know the lawful ordering of things only by a knowledge of a transfinite ordering, which corresponds to that lawful ordering, the creative lawful ordering. For example, as I indicate this extensively—and I think in what is a very happy mode of representation, of pedagogy—in my *In Defense of Common Sense,* only the principle which determines the ordering, implicitly, of the successive scientific revolutions, i.e., as I did with the A through E lattices, only that principle represents knowledge. Only that principle corresponds, even in approximation, to a lawful ordering of the universe.

Therefore, any knowledge of the universe as to the principle of ordering can only arise from the standpoint of the creative reason, i.e., the sovereign, creative reasoning powers of the individual: being conscious of those sovereign, creative reasoning powers and other creative phenomena which are analogous, shall we say, to what happens in creative reasoning.

That gives us the essential map of the universe in germ. To go further, we have to take the other principles into account. We have already demonstrated again, socially, that the efficiency of creative reason is, in first instance, represented by the nature of the connection of each isolated individual who does creative reasoning in our society in the present, with the past, present, and the future, as I have indicated earlier. That demonstrates that that causal relationship is the nature of the efficient relationship between creative reasoning and the universe. That is, the individual, creative reasoning, and the universe. This gives us the map.

Whenever we go away from this map, we are wrong. Whatever we build, there is a fundamental fallacy in it, if we depart from this map. Hence, the *Monadology* is perhaps the most essential document in all of physics.

You will note that Leibniz, in essence, says, in his own terms of reference, exactly what I say here—which is not entirely accidental; about the age of 13 to 14, I learned this from Leibniz directly. I wrestled with it then for over a year and I got it into my head; so today, I don't have it necessarily in the form I learned it from Leibniz, although I was stimulated to my discovery by him. I have learned it in my own way; but, I can go back now and find that what I am saying and what he is saying are really the same thing, in the sense we are talking about exactly the same phenomenon and are posing exactly the same questions.

IV | Plato's 'The One'

The human mind, as I have represented this in *In Defense of Common Sense* and other locations, is characterized by the creative processes of that mind, as those processes are developed. That is, the potential of the human mind is the form of essential behavior of that aspect of human mentation which sets man apart from the beasts and which distinguishes the function of the human mind as human, as distinct from bestial. Those are the characteristic features of the mind as a whole: the creative processes as they may be developed and show their potential.

This potential is a potential for mastery of the universe, in which the creative potential of the individual mind is sovereign. At the same time, this sovereign, individual creative potential is in what I have previously described as an unmediated, efficient relationship to the universal, by the following steps.

First of all, if every human mind is engaged in fruitful, creative activity, according to principles of creative reason, it is efficiently acting upon the past, the present, and the future of mankind. By acting upon mankind, i.e., mankind's practice, we are acting upon the universe as a whole, past, present, and future. This, as I have said, is the individual, the sovereign, creative power of reason in the individual, unmediated relationship to universality.

110

Thus, the practical relations of mankind, in terms of the individual, to the universe are so defined. So, the substance of the practical relationship between the individual, on the one hand, and the universe, and the human species in the universe, and all aspects of practice subsumed by the human species or impinging upon the human species, are in a relationship so defined; that is, defined in terms of this principle of practical reason.

That means that the universe is defined for us as composed of sovereign monads: human creative reason, in this kind of multiple relationship to the universe. The universe, taken as a whole, is thus *One*,[1] an unmediated *One*, as indicated, the essential *One*.

Otherwise, reason is related to other objects in the universe, other created objects, and so forth, in that universe. But, always in its relationship to other objects, the primary, unmediated relationship between the particular and the universal subsumes and is the substance, of all relations to other objects.

Let us pause at this point and imagine, that you think back and forth several times over what I have just said and its implications. What this means, among other things, is that *the idea of simple substance must be eliminated from physics,* if we are to have a correct physics. Simple substance, simple space, simple time, or even a simple form of space-time-matter, must be eradicated from our thinking, if we are to have a correct view of physics. That, of course, is a difficult thing to do, because we study physics in textbooks and classrooms, in which the deductive version of mathematical physics is the accepted classroom version. Therefore, for nearly all among us, nearly everything we know about physics, including our description of the experimental evidence, is couched in terms of this deductive classroom physics.

Yet, I have just said, on the other hand, that a true view of the universe rejects the most axiomatic features of mathematical physics of the classroom variety, on two counts.

First of all, generally, we must reject the deductive axioms, or deductive axiomatics, of a mathematical physics. To say the same thing in a more profound way, the idea of a simple matter, simple time, simple space is rejected; but, also a simple space-time-matter, is also rejected by this, shall we say, nonlinear characteristic of creative reason.

In Defense of Common Sense, for example, illustrates what we mean by that which is essential, that which is in relationship to the universe as a whole; everything to which that individuality is related within the universe, it is related to in terms of that nature of relationship between the individual, creative reason, and the universe as a whole. Therein lies the essence of the matter.

Let us proceed from that. Is this real knowledge or is this merely a form of knowledge? In other words, is it the case that because our mental apparatus is so organized, as I have just indicated (as the apparatus of knowledge), that the only form of universal physical knowledge we need to know, is in that form? That, whatever form universal physical law external to that form we might expect to be, we do not take into account? Or, does it mean that universal physical laws are *efficiently* in the form they must be properly represented by the mind, to accommodate to the imagery of unmediated relationship between particular and universal, as we just indicated? Yes! That latter is what we mean.

Now, let us look at the thing as to form. Let us assume hypothetically, that we are examining now the proposition, that whatever the form in which physical reality is ordered, external to our perception of it, we can only understand that form when it is translated into the form in which our thought must proceed, or *by virtue of,* or *in coherence with,* this notion of the unmediated relationship between the particular creative reason of the individual and the universality. That is the proposition to be examined.

That is where the fallacy lies in most thinking: to say that we have deductive, that we have geometric, that we have other forms, and so forth, and that in this way we may choose different forms of representation to represent the common reality or to distinguish, as in a more general way, between an objective realm, which is not directly known to our senses in its own form, and the perceived or subjective form in which that realm and its efficient relations are reflected upon the form in which we are capable of thinking. That is the obvious issue. Can we make that distinction?

We have to reject that distinction. In the process, by the nature of creative reason, we are not trying merely to represent or mirror what is happening in the universe; we are acting upon the universe, to such effect that creative reason itself is the cause of those changes which are effected. At least, those which are *significant* changes.

Therefore, creative reason itself, in the form in which we represent it, is a cause of existence in the universe: It is a characteristic of substance, of substantiality. Thus, there is no difference between the form, in the proper *form of reason of knowledge,* and the *subject of knowledge,* the *object of knowledge.* No difference in form whatsoever, except to the degree we have failed to perfect the quality of creative reason to know this latter.

So, knowledge is practice in this sense: not knowledge of practice in the pragmatic sense, but knowledge of universal practice. That is, the practice which has the universal effect, such as the scientific discovery. That is, the scientific discovery has a universal effect as it is transmitted and assimilated by the human species. It changes everything; that is practice.

Nothing is practice, except as it can be so represented, respecting universality, in these terms of reference.

V | Matter Is Not Simple

Up to now, we have indicated in general outline the scientific method flowing from our development of the *Monadology*. This can be contrasted, in all cases, with the Kant-Leibniz controversy on the *Monadology,* and, also, compared with what we have referenced earlier on Euler's error on infinitesimal division. This is not to say that we start with the idea of a predetermined discrete existence.

What we are referencing, geometrically, in these monads, are zones of what appear to be negative curvature. (See Appendix VIII.) That is, imposing negative curvature upon the surface of a Riemannian sphere, projectively, would be the kind of image that corresponds with these discrete existences. That is, they are not discrete in the sense that the deductive method teaches discreteness; rather, they are discrete in the sense of generated singularities which take the nature of these negative curvature indentations, so to speak, in a Riemannian spherical surface.

The relationship among the discrete realm, this area of discreteness, this singularity, and the rest of the surface, for example, is that of strong forces relative to weak forces. That is essentially the only physical distinction that we can make from the standpoint of geometry: various orders of magnitude of strong, relative to weak forces, in terms of curvatures and things like that. So, that is essentially the kind of space we are talking about, the standard physical space-time we are talking about. That is what has to be taken into consideration.

The problem here, which we have already started to reference, is that the elementary magnitudes, pertaining to substance, pertaining to action, are no longer linear ones. They couldn't be linear in any case. Just look at it from this standpoint: They are not linear.

That is, mass is not a linear magnitude, nothing else is a linear magnitude. Interaction is not a linear magnitude. There are no linear magnitudes, linear expressions in the system. We might be able to approximate some of the nonlinear ones, under special conditions, by linear approximations; but that does not mean, by virtue of approximation, that the elementary is simple. As we eliminate, by necessity, the notion of an elementarity as being of the quality of the simple, we reject the simple.

We must reject the simple in respect to the notion of substance, to the notion of discrete existence; we must reject the notion of the simple in terms of the so-called space-time relations, of interaction in space-time. So, *simple* is not a quality which we allow in our universe; we cannot allow it, for reasons already given.

That which is seemingly most simple, even if it does or does not, in itself, act as creative reason does on the universal, is in a similar relationship to the universal (as in the case of the lesser monads). That is, the fact that a singularity exists and that it does not act in a certain manner, or under certain circumstances, or under all circumstances, is itself the act of omission of that kind of action which we would expect from a creative magnitude, such as the creative human personality.

So that, in all cases in dealing with pair-wise or other, more localized interaction, we are dealing with something whose complexity is defined, implicitly, by the relationship of creative reason and the individual, as a process, to universality. We are looking at the pair-wise relationship in terms of its own relationship to that universality: the pair's action or lack of action upon that universality. Or, what they must do to act upon it, the condition they must satisfy to act upon it, or the condition they must satisfy not to act upon it, that is, not to alter it in some sense.

Since the primary action in the universe as a whole is itself nonlinear, elementary, but not simple, thus, the conditions which these relationships, or local relationships, must satisfy, in description, and ought to be consistent with, to be part of the universe, are functionally defined in the same nonlinear way as we define the relationship between the higher-ordered monad, the creative individual, creative action, and the universe as a whole.

VI | Reaction to a Query

The tendency is to take a reference point in what is called "credibility," classroom credibility.[1] You take a textbook point such as *isochronicity*, defined in a certain way, and start to reason from that to fill in things, rather than employ my method. Most people are really rather uncomfortable with a method which is rigorous as to axiomatics: what is called in German *streng*, for example, a rigorous Platonic dialectical method.

What people do often, is to adopt a definitional approach, to reference something which they think is unchallenged, and use that definition, to make a construction and to determine from a deductive, inductive standpoint of construction, whether that construction is *plausible* or not. That is poor physics, terrible mathematics. I recognize it is the generally accepted approach to these things, academically; but it is still rotten, because it misses everything. It misses the very thing you must do to make any significant discovery, at least a fundamental one. You cannot make fundamental discoveries, empirically, and then order their representation by that poor, shallow choice of method.

Spend a year of your life doing that kind of thing, and come up with a few important, although not fundamental, discoveries, which you spend most of your life refining. You are not going to make a really fundamental discovery by those methods, by that kind of thinking. To make a fundamental discovery, you must resort to a different way of thinking, which I have been emphasizing so far in this Project A series.

It appears on the one hand (the Kantian view), that a certain kind of geometric thinking is inherently, *a priori*, synthetic *a priori* geometry, even though we can't account for its derivation. That it is axiomatic why it should be that, rather than something else *a priori*; "It just sort of is." All these kinds of views are Kantian, in one sense or the other, or Kantian in this respect. That is not the way the real universe works.

The Isoperimetric Theorem

Think of an isoperimetric construction: People are always trying to correct my language on this and their corrections are wrong.

Most strictly, the so-called circular action should not be thought of as circular action ontologically. (See Appendix I.) It should be called, ontologically, *isoperimetric action*, or, simply, *action*. And the rate of action tends toward the notion of *power*. See, we don't have "energy" anywhere in this thing, because nowhere does energy legitimately arise, except by an arbitrary axiomatic addition based on Kelvin's and Clausius's misreading of the competency, or the scope of competency, of Sadi Carnot's work on heat and of the work of Fourier on heat (particularly Sadi Carnot's work on the thermometer scales and heat).

The isoperimetric theorem represents ontologically exactly what it does: It is the maximum *work* with the minimum *action;* that is all. The rate of that, of the maximum work from the minimum action, is *power*. Any other kind of action is related to the amount of work accomplished which is not worth more than the minimum action to accomplish the same work, or in the same time-frame, the same power, using the minimum action, minimum pathway of action. That is all that is involved. It is not any particular geometry; it is not the idea of circles or this or that; it happens to come out circular.

We don't mean this is a more elaborate way of interpreting a circle; rather, the circle is a way of representing this. Most people have it backwards. They say, "The isoperimetric theorem gives us a new interpretation of the construction of the circle." Bunk, no such thing. The isoperimetric theorem is fundamental; the circle merely is a representation of it. So, we are not discovering a property of the circle with the isoperimetric theorem; rather, we are discovering that all of our assumptions, which we called "circular" before that point, were more or less false. The circle is nothing more than a representation of what we have just discovered, when we explore more deeply the implications of the isoperimetric theorem.

So, from that, we can derive an entire geometry, up to a point. But you cannot, as Euler does, put indefinite divisibility in there. Nowhere in the construction did we have any basis for introducing the assumption of infinite divisibility, nor did we demonstrate it. So, how the devil does Euler *dare* insist it is obvious, that infinite divisibility is possible? No such thing; not obvious at all; not true, on top of it. But that doesn't mean that the universe is made, as Descartes indicates, of preexistent, self-evidently discrete particles; also not true.

That is the kind of problem we are dealing with here: People have difficulty in thinking in my terms of treatment of axiomatics.

They don't examine the assumptions. They say, in their method, "These are good rules for making definitions. All we are doing," they say, "is making a very elementary kind of definition, simple definition. We are following rules of representation which everybody accepts. Don't you see? This is a proof."

It is no proof at all. I'll take your proof, if you use that method; I'll tear your proof apart, show that what you have done, is build an edifice on quicksand. Underlying what you have done, are assumptions which are unproven, just as in the case Euler says, wrongly, that it is ridiculous to say that an angle is not infinitely divisible. Well, it is not ridiculous at all. Euler makes an arbitrary assumption; there is no proof and there could be no proof for it. He makes that the geometric basis for refuting Leibniz on the point of the monad. Whereas, as I have indicated earlier, the monad is as self-evident as anything; but that does not mean a self-evident, discrete particle in the simple sense of simple substance.

Just as a matter of reprise, here.

The problem inclusively being addressed, by this series of sections, is the tendency of people to slip back into an academic mode of thinking, a way of thinking which prevents certain questions from being addressed effectively; and which, worse, leads to the propagation of serious errors in approaching problems. That is, when you depart from the Socratic method, to the business of elaborating definitions based upon what are deemed non-controversial beginning-points or beginning-points "which ought not to be controversial among professionals," then you have sown the seeds of disaster; you have indulged in arbitrariness.

The essence of Socratic method and the essence of scientific method, as opposed to what is taught in the mathematical-physics classroom these days, is *absolute rigor. Nothing can be assumed on the basis of popular sense;* "common sense," professional, or otherwise.

We have referenced the case of the isoperimetric theorem in geometry and reported that the isoperimetric theorem is not an explanation of the circle; rather, the circle is nothing but an image, properly, of the isoperimetric theorem, and that every other understanding of the circle is wrong. That is, when you understand the circle as self-evidently something this or that in geometry and then say, that the isoperimetric theorem is a good explanation of it, you have it backwards. Rather, the circle (provided it means multiply-connected circular action), is a good representation of the isoperimetric theorem.

It is the isoperimetric theorem which is provable; the circle is not provable, it is merely a representation. Only to the extent that the circle is multiply-connected circular action, is elaborated in a manner consistent with a notion of isoperimetricity, as I have defined it, only then is geometric construction valid; and it is only valid to the extent that this isoperimetric principle and its implications, are applied to the notions of multiply-connected circular action in a manner which is truly consistent with a hereditary principle of construction based on nothing but what is directly implicit in the isoperimetric notion.

That is rigor.

Again, or deeper rigor, more specifically: that Euler's attack on Leibniz's *Monadology*, specifically, Euler's absurd insistence, implicitly, for example, that any angle of circular action, no matter how small, is divisible, is typical of an unscientific absurdity of the type we are attacking here.

For it can be shown, as I have indicated, that the possibility of all knowledge, human knowledge, depends upon the potentiality of a sovereign principle of creative reason, sovereign to the individual person, a principle which is implicitly in unmediated relationship, not only to all generations of humanity past, present, and future, but through humanity as a whole and its interaction with the universe as a whole, to the universe as a whole, past, present, and future. The fact is, that that is what is provable. The possibility of knowledge would not exist, unless that were the way the universe is arranged. Therefore, that is the starting point, rather than the isoperimetric theorem or any merely formal, topological construction.

In that physics context, however, the isoperimetric proof, the maximum-minimum, which is a derivative of Cusa's Maximum-Minimum principle, is the formal foundation of all mathematical physics, properly defined. Not as the implications of the circle, but the circle of multiply-connected circular action as a representation, an image, albeit a defective one, of the Maximum-Minimum principle in terms of the isoperimetric view. (See Appendix II.)

Remember, the Maximum-Minimum of Cusa, in terms of its scope and implications, is identical with what I said about the *Monadology;* and the unmediated relationship of the individual powers of creative reason, to the extent that creative reason is the active aspect we are considering of an individual, with not only the human species past, present, and future as a whole, but also the universe as a whole. Maximum-Minimum being thus reflections of one another, in the sense of *imago viva Dei,* the living image of God. That is the basis of everything: philosophy, statecraft, strategy, law, and physical science. (See Appendix XIII.)

It is only to the extent that one can begin with that and nothing but that, and trace a hereditary pattern, e.g., in physical science, that one has a rigorous notion of a physical science. A physical science premised on anything different than that, is an unrigorous notion of physical science, which can be no better at best, than a collecting and rationalizing of reconciliation of assorted elements of experimental evidence and related evidence, in the configurations which are subject to later interpretation, subject to later knowledge.

The typical situation in physical science, without the rigorous approach which I have indicated, is to list an array of constraints of added equations, added conditions, added constants, and so forth and so on, a list which may grow larger, larger, and larger. Obviously this list of equations is not science; it may be necessary work, but it is not scientific knowledge. Scientific knowledge occurs once this array of equations is reduced to a single principle, which is derived in a truly hereditary way from the only fundamental axiomatic sort of assumptions which are permitted, as I have indicated.

That is what I am trying to address again here with this series: to point out to you that I confront often among us, constantly, a lack of rigor. I have confronted this in a most exemplary way in the matters of physical science, where people say, "Start with." "Start with," famous last words. Or "Let us be practical." Or "It is well established that. . . ." Whooaa, nothing is well established, except the underlying fundamentals.

It is precisely the acid of criticism, of Socratic dialectical criticism, of bringing forth assumptions, tracing them to their ultimate roots, and overthrowing entire systems of thought, entire conceptions, on that basis; that and nothing less than that is true science.

It is more important to get that, than to solve any particular problem in physics; because, once we establish a science that is free of the Newtonian deductive heritage of mathematical physics, which is based on those principles we are defending here, then science will go forward at great speed. Whereas, we have come to the point that the clinging to deductive mathematics, the so-called accepted classroom mathematics, is the greatest impediment to physical science within the ranks of physicists, apart from extraneous things that such irrationalists as the environmentalists, the ecologists, and so forth, introduce from the outside.

On the True Nature of Substance

In the preceding section we referred to some basic principles. Let us review some material from a more advanced standpoint than we had previously, in light of what we have just said.

First of all, in Cusa's *De docta ignorantia* (*On Learned Ignorance*), for example, the circular action arises as a kind of metaphor, to represent the relationship between the maximum and minimum, i.e., between the Creator and the individual personality, not the other way round. Thus, the substance of the discussion is this relationship, the maximum-minimum relationship; the circle arises and various aspects of the circle arise, as a way of representing, symbolically, so to speak (a little more than symbolically, but symbolically in one sense), what we have discussed as the substance. Therefore, the circle is not the substance. The circle is a kind of mirror-image, symbolic mirror-image, of the substance; the substance is the relationship between the Creator and man, *imago viva Dei:* the maximum-minimum relationship.

Let us look at this circular action with that in view, saying, "We know the circular action, but not a linearity of the space or even space-time, when we speak of circular action. The circle, in itself, by which the circular action is being represented, is not substantial, it is not material." Let us define the materiality, in the sense of, "Let us discover, in the imagery of the circle, an idea of the circle, or circular action, which corresponds to the substantiality of the maximum-minimum relationship between the Creator and the individual person, *imago viva Dei.*"

We start very simply, obviously with *action*. We don't have circles, because circles don't exist; they come into being. Nothing exists as such; we have to account for the method by which it comes into existence, otherwise it does not exist. The proof of existence is to define that which is subject to this proof in terms of *becoming* existent. The *becoming* existent of the circle is isoperimetric, for example: circular action. It is a representation of it and what that connotes: coming into existence of the circle and circular action.

So we no longer speak of circles, as such; we speak of *circular action.* The circle, in itself, comes into existence as a result of the circular action, which is defined as a self-bounded area. *Self-bounded:* So the perimeter is included in the circular area, is a self-bounded existence, brought into being by peri-metric circular action, or that to which circular perimetric action pertains, or isoperimetric action pertains.

Therefore, we have an *action* in relationship to a result. The result is *work.* The self-bounded circular area is the *work* accomplished by circular *action. Action, work.* We put that into the context of a power relationship. We have *power* as the rate in time, at which the circular *action* creates *work.*

Now, for example, the number of cycles per second, in terms of circular *action* or isoperimetric *action* creating circles: That is one way of measuring *work, power* of *work, power* to do *work*. How much *work?* We have a unit circular area, self-bounded circular area, and the number of units per second accomplished by isoperimetric *action* is a notion of *power.*

We actually don't measure this in units of simple space. In all important functions, we have nonlinear functions. Why they have to be nonlinear, why elementary functions are nonlinear, is already implicitly indicated in the maximum-minimum relationship. You have this creative characteristic of the sovereign individual, *imago viva Dei,* as is indicated in *In Defense of Common Sense;* it is a nonlinear relationship. So, the elementary form of existence of the individual, the elementary form of existence of the universe as a whole, is immediately a nonlinear process, a very special kind of nonlinear process.

For reasons previously considered, all relations within the universe, other than those which are simply the direct relationship of the individual, *imago viva Dei,* to the universe as a whole, are also subsumed by that same nonlinear function. Thus, the *most elementary form of substance* in the universe, the most elementary form of action, is of this nonlinear form. That, its elementary substance, is of this nonlinear form.

Thus, we must look at this circular imagery in terms of the *action* itself being of that nonlinear character, and the work accomplished as being of that nonlinear character. Therefore, we are speaking of the power of a form of action which has that nonlinear character. Therefore, we are dealing with a slightly higher form, implicitly, of that nonlinear process.

Then, we find, that that higher form is itself subsumed by that; so, we simply have such a kind of reflexive relationship. Since we can conceptualize the transfinite arrangement, which includes functions of different power (that it is on that level higher than one order of magnitude, or one order higher than the notion of power), that transfinite level, at which, at minimum, human creative reason functions, at which substance in the relationship between the Creator and *imago viva Dei* individual is located, is the level at which all laws of the universe are located.

What this comes down to, in the simplest aspect, is that we count *power,* and we count *action,* in terms of singularities, meaning the kind of singularities which are generated by multiply-connected, self-similar action, derived from the self-similar isoperimetric action of the most elementary kind. In particular, in terms of *power,* functions of power, we are looking at different rates or variations of rates.

So, we are looking at *rates of increase* of the generation of singularities, as that power function. That means, we go one step beyond the ordinary Cantor function in this respect— the Cantor function which pertains to the implicit enumerability of the density of mathematical discontinuities within an arbitrarily small interval chosen. Now, take the same interval, as we indicated earlier, and increase the rate at which these singularities have been generated for that unit. Then, the notion of power, as of the second order, as the rate of increase of that rate of generation of these singularities, becomes the immediate notion on which we focus.

In that area, in the still-higher ordering subsuming that concept, lies, at least implicitly, the proper notion of substantiality. So, instead of looking at a circle as a self-bounded singularity on a plane sheet of paper, so to speak, without looking at that piece of paper itself, the *substance* is the still higher ordering of power relationships, that nonlinear function, which we have just referenced. That becomes then, the functional notion of *substantiality.*

Now, let us just reference this to the Planck length. This would mean that the number of singularities contained within a sphere or cross-sectional circular area, or something approximating that, of that Planck length in diameter, would b increasing in the density of singularities within it. We are looking at increasing density of singularities in that illustrative sense. So, that is the essence of the nature of substance.

That illustrates to us rigorously, from an axiomatic standpoint, why no linear system of simultaneous equations or inequalities can represent anything actually in our universe; why all deductive mathematics and mathematical physics is intrinsically, axiomatically absurd.

The Case of Classical Music Composition

Continuing as before, switching momentarily to music to introduce another point relevant to art in general, and, more broadly, to creative reason in general.

In the case of classical composition, in the case of counterpoint (not in the sense of schoolbook texts, but strictly in the sense of principles—provided that this is based on the proper tuning, of course), there is a very elementary kind of illustration of the creative principle, from the standpoint of the representation in my *In Defense of Common Sense*, for example: the simple singularities, which occur as harmonic or rhythmical dissonances, not arbitrarily, but *generated* from the lawful elaboration. These dissonances have to be resolved. These are not resolved in order to reestablish the theme as subject of the composition. Rather, the resolution of the dissonances in this form, in well-tempered polyphony, *is* the subject of the composition. (See Appendix X.)

That is, the composition exists for the purpose of defining and resolving the dissonances. The solution to that, as expressed in respect to what is chosen as the thematic material, so-called, employed to create the ironies, becomes the composition as a whole.

Thus, the elaboration of the irony, the dissonances to be resolved, the treatment of the material afresh from the standpoint of this development—these complete the statement of an idea and present us with a creative discovery which is precisely analogous in that respect to a fundamental, valid, scientific discovery. It is not merely analogous, but employs the same faculties of the mind, maybe in a different mode in some respects, but the same essential faculties of the mind.

In creative scientific discovery and in the proper composition, performance, and hearing of music so performed, there is a distilled expression of the quality of emotion which is called sacred love, as opposed to profane love: *agapē*. The function of music is expressed by the correlation in that way of this keener sensing of this emotion of sacred love (*agapē*) with the overall process of development of a composition to encompass one or more creative discoveries, a development which is itself the composition.

This applies to poetry, from which music is derived; it applies to drama, which is a branch of poetry, in another sense; it applies to classical visual art, where the same thing is done.

Exemplary is the case of the work of Leonardo da Vinci (1452-1519), in whose work this particular implication of classical method is made explicit, as we have discussed, for example, in the case of the "Virgin of the Grotto." It exemplifies that sort of thing.

So, art and science are derived, contrary to Kant, from this same faculty, this faculty of creative reason, with these qualities. That is the point to be emphasized, particularly with respect to music, and also with respect to science.

For example, how does the mind actually know that it is coming close to a creative scientific discovery? Or how does the mind of the composer, the great classical composer, know that he is on the right track, so to speak, to a major composition, or toward something of the quality of a major classical composition? Or in any other classical work of art?

We find that even the successful composers and scientists are dreadfully lacking in certain kinds of what plausibly is the required knowledge to solve the problem they are solving. They solve it nonetheless. From the outside, people say, "Well, that is insight," as if insight were a magical quality, some "unerring instinct," so to speak, which guides them to a solution for which they have no explicit, formal basis for their solution as a whole. Something is added to the material they know, to cause them to leap, as it were, it appears, to the right solution.

We find that, particularly the great performer of classical works, is guided to the right interpretation, under the influence of a strong sense of sacred love. Whereas, the romantic is driven, as in the case exhibited most boldly by Wagner's famous "Liebestod," by nothing but the erotic emotion. The erotic is equivalent to linearity, to entropy; whereas, the sacred is, in a sense, explicitly equivalent to negentropy. It is by *following the pathway of negentropy,* to give the sacred love another descriptive form, that the discoverer is led to the solution.

It is more than just being led by following a trace, as of the trace of sacred love; along this track one finds sacred love. The driving motive of creative discovery, the motive which supplies the potential of the concentration span required, is the same quality of emotion. Thus, we see something here. The idea of beauty, as we associate it with great classical art, emphasizes an aspect of the creative processes of mind, which is otherwise essential to creative scientific work; this emotion we can associate with the word *agapē*.

So, we see, in even this aspect of life, in the relationship between the artistic and the scientific experience of the scientific worker, that the scientist *requires* classical art, including classical music, in order to be a better scientist. The experiencing of a form of creative activity, which generates beauty as the classical form of experiencing a stronger impulse of *agapē*, in the development aspects of the composition, is a strengthening, a well-source, so to speak, for continued, creative scientific work as such. Not only are the two based on the same principle; but the one is necessary to the other. A scientific sense, whether in the scientist or not, is necessary for classical musical composition, for example, as obvious for the case of classical arts, as Leonardo da Vinci and others exemplified this. The more essential thing to bear in mind, is that classical art is essential for the moral development of the scientific creative potential of the scientist.

This is not restricted to that. In every aspect of life, classical art is essential to enhance the experience and command of that which separates man from the beast. Thus, we give to this combination of classical art and this emotion the name *beauty*. In the truthfulness of this classical art, insofar as this art imitates creative scientific work by means of beauty, we have the equivalence of truth and beauty, and beauty and truth. So the function of classical art is essentially to give mankind an experience of truth and beauty, and beauty and truth in this way: to give mankind the light of this beauty, to illuminate scientific thinking, scientific potential, and, indeed, every aspect of life. So bury Kant.

On Natural Law and the Rights of Man

Let us go to the question of natural law as such. We have covered some introductory, axiomatic features of the basis for a hereditary, constructive approach, to a constructive physical geometry, consistent with Leibniz's and my own definition of monads. Now let us look at natural law in a broader sense, as it applies to political or historical processes, and see it correlate to that.

We have also considered art, an inclusion which gives us, in total, a general social setting of the individual.

This historical question brings us right smack into the middle of the principal topic of *In Defense of Common Sense*. That is, the significance of the individual's behavior, is the impact of that behavior on the enhancement of the survival of not only present and future generations of mankind as a whole, but also, past generations.

But first, to get that little irony out of the way: past generations? How so?

We are the past of our future. The question which ought to occupy our attention, particularly in light of the current and recent behavior of President George Bush, and some others, is whether the United States in the future, will survive. In other words, will the outcome of our having lived and acted survive?

In some degree, that question is left to the future, to decide whether we, in the past, their past, have survived or not. So, similarly, today, look back at the Founding Fathers of the United States. Did they survive? Did their principal work, the United States, a federal republic based on constitutional law (informed, poorly, but nonetheless definitely, by natural law, in the Augustinian, not Grotius's sense), survive? Well, of course, they died; but did they survive? Did their actions lead to a survival of that cause for which they acted? Were they fulfilled in the future? And for how long in the future? This is the meaning of, "Did the United States, for how long, survive?" The answer to that question might very well be "no" at this point.

So, we in the present, bear now and for the future, the responsibility for the survival of those our forebears. Clear?

You come from one or more varieties of any, say, ethnic extractions, from many parts of the world. Let us take the American Indian. Now, did the American Indian survive? Very interesting question. Do American Indians today, or any particular American Indian, play any important part in the survival of the human species? Are they essential to the survival of the human species? Well, there is some doubt of that; obviously, some Indians have; but, in general, the great majority of Indians today, those who are confined to reservations, are denied the right to the survival of their ancestors. That is, after all the killing and the starving and the dying, and all these kinds of things that went on with all these people who once roamed the forests and plains and so forth of this nation, this area; did anything good come out of it all? Well, that is placed in doubt, isn't it? Shall we say, to make a pun, which is a rather cruel pun, perhaps, but appropriate in the circumstances: Did the American Indian survive? Did it all amount to anything? We could say, "One must have reservations on that subject."

It is a very important question. Not only is it a practical question, but as illustrated by the case of the American Indian, it is a very poignant question. Not only did the United States survive, but did the entire American Indian population, as an American Indian population, survive? Did it produce something of lasting value, as the Cherokee nation tried to produce before that great Democrat, Andrew Jackson, committed his genocide, his Nazi-like crimes, as the Cherokees would rightly view him? When the American Indian encountered European culture, did they assimilate its best component, did they rid themselves of barbarism, to bring out that which is the best in them, in conjunction with the European culture they encountered? And did these American Indians thus go on to play, at least in proportion to their numbers, an essential role in ensuring the future success of the United States and the survival of the human race? They mainly did not; they were denied that.

Ah! Therein lies the essence of a human right. And therein lies more than a right for the American Indian, descendant of those forebears—and responsibility thereof as well. Therein lies the key to the whole question. *Are you given the right to be fit to survive?* Are you given the right to do something which will contribute to humanity's survival, in the present, and the future, and the past?

This is not giving a loaf of bread, as such; this is not producing something. This is contributing something culturally, to the advancement of culture, in some way. Even the raising of a child, who might become creative; or whose children in turn, their grandchildren, might become creative. Even that is a contribution. Were you given the right to do this? Have you done it, if you were given the right to do so? What does it mean, to be given the right to do so? Doesn't it mean an education, doesn't it mean the social environment which is at least somewhat conducive to that? I don't mean a privileged social environment, I mean one in which you are not yelled at so constantly that you can't think; not living in a neighborhood where it is so noisy with screaming and screeching and yelling all the time, that you have no rights to think: the typical victimization of the black or Hispanic ghetto, our slums today. Give them every material right, in one sense, but let the yelling and the screaming and the howling, the noise-making go on; nothing good much can come out of that.[2]

So, these are the kinds of questions we have to consider in general.

Now, let us look at this as a matter of principle, as we do in *In Defense of Common Sense*.

The test of the rightness of an opinion is that it must be *more* than an opinion. A mere opinion is worthless. Any man's opinion, insofar as it is merely an opinion, is worthless.

We see this illustrated today by Project Democracy. Project Democracy is a fascist movement. It was called, in the early 1970s, "fascism with a democratic face"; or "fascism with a smiling face"; or "fascism with a liberal face." It is a "democratic" form of fascism. The content of Project Democracy's policy is fascism, in the sense that we use the term fascism for the policies of Adolf Hitler. It is based axiomatically on the theory of opinion, that there is no right or wrong, which is how fascism crawls in through all the windows and doors—"because there is no right or wrong, don't you see?"

Jeffrey Sachs, who is, in fact, a fascist, who teaches at Harvard, and who is imposing fascism on the Poles, can argue, under liberalism (under liberal democracy), there is no right or wrong; there is only opinion. There is majority opinion; there is authoritative opinion; there is a consensus. A consensus doesn't say something is wrong; it is not wrong, you see: *"Alles ist erlaubt"*: "All is permitted." A game of power. A Nietzschean game of power.

Who has this power? The bankers; the government that works for the bankers; the thieves; and those who work for them, and so on: fascism. But on the surface, it is democratic.

Do you wish to express publicly an opinion that there is a fascist government in Washington or fascist policies of the government in Washington? If you do, you'll be victimized by the Department of Justice.

Now, express your opinion. All say what your opinion is; stand up and say one after the other, what your opinion is; what is the majority opinion here? The majority opinion: It were discreet to support the present fascist policies and method. Ah! We have, at last, achieved a democratic fascism.

In the longer run of things, a nation which does as the United States under Bush is doing today, will not survive. A wrong opinion, if it prevails, ensures that sooner or later, that nation will be exterminated. The great debate today is which of the two superpowers, the Anglo-American or the Russian, is Sodom, and which is Gomorrah? At present, if we project the outcome of their present policies, their present cultural policies, as well as their economic policies, their political policies, including legal policies, we project that the United States and Soviet Union, the Anglo-American power and the Soviet power, will not survive. And therefore, all who contribute to the present policies or the present administration, are persons who have rendered themselves morally unfit to survive, by virtue of the fact that the net effect of their existence, is to render the nation unfit to survive and to cause it not to survive.

That is the essence of natural law. Natural law pertains to the sacredness of individual life, by virtue of nothing else but the sovereign individual potential for sovereign, creative reason.

The capacity of the individual for opinion is not sacred. It is not worth a damn. Experience of the individual, as knowledge, as mere experience, is not worth a damn. Democracy is not worth a damn, at least as it is fabricated by the Anti-Defamation League's Carl Gershman, the nominal head of that fascist Project Democracy, which gave us the drug-running Contras, among other things.

What is sacred is creative reason, as a sovereign potentiality of the individual person. The worth of this person is the degree to which he or she develops that reason. That reason is expressed, in practice, by its production of the means for the survival for the entire society, past as well as present and future. Thus are right and wrong and law defined from this standpoint.

That is natural law. That is the law of the Creator, which we know, not because it has been dictated to us or been revealed to us in a dream by our Creator; but, because it is written on the face of the universe, that whoever violates that law shall bring about, by means of the law they violate, their own destruction.

VII | Self-Conscious Reasoning

On the subject of creativity per se: The great difficulty which I observe in discussing this subject is, that most people lack a conscious referent for it. I have discussed many aspects of the creative experience, that is, an empirical experience against which these kinds of concepts can be contrasted, as were it an experimental method of approach to the subject.

If you describe *agapē*, the kind of emotion involved, and recognize it in some sense, but only as a member of a listening audience to music, not as a performer or composer of music, the focus is on trying to experience the feeling, a focus which may lead sometimes to manic excursions, trying to intensify the erotic, trying to turn a surfeit of profane love into sacred love. It is quite something to watch, something I prefer not to watch.

But the obvious point is, that one must set up experiments, which define the difference between the two states. I can indicate from experience some of the correlatives, the preconditions, the circumstances, the conditions which one must more or less consciously, explicitly, impose upon oneself in order to generate creative thinking. That is, generally creative thinking, as distinct from the deductive and other banal types.

To actually experience it, however, and to be able to look at it self-consciously, as I shall indicate, is another thing than to describe it. Before one really knows what it is about, one should experience it wittingly, consciously, rather than merely attempt to describe it as in a faithful classroom academic exercise, describing accurately something one does not really know: typical university occupation, even in my days and more abundantly so since.

There are several things to be considered. Let me address the emotion, the sacred love, the intense feeling of sacred love, which is always associated with the creative act. It is, as a matter of fact, the emotional state one must muster, or must be found to have mustered, before one is going to go any place with creative insight. It comes sort of intensely; it is sometimes described as a light turning on in one's head—the sort of emotion that goes with that.

People will sometimes attribute that mistakenly to different kinds of experience; but that is a fair picture of this agapic emotional state, sacred love, as distinct from profane: the most intense experience in music, the non-erotic, the non-Wagnerian, non-romantic experiencing of music.

The other thing to bear in mind is, this doesn't work without *self-consciousness*. As some will recall, a couple of decades ago and earlier, I placed great emphasis on this business of self-consciousness. I addressed how this self-consciousness might be achieved, how one could enforce it; some of us conducted experiments in group discussions, as part of our effort to try to understand these matters, and there was some comprehension realized in this way.

The essence of the creative method essentially is non-linear.

You conceptualize your own state of mind and you conceptualize it in a Socratic way, such that you don't simply admire, accept, the state of mind of yourself that you are observing, as if it were a hero in a drama, some silly soap opera or something.

You look at it critically. See your own follies; see the assumptions you make, as in tragedy. One might say, "Have consciousness of your ordinary conscious states, as you would of the progress of a tragedy; looking at your everyday self, your ordinary self of the classroom, or whatever, as you would look at Hamlet, for example." Then you are trying to be conscious, to have insight into Hamlet's mind, as you are observing it. You, being Hamlet, and using the knowledge that you have, as being yourself, and therefore having access empirically to everything that is happening to the Hamlet inside you, the one who ordinarily speaks, you can look at some of the assumptions that Hamlet is making, in order to behave the way he does, or to justify, or to perpetuate the way he behaves as he does. You can look then and see what the alternatives are, as to how you might change Hamlet's assumptions to cause him to behave differently.

In that kind of simple self-consciousness, two things happen.

The location of consciousness shifts from ordinary consciousness, the reacting, as the student in the university classroom, for example, reacts in answering an examination. Usually, the student is reacting, is generally not thinking, but reacting at a lower level. Now, put yourself up to a higher level, look down upon yourself being that student, and describe to me what is going on in that student's mind, why the student is reacting the way he is, and what would cause his mind to behave differently.

Simple, very simple kind of thing. The important thing is not to get completely distant from the subject, the student as the subject of the examination; the important thing is to change the subject, into becoming us, the conscious self that is looking at the student's conscious self. The important thing, then, is to place the importance upon achieving the corrections to be made: *simple self-consciousness.*

It is only in that state that any creative work can be done.

For example, to look at this experimentally: Reference my *In Defense of Common Sense,* the way I structure the argument there. If you can look at the student's mind as the mind of Kant, as I do, in *In Defense of Common Sense,* now you are criticizing the fallacy of Kant, you are criticizing the fallacies of the student. In that way, you shift the "I" from the student to the one who is looking at the student, looking at the Kantian; and what I described in *In Defense of Common Sense* as the problem to be solved, to be addressed there, is exactly what you must do in self-consciousness.

Now, with a certain quality of zeal and a determination to persist, without losing track of what one is doing, the result is a movement toward the kind of concentration which, extended over days, weeks, whatever, leads to creative discoveries.

I rather think that people have not only to do creative things, but have to accomplish them with aid of the viewpoint that I have just indicated, before they really know and understand what I reference as the creative processes. Then one sees, or you should be able to see, from what I have said so far, if you meet all the conditions that I have indicated, that you have an experimental setup, so to speak, in which you can begin to isolate the critter, with which you can begin to look at the creative process.

You also can see, for reasons already given in *In Defense of Common Sense,* and re-emphasized from a different, fresh standpoint here, that creative thinking is intrinsically nonlinear, as I have described nonlinearity, in describing the laws of the universe, here; that this indicates that the self-consciousness involved is nonlinear. And since it is only from this standpoint, that the laws of the universe can be comprehended, for reasons already given, then the laws of the universe are elementarily not simple, but nonlinear of this negentropic form.

You have essentially, therefore, a universe which is not entirely dissimilar from Kepler's (1571-1630); in which the characteristic of the universe, as Kepler's model implicitly states, is negentropy, rather than entropy, and in which the Second Law of Thermodynamics is not tolerated, except in the loony bins of society.

VIII | Is There an American (Protestant) Ideology?

I note the fact that this section is drafted on Sunday with a certain irony, because I refer inclusively to what is sometimes named the problem of the American Protestant ideology.

Contrary to the admirers of Adam Smith at the University of Chicago, such as Thorstein Veblen, the Rockefellers, Teddy Roosevelt, and so forth, the American Protestant ideology is not the hallmark of economic success which Teddy Roosevelt's New Age cult professes it to be. Let us look at the aspect of this which is relevant to this Project A.

It has been my not-uncommon experience in past years, in speaking to some Americans, to speak of the fact that the financial system is collapsing or to offer a list of catastrophes to indicate the way and the approximate time-frame in which they may be expected if the United States continues its present course of action. During this, in some of these cases, some of these fellows will interrupt me to say, "Yes, I agree; yes, I agree; yes, I agree."

I say, "If you agree, what are you going to do about it?"

"Oh, I'm going to wait for the Rapture, I don't have to worry about all this. What you say is all true, it is all happening; but I don't have to worry about it; I'm going to be raptured."

That is one aspect; it is an extreme, if actual case, also a very illuminating one. That illustration goes to the core of the matter; it goes to the core of the worst, radical version of Calvinist dogma, and of related, radical forms appearing within Protestant dogma. The worst version of Lutheran dogma is a variant on the radical Calvinist dogma and is the same thing, in effect. The typical such American radical Protestant follows Adam Smith, as does the Quaker who refuses military service: "The larger matters and the larger consequences of my behavior, I leave entirely to the ministrations of God; they are beyond me. And God will decide that, I have nothing to do with it, I have no responsibility for that. I," he says or she says, "am responsible only for my immediate personal affairs: my happiness and that of my family and friends, my wealth, security, and so forth; and my personal dealings."

The essential character of the American Protestant variety of predestinationism, of the rapture variety, of the radical Calvinist variety, or of the Quaker variety, or of the Lutheran variety, is an essential, underlying immorality: a refusal to recognize that individual behavior has something to do with the ultimate consequences of the present for the future society, and that acts of omission are as much acts, as acts of commission, at least in many respects.

So, the failure to recognize this is typical of the American ideology.

Therefore, I must say to my interlocutors, "You are each responsible for the outcome of your nation's future."

They retort, "What kind of nonsense is that? I reject that," they will say. "I reject that. That is your opinion. I reject that. I take care of my personal affairs, I'm a moral person, and these are the matters I have no control over. I'm not responsible, I'm not responsible, I'm not responsible."

Yes, they are very irresponsible. This irresponsibility is the characteristic of the so-called Rockefeller variety of American Protestant ideology, and its secularized expressions. This is not necessarily an irredeemable feature of the American, but it is a widely prevailing viewpoint which imposes itself upon many Americans and which, because it is *popular,* is deemed acceptable and authoritative widely.

The key to the weakness and the stupidity of the Americans is the term "popular," and the equation of the term "popular" with democracy, "consensus politics," and so forth; that truth and falsehood are rejected and moral value is put on the "consensus," "popular," "majority," and so forth, even though popular opinion usually happens to be wicked or merely stupid.

Thus, the essential thing which keeps nations going in times of crisis or great travail, is the role of the individual or small group of leaders in taking the leading position—the others, recognizing that they are morally obliged to act similarly. That is lacking, in general, among Americans, at least most of the time.

So, the essential part of history is rarely understood by Americans.

At the same time, the importance of the individual is not understood. The so-called Rockefeller type of American Protestant, is not really a Christian.

Christianity, if we take the Gospel version, for example, or the New Testament Epistles, quite explicitly casts man in the image of the living God and does this in respect of creative reason. The Christian is responsible, in the sense that Cusa describes the relationship between the maximum and minimum, and so forth, with corresponding implications. The essence of Christianity is that "I am responsible; I have potentialities, which I am obliged to develop, to the degree of need about me; and, I am responsible to apply those developed potentialities to better the condition of mankind. I am an instrument; I am responsible; I am the agency."

For example, another expression of this cited pathology: "We must meet our responsibilities; yes, we must *pray* for the right outcome." Pray for the right outcome? By what means do we propose that prayer will prevail in inducing this right outcome? The Christian retorts, "Prayer must, among other things, summon in me the strength to become the instrument, the solution."

So implicitly, the misled American Protestant of the type I have described, does not accept the implications of the divine spark of reason, of *imago viva Dei*. They may accept it, in one sense, in one degree; but they haven't made the connection to individual responsibility, the universal responsibility of the individual, and, thus, the universality of the individual.

Thus, in both of these cited varieties of regrettable tendencies, we find the inclination to a false, anthropomorphic theology: God as an anthropomorphic being; and He is portrayed in what is sometimes called an "Old Testament fashion," in the sense of being some kind of a capricious Mesopotamian potentate, whose laws are known to us by dictate and are arbitrary: "It is not for us to know; it is for us to accept revealed instruction: not to accept knowledge, the responsibility of knowing."

Thus, you have the American populist. The populism and the Protestantism of the type I have described, interface.

This is not to imply that all American Catholics are virtuous heroes; Pope Leo XIII referred to "the American heresy"; this problem has been referred to from Rome many times. All too frequently, the professed American Catholic is not necessarily a Christian, even though many of them would like to be called such.

So, we are not just picking on the Protestants; we are looking at a phenomenon; we find the same phenomenon is characteristic of what Rome has often described as defective American Catholic behavior. Such errant American Catholics are defective and tend to be gnostic and heretic, precisely to the degree that they imitate, all too often, the New York City Episcopalian of the present New York City Cathedral of St. John the Divine. It is to that extent the American Catholic tends to be not a Christian, as the Bishop Paul Moore type of

Episcopalian is not a Christian, but, is, rather, a gnostic and sometimes veers, as the Lindisfarne crypt of St. John the Divine does, toward outright satanism.

So, this is the problem we have to face in ourselves axiomatically, as those of us who are exposed to it in the United States; and also in Europe, for example, in those of a Kantian inclination. For both, it is the same problem. The Kantians are immoral; they are professedly irresponsible, as the overtly anti-Christian Adam Smith makes a point of it; and as Jeremy Bentham (1748-1832), after Smith, makes the point much more clearly and much more nakedly.

These are the problems we face in pedagogy, even in a preliminary way, in approaching the subject of natural law in the United States. You are talking to Protestants and to Catholics who are Protestantized, and so forth. Among many Jews, the same thing or even worse (cabalist lunacy). That is what we are dealing with.

Now, you say, "These ideas of LaRouche are not popular"; recently, truth itself is not popular in the United States. Everyone says, "Well, I'm telling the truth"; but most of them don't know what the truth *is*, so how can they be telling the truth?

Worse, not only do they not know what the truth is, but they are not truthful. That is, their errors do not flow from a method which is seeking truth. They may think they are seeking truth, but they are not looking for truth in fact; they don't accept truth in fact. Instead of truth seekers, they are poor pragmatists, who would rather seek ideas that are popular or presumably will be popular; they test the merit of ideas by their actual or implicit potential popularity.

The lower, animal type of belief, which is the typical American popular-opinion level of the mass media, the bite-sized opinions expressed by the mass media, the buzz words, all that nonsense, that level of animal-like thinking, is characteristic of most Americans most of the time. Equate us with another type of thinking, which is truth-seeking, Socratic, critical, in the sense we described earlier; what we represent is generally rejected among liberals. So, when we are trying sometimes to be "popular," or we are being instructed how to become popular, how to become influential through becoming popular, we destroy ourselves; we become less than ourselves and we fail.

The strength of my friends' association lies, regrettably, but unavoidably, in our being often unpopular, because of our adhering to truth and truthfulness. We follow the truth where it leads and we find that society is sick. We find the society riddled by qualities which are rightfully subjects of scandal. We find people who call themselves liberals, who, in point of fact, through economic policies, are greater mass murderers than Adolf Hitler. And so forth and so on.

This is the kind of society in which we live; and, we, to the extent we follow the truth, and, by the path of truthfulness, make ourselves unpopular. But, by making ourselves unpopular in that way, in the service of truth, we touch that aspect of our fellow human, even of our adversary, which is human, which is *imago viva Dei,* which really seeks the truth, which seeks the path of truthfulness and which knows that it must combat the degrading impulse to be popular, as a whore is popular.

One must fight the whore in oneself to face the sometimes dangerous pathway of truthfulness, and virtue—*virtù,* in the classical Italian sense.

So, that intermezzo is added.

IX | Determinism and Matter

Thus far, I have defined the universe as a whole, repeatedly, as *elementarily* "nonlinear," and, *yet not simple. Not simple:* That is the essence of the matter.

The nonlinearity exists *primitively,* that is, *elementarily,* only in the whole; that is, in the universe as a whole. This character of the universe as a whole is expressed for our knowledge and in practice, in what I have referred to for purposes of metaphor, as the *unmediated* relationship between the sovereign individual's creative reason and the universe taken directly *in its entirety:* not part by part and not as a sort of a philosophical gas system.

In other words, the meaning of "elementary": The process of division into ever smaller parts (again, the Euler problem), does not signify that we are approaching elementarity. The monad is not elementary because it is small; it is not elementary in the sense of being a building-block. The character of the monad lies in its relationship, its direct relationship, to the universe as a whole. Therefore, the little monad is as big in this respect, *in this relationship,* as the universe as a whole.

Elementarity, the elementary, indivisible building-block of the universe *as a whole,* is the universe as a *whole.* Relationship in the universe, is defined elementarily by the relationship between the creative processes of mind, as in valid, scientific discovery, and the universe as a whole, through the action of such discovery upon the past as well as the present and future generations of all mankind, and through the totality of human existence, so represented upon the universe as a whole. Thus, also, the line is related to the universe as a whole, since that which is adduced by creative reason, is the ordering of the universe as a whole.

This lawfulness of the universe, taken as an essentially indivisible *oneness,* must include all of the changes in the universe of which mankind's creative powers shall ever become capable.

Let us look at Kepler's construction of the solar system, as opposed to the unworkable, and obviously fraudulent, Newtonian construction. (See Appendix VI.)

In Newton, we have the three-body problem. Why do we have the three-body problem? Because the relationship among bodies is determined, in reality, by the curvature of physical space-time and not by the relations among bodies pair-wise, as in Cartesian notions of matter, space, and time. Therefore, for that reason, the three-body problem rightly does not exist, in the sense that there is no solution to it, because *the solution requires another consideration not advanced by Descartes or Newton, which is the curvature of physical space-time.*

On the basis of the evidence developed by Leonardo da Vinci et al., Kepler strikes upon the nature of the curvature of space-time and shows, that all physical laws in the universe are derived from physical space-time. In modern language, that is what Kepler is saying implicitly. He says similarly: Because of the relationship between the creative powers of mind (as an example of the living processes made self-conscious, efficiently self-conscious) and the Creator, the universe must necessarily be founded on a principle of least-action, consistent with what we would call in modern language, *negentropy,* negentropy corresponding to the harmonic orderings congruent with the Golden Section, living processes. (See Appendix III.)

We see, for similar reasons, that creative mental processes, in the sense of any hereditary construction principle, will be ordered, in respect to that construction principle, in terms of a similar Golden Section harmonic ordering. Or, at least, we can show in respect to this the necessary effects of the realization of such creative discoveries.

So, in this respect, mankind is not only acting upon the universe in a practical way, through scientific discoveries, in changing the mode of behavior, as behavior on nature; but man is also acting upon nature by understanding the laws of nature. (See Appendix XIV.) To understand the laws of nature, even though the practice which we referenced is human practice, nonetheless, what we are referencing directly by means of human practice, by the reflection of human practice, is the laws of the universe as a whole. Directly. So, man's mind, the creative processes of mind, are directly related to the universe as a whole and not *only* through the action of mankind as a whole upon the universe as a whole.

These are the kinds of distinctions.

Then, again, as we said before, to the same effect: Given, let us say, a monad, which is not an intelligent monad floating around in this process, we do not substitute, suddenly, a pair-wise relationship among monads of this sort, to account for their behavior. This is not a situation where we have on the one side higher monads, which are directly related to the universal, whereas there are the lower monads which are not, because they lack this creative quality. Rather, the universe as a whole is so constructed, that the pair-wise relationship of these lesser entities must be congruent with the nonlinear lawfulness which characterized the universe as a non-simple elementarity: i.e., universal space-time curvature.

This is obvious in the case of Kepler. Kepler discovered, wittingly, a law of gravity, which he regarded as, probably, an electromagnetic principle. We can understand that today; we may not have solved all the problems of correlating the strong forces of gravitation with the relatively weak forces of other electromagnetic aspects of the matter, except as we introduce negative curvature. Then, suddenly, we are required to get into strong forces, relative to what we call weaker electromagnetic forces and, therefore, we see a necessary geometry, even if we have not resolved this satisfactorily, experimentally. We can see a direction in which to go. But Kepler, identifying the electromagnetic principle as the relevant one to this phenomenon of gravity, caused by the curvature of space-time, was on the right track. He did not, at that point, tackle the difference between relatively strong and relatively weak forces or things of that sort.

When you turn Kepler inside out, as, most probably, Hooke and others did, in respect to the work of the reductionists of the seventeenth and eighteenth centuries, you see, as various fellows understood this, including Planck, that you can directly derive from Kepler's Laws all the expressions used in Newtonian physics, simply by an algebraic manipulation. But how did Kepler develop that from which this Newtonian schema is derived by a reductionist manipulation, algebraic manipulation?

Kepler derived it from a principle which is consistent or coherent with what I'm arguing in respect to elementarity, which is ontologically nonlinear, not simple.

So, what we are seeing with the Newtonian ratios are nothing but the distorted shadows of actual knowledge, the actual knowledge being the Keplerian form, and the Newtonian merely a shadow. (See Appendix V.)

We see the same thing in Galileo Galilei (1564-1642). Galileo was informed of Kepler's work and parodied it, with corruption, to assert things which he did not actually, empirically, prove; but simply to show that, in effect, he could have claimed to have discovered empirically what he did not discover empirically, and, thus, show that Kepler's method was not necessary; was, in other words, superfluous.

X | What Is Change?

It is most useful to consider an apparent anomaly at this point. The anomaly is: *the action which results in no action*.

Again, let us reference *In Defense of Common Sense*. Let us take any of the axiomatic systems of hereditary principles A, B, C, D, E, and so forth, respectively. To any among these, if we supply any action to be interpreted by, say, hereditary axiomatic system A, there will be no theorem generated by that action which is not consistent with the axiomatics of A, the hereditary principle of A. Similarly, for B, C, D, E, and F.

From the standpoint of A, that example, an anomalous aspect of an event which differs from an acceptable axiom of A, or is inconsistent with A, will simply be disregarded as an erroneous, non-real occurrence.

In the practice of science, this treatment of anomalous reality appears all the time, or nearly so. People say, "Well, it didn't occur, it couldn't have occurred, because. . . ." When one sees several times, that something anomalous did occur, instead of rejecting the event because it is not consistent with the hereditary principle, one might, rather, realize that the hereditary principle is flawed, by virtue of the recurrence of the anomalous event.

One tests this, very simply, by proving that the alternative system, the alternative axiomatic system, generated by accepting the actuality of the anomalous event, generates a network of theorems which is consistent with the physical evidence, more consistent than the replaced or superseded axiomatic system *A*.

So, in the first instance, when we reject the aspect of the event which cannot be rendered consistent, we have *no-change:* We have no acknowledged result.

There is another aspect to this, a higher form of no-change or *change that is no-change, but is also change.*

Take the same array, *A, B, C, D, E.* The event that causes the scientist to generate *B* as a successor to *A,* is of a very precise form. That is, even though there is no point of consistency between *A* and *B,* we can define the inconsistency. We can define this geometrically; we can provide a locus definition, which gives us an adumbrated algebraic definition, and so forth and so on. So, no event but one which is consistent with that difference will carry us from *A* to *B,* that is, will generate *B* out of *A.* Anything inconsistent with that inconsistency, would either lead not to *B,* or, if it is required to lead to *B,* will tend to be ignored. If the latter is not ignored, it will lead toward a completely different axiomatic system, which then comes under the same test.

Now, let us apply this principle to political and social processes and events.

We have this all the time; we have these kinds of envelopes all the time. Within limits, once something that can be represented as axiomatically determined in the course of events in process, any event, within certain bounds, introduced as a novel event to that system, will lead to the same general result as any other such event. It makes no difference what the choices are *within those bounds.* We will still end up with the same general outcome.

Now, for example, let us take simple economic forecasting. In the recent period, at every point we were looking ahead, someone said to us, "When is this going to happen?" In response to that query, we could list an array of events which will be the probable, mutually exclusive alternatives.

Now, in each of these cases, the event is a crisis, which takes different forms; but all of the forms add up generally, within certain limits, to the same result, even though they are different in detail.

It is one of the difficulties of forecasting, that it is more difficult to cause the layman, even the informed layman, to understand such a forecast and its significance, than it is to construct such a forecast (at least for me, an old hand at this sort of thing).

They rebuke me, "But that is no forecast. Which of those is it?"

I say, "It could be any number of them. But they all add up to the same thing." And whatever that is, when it happens, will cause another series of complementary events, which, whichever route is taken in detail, will add up to the same general thing in turn.

So, for the most part, we have systems which don't change. They change, but they don't change. And even the change may change, but without changing. *What we can forecast is that which does not change, the invariant, common feature of a variety of alternative sequelae.* That common feature is forecastable. If the common feature is disaster, then we can forecast disaster. But we cannot forecast in exactly which form the disaster will occur, because we don't know, in advance, which of the alternative routes will be taken, willfully. But once we forecast the disaster, we can examine the disaster, in all aspects and find how the characteristic, which is disaster, will determine a characteristic sequel.

We can determine also something else, which takes us to *how to change the no-change.*

By looking at that which must be done to get us out of this kind of limitation, these kinds of boundaries, this trap, we select a course of action which takes us into new dimensions, which changes the characteristic of the event. Either we wish the disaster, in which case we don't try to change the characteristic of the event, we don't try to change that sequence; or we don't wish that latter outcome, that characteristic; in which latter case, we must select only events available to us which will cause a different characteristic to emerge.

So, then we have the boundary conditions within which certain events lie. These events mean, effectively, no-change, which is of one order or another. Any of these events is somewhat interchangeable; not entirely, but somewhat, at least in terms of that general thing which may be most significant to us in the result.

But there are also events which lie outside that narrow domain, outside these more restricted bounds, which can produce a different common characteristic of an alternative set of events, than the first case. That is the way we have to look at not only political processes, but, that is the way in which we have to look at physical processes.

XI | Change and No-Change

The subject is a further examination of the concept of "change/no-change."

The question of change arises on, immediately, two levels.

First, there is simple change in experience. On this level, any event, or the lack of an event when it is to be expected, is a change. What we are looking at, at a higher level, is changes in the way we think.

This is most simply illustrated, from a deductive standpoint, by the fact that a logically consistent, deductive form of thinking, is always based upon a set of underlying axioms and postulates, both stated and conscious and explicit, as well as unstated or unconscious. Thus, the events which are of significance, as in the case of scientific discovery, are events which are not consistent with a theorem which might be derived from a given set of axioms and postulates.

So the practice of science and serious statecraft, i.e., something higher than the politics practiced in Washington, is ordered.

In both science and politics, the object is to increase the per capita power of society to exist and develop. This is the proper object of science. The essential difference between one set of underlying beliefs about science and another, from a practical standpoint, is which set of implicit axioms and postulates guides us to greater practical power over nature, per capita.

The same thing is true in politics: Which set of underlying political principles, notions of the nature of God, man, and nature in general, guides us to form some practice which corresponds to an increasing power of mankind over nature and the increasing security of a society?

So, it is on the second level, of changes in axioms and postulates, at least implicitly, to the purpose of increasing the per capita power of man and society, that our attention ought to be primarily focused, rather than on the inferior level of the simple response to judging of simple experience. Hence, those aspects of experience which do not challenge the existing set of axioms and postulates, belong to the area of no-change, even though there may be change involved, of course.

This involves also, as we have already indicated, the case in which an event has occurred, which is anomalous, and which thereby would tend to require an overthrowing of existing sets of axioms and postulates, at least implicitly so. But, we refuse to recognize that event; or, we refuse to recognize the aspect of the event which represents this challenge. We do so in order to defend the system of axioms and postulates in use, against the threat which is represented by this anomalous event, or the anomalous aspect of an event (which is otherwise tolerated). *So, even though a change might seem to be required by the anomalous event, no-change occurs,* because the mind refuses to acknowledge the anomalous aspect of the occurrence, or relegates it to some mystical realm, for which the conflict between the event and the axiomatic assumptions is reduced, as in the case of Descartes, who simply takes everything which is disagreeable to a radically reductionist standpoint and relegates it to the mystical domain of *deus ex machina*. Thus, the problem is defined; and thus the importance of this subject of change/no-change in these discussions.

XII | Self-Consciousness Again

This is on the subject of *self-consciousness*.

We have outlined a schema, as we have in my *In Defense of Common Sense*. We have indicated some transfinite levels, *transfinites* in the sense of the term used by Georg Cantor, the famous mathematician of the nineteenth century, but also in a broader sense, although not inconsistent with Cantor's usage.

We have, first of all, the level of simple experience: Consciousness of simple experience is one level of transfinite: consciousness of events happening to one, not merely perceptual consciousness, but consciousness of a theorem-ordering, that something is happening. One thinks in terms of causalities, simple causalities respecting day-to-day experience, or assumed causalities. That is the simplest level of true consciousness, apart from mere awareness/perception.

Then, we have the second level, which we introduce forcefully by aid of a negative view of consistent deduction. That second level is the Kantian level: the fact that all mutually consistent theorems, i.e., perfectly consistent deductive mathematical physics, can be reduced to a set of underlying axioms and postulates, which axioms and postulates combine to represent what is called the hereditary principle. That is, that no theorem can be constructed in deduction, by deductive means or otherwise, which is not simply an elaboration of something already asserted implicitly in the hereditary principle in the underlying set of axioms and postulates.

Thinking about the changes from one such set of axioms and postulates to another, is the second or next higher order of consciousness. By thinking about that, we mean thinking about some notion of an ordering of change from one set of axioms and postulates to another set of axioms and postulates, and thinking at the same time of the changes in our notions of causality, on the simple level, simple causality, which are accomplished by these changes in choice of set of axioms and postulates. We also are thinking about the evidence in the empirical realm, which might be called *crucial experimental evidence,* which compels us to see a flaw in assumption within a set of axioms and postulates and thus forces us to reform our axioms and postulates, to generate a new set of axioms and postulates. So that is the second level of consciousness, an awareness of this.

The second level of consciousness also includes the notion that there is an inherent ordering, which ranks one set of axioms and postulates as higher in rank and order than another. This notion of rank is inseparable from the notion of power, which is why and whence my work in Physical Economy comes directly into play as reflecting the essence of philosophy in this matter.

The notion of *rank* and *power* is associated with an increase of the power per capita of the human species to survive successfully, which means to continue the development of that power.

This takes us to a *third level of self-consciousness, which is looking down on the level of succession of the sets of axioms and postulates.* On the third level, we are into the realm of true Socratic thinking, in which we are not merely negating the errors, obvious errors, or reducible errors in a set of axioms and postulates; we are now looking at the ordering principles, the choice of ordering principles, by means of which we might order progress among alternative sets of deductive axiom and postulate arrays.

So there is the third level of self-consciousness.

In each of these cases, what is involved is consciousness; that is, our actual human consciousness, as an individual, taking our consciousness on the relatively lower level, as a *subject* of consciousness, as an *object* of consciousness, and thinking about our thinking. That is, going to a higher level, to thinking about our thinking on a relatively lower level.

So we go from the level of thinking in terms of simple causalities, subsumed by only one set of axioms and postulates, to thinking about the differences in notions of simple causality associated with an ordered change in choice among implicitly alternative sets of axioms and postulates. That is the second level.

On the third level, we take the activity on the second level as the object or subject of consciousness; and we might also think about, on the third level, our consciousness on the third level.

Thus, as long as we are able to do that, to achieve these three levels of consciousness, as consciousness, by no mysterious means, no mystical means, nothing more than precisely what I have described in essence, we have two results. We can master our fate to a large degree, as we are not compelled to follow blindly the current consensus of the Bush administration combination. We can choose sanity, we don't have to put butterfly nets around ourselves, as most of the Bush men should be doing. We can also conceptualize creative reason as a consciously comprehensible form of human thought and activity. We can do what Immanuel Kant, Descartes, and Aristotle could never succeed in doing and which they denied could be done; but, we can do it.

This latter is obviously what was done by all the greatest scientific discoverers and greatest artistic composers. Whether or not they were fully aware, in the terms I have just referenced, of what I have described, they practiced consciously what I have described.

This is related to something which can be called spiritual; that we all have, in a sense, two natures.

We have one nature, which is essentially below the belt, including treating the mouth and sense of smell, in terms of aesthetic aspects of the mouth and sense of smell, as upward extensions of the gut. That is the lower level. That is the level on which man is closest to the nature of a beast. He is a little bit brutish, bestialized, shall we say; he is egotistical in the narrow sense; he is a *pragmatist,* which is a form of bestiality.

Or, man is on a higher level. On the higher level, man is simply thinking and locating his or her self-interest in terms already referenced. Man is locating himself or herself as a sacred individual, as in the image of the living God, as the embodiment of a sovereign quality of potential for creative reason, in which self-interest is associated with the discontinuous development of that potential. The development of that potential is associated, not merely with the progress which enables mankind to increase power for the survival of the human species, but is located, as we have indicated, in a conscience-strickenness respecting one's debt to past, present, and future humanity as a whole, and respecting man's role as a species, as a servant of the Creator in respect to Creation as a whole; that we are responsible to the Creator to assist in the process of continuing upward Creation.

Once we locate the meaning of our individual lives' soul, then we look at what we are thinking, as well as what we are doing. From that critical standpoint, that enables us to say, "Is our belief correct or is our belief absurd?" as opposed to the person who says, "I was raised that way and I'm going to believe that way until I die," which is not very intelligent, is it?

One says, "I believe that way, not merely because I was raised that way, which was an advantage to me; but I have come to understand why this choice of Christian civilization, for example, was the right one; why anything else would be a mistake; and why this is not merely our civilization; it is something, the best of which we hold in trust for all mankind."

That is the beginning of the *emotional,* intellectual ability to rise above the relatively bestial level of thinking in terms of simple causalities. Then, when you think further and become a philosopher, in the sense that Plato and Plato's Socrates identify this, a philosopher-king, a true statesman, then one must think and say, "What *is* creative reason?" And we think of what is creative reason in terms of what we must do, what we yearn to accomplish; the yearning for atonement, so to speak, with humanity as a whole, and with Creation as a whole.

We are sensible of the fact that we have short lives, mortal ones, in which all the sensual pleasures of modern life go into the grave with us; and so we sense our immortality, not in respect to the survival of our mortal flesh, but rather, in terms of the mission which makes us useful to past generations, as well as present and future ones. Thus, we are able, in thinking in those terms, to reach out and see ourselves in respect to a necessary existence in the service of Creation as a whole.

We seek that quality; we seek to find that identity; and we find it within ourselves in creative reason. We yearn for it; we yearn to distance ourselves from that which denies us the development of that quality in ourselves. We have precious little time to do it, because we are going to die soon. It may be years, it may be decades; but we are going to die. That is a short time to get the job done, with limited opportunities available to us. So, we yearn for it.

People who are of that cast of mind and who have achieved a certain amount of rigor (which takes time in achieving that cast of mind), can go more or less readily to the second and third levels of transfinite as I have described. Thus, what I have talked about, as the problem of change and change/no-change, in the preceding, should be understandable, should be comprehensible, in practical terms of reference.

What do we care, in the long run, of these little things that most people care about? We care about them; they have to be taken care of. But we do not obsess ourselves with following the simple causalities, which are seemingly given to us, by an established way of looking at these things. What we have to do is to outflank the problem. We go to a higher level, the second transfinite level of consciousness, and look at ourselves engaged in this play. We look at ourselves as Swift's Gulliver might look at the Lilliputians and look down on them and say, "There am I; I'm that little Lilliputian over there, I'm looking down on myself. What am I doing? What kind of silly fool am I, playing this game?" Or, as a playwright, putting a great tragedy on stage, in which he may put something of himself or something of somebody else; where you recognize yourself on that stage of that tragedy. You say to yourself, "That's me up there. What am I doing? What am I doing?"

That brings you to the second transfinite level. A choice of the set of axioms and postulates, so to speak, which govern causality, particularly one's own role in causality. But that is not enough. That is useful, it is necessary, but it is not enough. One is driven, thus, to find, "Well, what is truth? This is true, this is more true than that; that is false. We have proven that." "But what is true?" That requires going to the third level of transfiniteness, in which we understand the ordering principle and understand a relative absoluteness. We recognize this as the Good. And when we reach that level and when we think in those terms, we are good. When we think in lesser levels, we are not good.

Thus, those of us who would be good, must be, from the standpoint of outsiders, from the Lilliputians looking up at us, as Gullivers: We must be preoccupied with these three levels and the problems of change and change/no-change.

XIII | A Self-Conscious Scientific Method

In light of what we have just said about self-consciousness's role in this process, several points should be made respecting the nature of an adequate, i.e., rigorous, scientific knowledge.

First of all, it should be apparent that it is on the third of the three levels indicated, that creative reason is consciously located as both subject and object of consciousness.

The first level is the notions of causality associated with experience, or a causal notion of experience, as opposed to, simply, a perceptual one (which we deal with as somewhat below the dignity of the term consciousness). This includes the consciousness of the existence of *self,* as an actor in the causal sequence. All of that lies on the lowest level of self-consciousness.

The second level of consciousness is better called the simple Socratic consciousness, in the sense that we are aware that a deductive or formal reasoning (or linear reasoning, which is the same thing as deductive reasoning) is always governed by one of a possible set of axioms and postulates, such that these axioms and postulates, taken as a set, anticipate every theorem which might be attached to a lattice based uniquely upon that set.

This relationship is called the *hereditary principle,* and no theorem respecting experience, that is, no notion of causality, including the relationship of self within the causal process of simple consciousness, can be reached except as in terms of expressions of theorems consistent with, in the simplest case, a specific set of axioms and postulates or with a specific theorem of that set.

Thus, the second level: the notion of an array: that we can go from one set of axioms and postulates, to another set of axioms and postulates, and that the theorems generated by, say, set *A*, are never consistent with any of *B*, and so forth. But, we are aware there is a connection between *A* and *B*; we are aware of a kind of mathematical discontinuity, separating one set absolutely from the other in terms of being mutually inconsistent.

This enables us to see the third layer in the ascending rank of consciousness. In this, we are focused not on a successive layer of mathematical sets; but we are, rather, concentrated on the process by which possible such sets may be ordered to represent an ordered series: an enumerable series, in terms of a generating principle, such that the sets proceed from relatively lower or higher order, when the measurement of lower to higher is the increase of the per capita reproductive potential of the human species.

It is on that third level that we locate the action which constitutes creative reason, as an object and subject of conscious thought. In general, except as we imply a fourth level, which is the consciousness of this, such as the notion of universality, this is the nature of possible conscious human thought.

The question arises: To what degree is this subjective? That is, to what degree does the thinking, as in scientific knowledge, defined so, in terms of these three levels, by human beings, constitute a true science? An interesting proposition. *To what degree would a different species, presumably with a comparable intelligence, think quite differently?*

In general, we would have to say, with respect to the third level, not necessarily the first level: "They could think no differently: Otherwise, they would not be equal." The human species has an indefinite potential for increasing its equivalent of its reproductive power. That does not always mean that this increases the total number of persons; but it means that the equivalent of the power to increase the total number of persons is always there. It may be converted into some other expression; but it is there. So reproductive *power* refers, not to the reproduction of the *number* of persons of the human species, although that is implied; but rather, to the condition of the species as a whole, with respect to the universe as a whole. Both productive power and reproductive power are subsumed notions of this *power*.

But in terms of the creative principle, if we can postulate or hypothesize different species which have intelligence comparable to the human species, but might have all kinds of other differences, they might differ, in respect to the first level of consciousness, but they could not differ, essentially, with respect to what we have indicated as the third level of consciousness.

So much for that preliminary observation.

Now let us see what we are really saying.

First of all, the general condition we are referencing, as outlined in *In Defense of Common Sense,* indicates that simple empirical knowledge is not knowledge, nor is it scientific knowledge. That does not mean that simple empirical knowledge is irrelevant; it means it is not scientific knowledge; it is merely a device which plays a part in the development of scientific knowledge.

The interesting part, which goes back to the change/no-change proposition, is: The most important thing about empirical knowledge is the extent to which it is or is not, in Kant's terms, possibly anticipated as synthetic *a priori* knowledge.

For example, anything outside an accepted theorem-lattice which could be predicted by a mathematical physics without experiment, would be analogous to something synthetic *a priori* in a Kantian system. Then, what would be of interest to us in such a mathematical physics, for example? In all cases, from a scientific standpoint, we would only be interested in determining, given any array of events or individual events, which of these arrays or individual events, conformed, predictably, to synthetic *a priori* extrapolations from a given set of axioms and postulates; and which did not.

The only thing of very much interest to us, would be the situation in which some of the events did not correspond: were, in those terms, anomalous.

So, *the variable rate of occurrence of anomalous events, with respect to all events,* including the non-anomalous, is the kind of event in which the well-advised scientist is interested primarily.

So, in that respect, empirical experiment plays an essential part in scientific knowledge; but it is not the substance, directly the subject or substance of scientific knowledge.

What we do, properly, in design of experiments, is to design experiments to bring forth, predictably, the anomalous kind of event, with respect to existing mathematical physics, to demonstrate that this predictability is accomplished, by allowing a different axiom or hypothesis, for example, than exists in the generally accepted mathematical physics. The occurrence of that which is absolutely anomalous, with respect to currently accepted mathematical physics, but which is allowed by a different hypothesis, constitutes what we call, sometimes, a *crucial experiment*. And thus, science is based, essentially, on a Socratic doctrine of hypothesis, or at least that is the proper representation of scientific activity, whether some scientists recognize it or not. (See Appendix XIV.)

This brings us directly into the main subject matter of *In Defense of Common Sense*. Science is concerned, in terms of reference modeled upon the idea of crucial experiments, to discover the discontinuities which compel us to overturn axiom and theorem-lattice *A,* in favor of axiom and theorem-lattice *B,* and so forth and so on.

Thus, we are forced to level three in consciousness.

Rather than just saying that we have to change from *A* to *B* in some undetermined fashion, we say what we are concerned about fundamentally is that which is crucial-experimentally right, which confirms an ordering principle which will enable us to say, with crucial-experimental authority, that *B* is greater than *A,* and *C* is greater than *B,* and so forth and so on. The concept of that ordering principle, as itself the only axiomatic of mathematical physics, would be mathematical physics on the third level of consciousness. An awareness of that may be seen as analogous to a fourth level, which is the kind of thinking we are reflecting, or Cusa is reflecting, and so forth, in dealing with these kinds of matters that we are addressing now.

Thus, we come to the next point.

So, the human mind is incapable of scientific thought, or actually classical artistic creative thought, *except* in these terms of reference.

Our definition of an object, the ontological features of axiomatics of our knowledge, are all referenced to this level three of consciousness, as we have defined it, immediately here, or just loosely described it here. Therefore, first of all, this is the only apparatus by which we could have scientific knowledge of our universe. Only from this Socratic standpoint is a rigorous mathematical physics possible, for example.

The question is, then: "Is this merely a projection, a stereographic projection, so to speak, from one geometry into the geometry of the brain; the geometry of the other, to the geometry of the brain? If we have a different geometry of the brain, would a different perceiver, having that different kind of geometry of the brain, get a different stereographic projection of reality than we do?"

Not really. Not in terms of the third level of consciousness, he couldn't. Because the crucial-experimental approach associated with level of consciousness three, is crucial-experimental with respect to the real universe. So, in terms of the ordering principle, it is only on level three of consciousness,

that the ordering principle of the mind and the ordering principle of the physical universe come into agreement. And there,
the agreement is not merely the stereographic correspondence;
there, the agreement is actually an essential identity, so that
any other species of creature which is intelligent, in the sense
of the human species being intelligent, would, in terms of this
third level of consciousness, have a mind exactly like that of
our best scientists, our best musicians, and so forth.

Furthermore, that being the case, that implies that our
mind, in these terms of reference (not in terms of simple
perception), is a representation of the lawful ordering of the
universe; that the laws of the mind, when seen in this frame
of reference, are essentially the laws of the universe. Not
perfected laws of the universe, but unperfected laws of the
universe. But the laws of the mind, insofar as they govern our
mental processes, on the third level of consciousness, *are* the
laws of the universe. Even though what happens on our third
level of consciousness in terms of particulars and its derivatives, may not be perfect, yet the principle which governs that
process in the mind, is a perfect principle.

Similarly, in the universe. The principle which governs
the development of the universe, the negentropic development
of the universe, is a perfect principle. And these two perfect
principles are in agreement. And that agreement pertains to
the notion of *imago viva Dei*. That is the best of all possible
worlds.

XIV | The Uses of Deduction

L et us continue the line we have just been exploring. Let us compare what we have said in the previous section with our earlier references to the importance of the *Monadology* of Leibniz and to the refutation of the attack on the *Monadology* by Leonhard Euler. We shall then see how what we have just said pertains to mathematical physics, for example, concretely.

We repeat: It is the generally accepted view, among educated mathematicians and mathematical physicists today, that the only acceptable argument in physics is that form of argument which is couched in the accepted terms of reference of a deductive/inductive form of commonly used classroom mathematics.

I object: "That commonly used classroom mathematics is faulty and cannot possibly represent the real universe." This was first emphasized, as I referenced this problem in the text of *In Defense of Common Sense,* in connection with Newton's *Principia.*

Newton was astute enough to recognize that what we call today the Second Law of Thermodynamics had made its ugly appearance, implicitly, in his text. He pointed that out to the reader and said, in effect, "This is absurd. That is not the way the universe functions and it is not my intent to convey that impression to you. However, I was compelled to show that, because of my choice of mathematics."

Now, what Newton was saying, effectively, is that the only mathematics which he considered acceptable at the time is a deductive/inductive form of mathematics, of the type which coheres, in most respects, with the doctrines and dogmas of Aristotle, Descartes, and Kant, the things we have refuted in *In Defense of Common Sense,* and in a number of earlier published titles, as well as here: That in any deductive mathematics or linear mathematics (the same thing), there is automatically introduced, to the physical evidence, superimposed upon the physical evidence, the appearance of a universal entropy. That is, a kind of averaging down of a statistical-gas-system process toward the point that there are no heat differences in the universe, and therefore no potential, in terms of the kinetic theory of gases or things of that sort, from which to generate, spontaneously, any work in the universe. So, the universe is seen to run down into heat death, through this so-called ergodic process or something analogous.

In point of fact, the universe is of quite a different order. The universe is a positively evolving universe, evolving to higher states. The universe is characteristically negentropic.

Therefore, we must reexamine this mathematics, this deductive notion of mathematical physics. It does not correspond to the physical universe, but mathematical physics based on that kind of mathematics does superimpose the appearance of things like a pseudo-law of physics, a Second Law of Thermodynamics, upon physics. It gives us a false physics.

However, we know from the standpoint just argued in the previous section, for example, and from earlier references to the Euler problem, relative to the *Monadology,* that a proper physics can be constructed free of this, if we are willing to forego the habit of deductive/inductive formalism.

What does that require?

We have to reject the deductive formalism, essentially, as we would depict it in a context we have been developing here, because we have shown that scientific progress, the essential feature of man's mastery of nature, is associated with a succession of scientific world outlooks. Usually, the successor, in this ordering, is superior to the predecessor. Crucial experiments, which overthrow or show the fallacy inherent in the axiomatic structure of an implicit or explicit set of axioms and postulates, lead to the generation of a new set of axioms and postulates, such that there is an unbridgeable gulf between any two successive sets. We can portray, at least if we use a deductive mathematics, the progress of science, in terms of this succession of sets of axioms and postulates—the deductive systems.

The deductive systems do not represent science; but they represent our attempts to *approximate* a consistently deductive representation of the possible theorems which might be advanced from the practice of physical science as we know it empirically at that point.

We have shown that creative reason cannot be encompassed by this; creative reason lies in what we have indicated to be the third level of self-consciousness. Therefore, we must have a mathematics which represents that. Obviously, a constructive geometry consistent with the third level of self-consciousness would be adequate for this purpose.

Let us just mention again the problem of geometry, to make sure we are absolutely clear.

We cannot, obviously, use any form of arithmetic or deductive algebraic schema as an acceptable mathematics for representing a competent mathematical physics. We have to throw them all out. Obviously, for similar reasons, we would also throw out a deductive geometry, such as a formal Euclidean geometry. For the same reason, we would throw out most of the formalistic versions of so-called non-Euclidean geometries, because these are actually, simply, neo-Euclidean geometries, that is, Euclidean geometries, altered by tampering with some among the axioms and postulates of an existing formal system.

We require, therefore, a purely constructive geometry, which depends upon *no* axioms and postulates. Otherwise, we can't be rid of this deductive curse. The question is, what are the specifications of that constructive geometry which are required for this purpose?

Obviously, it must be a constructive geometry which is based on the isoperimetric proof. It must be a projective geometry which is a multiply-connected form of action of this isoperimetric form. It must elaborate itself simultaneously as multiply-connected, in the sense of a double-conical geometry, for example. And it must correlate that with a simultaneous expansion in another kind of ply, the simple *Rouladen* (non-algebraic curvatures, which are generated by rotations).

Our geometry must also satisfy another specification. It must be based ontologically on the notion of monads. That is, we must think of a continuum, in which the continuum, in an evolutionary way characteristic of the system, generates monads.

Without getting immediately into the question of the higher monads, which we are, just look at ordinary good monads, good singularities. Let us concentrate on those kinds of singularities which correspond to negative curvatures, denting, so to speak, a Riemannian surface. So, let us call that, as I have proposed earlier, a *Riemann-Beltrami* surface. (See Appendix IX.)

So, those geometries which generate, in a lawful way, the characteristics of a Riemann-Beltrami surface function, are a minimal condition for a good mathematical physics.

This bears upon one of our big problems in physics today. Let us look at some of the implications. Let us take the case of Kepler versus Galileo, Descartes, and Newton, for example.

Kepler's physics is correct, at least as far as he makes any claims for it. That was proven during his time, and through the time of Carl Friedrich Gauss (1777-1855); Gauss's work on the implications of the asteroids proves in a crucial and unique way, that Kepler's astrophysics is correct, relative to every contrary claim of the incorrect Newton and Newton's supporters. (See Appendix V.)

The negentropic curvature of space-time associated with the harmonic orderings of the Golden Section, is the basis for the construction of Kepler's system, to a large degree *a priori*, as Kant would say, synthetic *a priori*. But, in a sense, it is not *a priori*, because Kepler shows two things.

First, on an empirical line of development, associated with the contributions of Leonardo da Vinci and others, in connection with the Golden Section's significance, Kepler had crucial empirical proof that the universe was negentropic, as we would say, that is, relative to entropic: It is fundamentally negentropic, that is, a developing system; and, with a curvature of physical space-time, consonant and congruent with, coherent with, the harmonic orderings consistent with the Golden Section. That is the instruction of Kepler's system.

He proves that, by finding that the empirical values correspond to and give scaling to such a geometry. And, thus, we have his system: the Keplerian system of harmonics, which he correlates with musical harmony, and quite rightly so.

So thus we have the two intersections. The geometry gives us a seemingly *a priori*, synthetic *a priori* view of universal physics, i.e., Kepler's physics. But this physics cannot be perfected without reference to those crucial empirical data which enable us to scale the system. That is also true in music.

For example: We can show, in a similar way, that classical music must be based on well-tempered harmonics, in which the harmonics is ordered in congruence with the Golden Section; but that doesn't prove middle C should be approximately 256 cycles per second. It may suggest it, but it doesn't prove it. What proves it is something else.

We look at the human voice, the well-trained human singing voice—and of all species, as we identify species of singing voice. We find, first of all, the human singing voice follows harmonics that are consistent with the Golden Section harmonics.

So far, so good.

However, we find that the singing voices, so tuned, have register shifts within them. These register shifts are consistent with the species of singing voice. And, therefore, we must scale the musical system to fit this empirical datum of the register shifts, which is historically pretty much how the well-tempered system developed, through Bach, Mozart, Beethoven, and the other classical composers, such as Chopin, Schumann, and Brahms, as opposed to the Romantics, such as Liszt, Wagner, what-not, who all went, of course, for the higher, elevated tuning. (See Appendix X.)

That is the general nature of the thing.

So, what we must do, always, is to guide the mind by such a constructive geometry. Use that guidance, relative to existing physical knowledge, to define new crucial experiments, which enable us to do two things: to demonstrate the appropriateness of our construction to physics, empirical physics; and to provide us a scaling of those functions, as we have indicated by the two examples: the scaling of the solar system by Kepler, relative to a geometrical construction or a method of geometrical construction, not a complete *a priori* one, but a method of construction; and the case of the well-tempered system.

Why do we get C approximately 256? Well, we get it from this evidence, in terms of the natural harmonics of the human singing voice. That is the essence of the matter. (See Appendix X.)

Thus, from this discussion, we see into some of the ways in which the third level of self-consciousness, and the organization of thought on that level, defines a necessary form of, for example, physical science, the way we can comprehend consciously, empirically, the lawful ordering of the universe. All we must include in that, as we specified, beyond the correct geometry as such, is to recognize that the geometry must be a monadology—that no constructive geometry will allow us to assume the infinite divisibility of any portion of physical space-time, but requires a monad at every point of singularity.

Of course, again, these monads are not self-evident, discrete particles, not discrete bodies in any sense. Rather, they are the generated singularities, like the singularities of a Riemann-Beltrami surface function, which are lawfully generated, and *necessary* in the continued elaboration of a Riemann-Beltrami surface function.

The monads define the special features of the proper choice of constructive geometry. Hence we have a continuous constructive geometry, which also has discreteness and yet, on a higher order, is continuous, nonlinearly, so to speak, despite the appearances of these singularities, which are discreteness.

XV | Religion and Creative Reason

L et us turn to the question of religion and examine issues of religion from the standpoint which we have elaborated thus far.

To begin, let us take the case of the long-standing split between the Russian (Muscovite) Orthodox Church, and the Western Christian churches, over the issue called the *Filioque*.

For those who are not informed already, the *Filioque* signifies that in the Latin Credo, following St. Augustine's writings, the Latin term *Filioque* was introduced to say that the Holy Spirit flows from the Father and the Son. This was adopted by Isidore of Seville, and so incorporated into the Credo there in Spain. This editing of the Credo went by various routes into all parts of Christianity and became formally a universal part of the Western Christians' Creed.

It was adopted by both the Eastern and Western churches in the ecumenical unification which occurred in 1439-40, in the Council of Florence. There, the Eastern Orthodox Church recognized, on the basis of evidence from their own writings' original intent, as presented by the later-Cardinal Nicolaus of Cusa, that the original intent of the Nicene Creed had been to incorporate the conception which is otherwise known as the *Filioque*. So, in the 1439 Council of Florence decisions, the Eastern Orthodox Church recognized that the *Filioque* of the Latin Church was a proper and essential part of the Christian Creed for all persons and was not simply a Western innovation.

This Council decision was opposed by certain people at Mount Athos (Holy Mountain), including a fellow who became the "Quisling" of Greece, later known as the Patriarch Gennadios. (Gennadios helped in betraying Constantinople to the Ottoman conquest and was rewarded for his treason, by appointment as patriarch of all the Christians of the Ottoman Empire.)

Gennadios, who represented a faction at Mount Athos, was supported chiefly by a gnostic faction in Venice related to the Bogomils and Cathars and so forth; he was also supported, notably, by the princes of Muscovy, who practiced a heathen variety of Christian doctrine in the gnostic form. The Muscovite form was derived from what is called hesychasm, that is, the bellybutton contemplation of oriental pagan mystics.

That is one split.

In Protestantism today, we have a split between orthodox Western Christianity and certain among the Protestant cults, on the same substantive issue. For example, radical Calvinism is a form of gnosticism, *in effect,* which denies the *Filioque,* denies the divine spark of reason in humanity.

You have also those Lutheran radicals, who implicitly join with the Calvinists on this, as do radical Pietists. For example, Immanuel Kant's Pietism was a significant factor in shaping his gnostic philosophical views. This connection was expressed in his famous *Critiques,* for example, as a follower of the gnostic, virtually satanic, David Hume and Adam Smith.

These issues come up more broadly today.

They are presented, ordinarily, as theological issues. In the United States today, at many divinity schools and theological seminaries, they would tend to be argued from the standpoint of William James's *Varieties of Religious Experience.* William James, the famous Harvard psychologist and pragmatist, was virtually a satanist or, at least worst, a gnostic, certainly no Christian.

Around the world, people would argue, "These are merely doctrinaire matters; and it is merely a matter of opinion, of one sect against another." They would argue, "The only thing that is fundamental, is the religious experience as William James defined it"; "These are matters of revealed religion, revealed doctrine, or allegedly revealed doctrine, as opposed to anything which can be settled by means of reason."

Unfortunately, many advocates of these various positions in theology, will argue only from the standpoint of revealed doctrine. For example, many Protestants will say, "Well, 'such-and-such' is revealed doctrine in the Sacred Word of God from the Old Testament."

The Old Testament as a whole is not pure and this is provably the case. Some of the Jewish texts, for example, were known to have been corrupted by the Babylonians in the seventh century B.C. by the scribes. These scribes imposed upon the Jewish texts, the satanic, Chaldean cult of Ishtar. The latter was superimposed, in part, upon the Hebrew texts, to bring them into conformity, by corruption, with the imperial pantheon of the Babylonian Chaldeans.

There was a second revision of the Jewish texts, in a similar way. The scribes under the Achaemenid occupation also created a pantheon, like the later Roman pantheon; the Hebrew religion, in order to be tolerated, had to conform in letter and in practice to the terms of membership in this polytheistic pantheon.

A good deal of the pseudo-Christianity and pseudo-Judaism come from this particular corruption.

Christian gnosticism comes chiefly from the Mithra cult of Simon Magus. Similarly, Jewish cabalism comes, in part, out of the same Mithra cult. The Mithra cult was explicitly the author of the Nietzschean Adolf Hitler, at least ultimately.

So, people will argue these issues, typically, from the

standpoint of scripture, revealed religion; they will do so even when it is provably the case that these scriptures are largely corrupted, as the Old Testament is extensively corrupted in the manner we have indicated above.

There are certain aspects of the Old Testament which we know to be valid from a Christian standpoint, because of their coherence with the New Testament doctrine. We also have historical access to proofs, based on knowledge of the cultural and religious beliefs of the relevant period, the time of Moses. We know what the Chaldean cults were, as opposed to Egyptian culture. The better part of Egyptian culture, not Mesopotamian, of course, is incorporated in the cleanliness code of Judaism. Anything that is paganism, we know to be corruption. For example, there is a certain amount of corruption in favor of the Canaanite Hiram of Tyre.

But, these are matters of background.

How should we deal with these issues?

Someone quotes his text, his interpretation of a text, and so forth, against somebody else's text or interpretation of a text; this gets us nowhere. This fails and leads Christianity, in particular, precisely into the trap of irrational formalism.

On the subject of the *Filioque:* We could know the truth if there were no text. If there were no Latin Creed with the *Filioque* in it, the *Filioque,* even without its incorporation in the Latin Creed, would still be true and we would be able to prove, that that were true.

Why?

For example: In the way we have indicated before, it is *provable,* by reason, that the human being, as a species, is distinguished, set apart from and above, all other species, including all animal species, qualitatively, by virtue of the divine spark of reason, that potentiality. That separates the human species absolutely from an animal species. Man is not an animal; and animal behavioral experiments tell us almost nothing about man, except the lower part of man, below the belt, so to speak.

It is provable, that creative reason is a creative principle, as we have described it. *It is provable,* that you cannot define Creation or the Creator, except from this standpoint of the definition of creative reason. *It is provable,* that man, by virtue of his potential, is *imago viva Dei. It is provable,* that Christianity presents Jesus Christ as the mediation between the Creator and Man, or the aspect of the Creator which mediates between Creator and Man, which brings man out of a state of taking orders from God as a potentate, to man who, out of love of God, a love based on *imago viva Dei,* acts out of the commandment of love, not the commandment of fear. That is all provable.

It is also provable, that this divine spark of reason is not a collective property of the species, in the sense that the Muscovite Russians would argue, but is, rather, a sovereign potentiality, a sovereign power, of the individual as an individual: a *monad.*

It is also provable, that this distinction we have just identified and outlined defines a different kind of ordering of society, as against barbarian or pagan society, and that this form of society is superior to and natural relative to, all other forms of society. That Christian civilization, as defined from this standpoint, not an arbitrary standpoint, is the highest form of civilization which man could achieve and every other form of civilization is inferior to it—that is provable. (See Appendix XIII.)

It is also provable, that any contrary notion of religiosity is false. So, why do we get into doctrinal arguments about text and interpretation of text, where reason guides us to the right answer?

The text is not to be despised by any means on this account. For example, the Gospel texts, the texts of the Epistles: These are historical statements of Christianity. They contain statements which are true, which may not have been known to be true by virtue of the action of reason in an ordinary sense—in the action of scholarship or science—at that time. However, we can know them to be true. They are accessible to reason, and we are gratified to find that the truth has been told; but we can prove it.

This takes us to the verge of the matter. There are certain mysteries of Christianity, but they are very limited. Virtually everything people would normally argue about, except this one- or twofold mystery, is subject to reason. Be informed by texts, perhaps helped by texts, by biblical texts, but not dependent upon them. *It is provable* by reason.

Let us take an example of this: I Corinthians 13, of Paul, the famous one.

What is stated there is provable, even if Paul had never written that; but he did write it. It is beautiful, in the center of a number of chapters of the same Epistle which converge on the same point.

Paul instructed the Corinthians on this point, and instructed others. Does it detract from Paul's conveying that, that this argument he makes were provable? No, it is like a hypothesis. Paul has stated a theorem. It is up to us to prove the theorem. But Paul stating the theorem was the essential act—that this was said, even though it were scientifically provable, without the Epistle. Would it have been understood as widely, would it have been applied, if that had not been done, if that Epistle had not been written? The implication is fairly obvious.

The point I wanted to stress here, in this kind of intermezzo, is that, as members of an ecumenical association, we must oppose arbitrary, doctrinal, textual argument in religion and say, "These matters which are of importance can all be reduced to reason; and whatever the text is assumed to say or is interpreted to say, is irrelevant in that sense. Where is the proof? Where is the proof?" (Except in that which is identified as a mystery.)

Now, on the Russian part, what do you get? Then you get the holiness, the holiness as defined by oriental paganism, brought into pseudo-Christianity as gnosticism, beginning at least the time of Constantine, who promoted gnosticism with Arius and the Sinai Desert monks (St. Catherine's of Sinai), as the latter hesychasts, or bellybutton worshipers, which were characteristic of the troglodytes of, for instance, Mount Athos's Holy Mountain, later.

This is sickness; but, this is the essence which separates the so-called Russian holy man of the Muscovite model, from the Christian; which defines Russian culture as really barbarism, with a facade of Christian terminology. It is not Christianity. The fact that the Russians would like to call themselves Christians may be commendable; it is not to be discouraged; but what they have got is not the true article.

Finally, to the Protestants.

We see that the Presbyterian Church, at least the Church of Scotland's leadership, is being destroyed from the top. It has gone outside Christianity, toward satanism, by way of paganism. That is what it was doing officially, with the motion on sexual practice set before it. This was done, in conjunction with the Muscovite Russian Church, with the ecumenical gestures which were taken during 1989, to promote precisely that. This has been the role of Archbishop Runcie, within the Church of England, who did the same kind of terrible thing. This is typical of the satanist, gnostic Cathedral of St. John the Divine, the Episcopal Cathedral in New York City, and its Lindisfarne attribute.

But, the essence of the matter here is the danger of the radical Calvinism, of Adam Smith's *The Wealth of Nations,* which has spread widely throughout Protestant Christianity in the United States. The danger is, the separation of faith from works: faith without works, which is the characteristic of radical Calvinism, the characteristic of the worst part of radical Lutheranism. Those aspects of Protestantism are what must be fought and combatted.

It is not a theological matter, as such. It is a matter of reason. We are obliged—contrary to the Quaker, who says he must not participate in military affairs—to be accountable for the condition of mankind. We are obliged to that by determinable, knowable, moral standards respecting past, present, and future generations in entirety.

This, the radical Calvinist rejects; this, the gnostic rejects; this, the radical Lutheran rejects. This must be combatted. It is not a matter of interpretation of the Bible, even though the Lutheran version of argument on this is false, as Calvin's is, even from the standpoint of the Bible. But that is not the hard proof.

The hard proof is: This is insane; and Christ and the Creator are not insane. That is the point to be made.

XVI | The Bankruptcy of 'Standard Theory'

So far, we have outlined material which should make clear the general significance of the following statement.

What is called, in conversation among physicists and in classrooms, "standard theory," is inherently fallacious, even in its treatment of so-called entities of so-called classical and quantum mechanics, including relativistic physics: to the degree that an ontological entity's existence and function depend, in any significant degree, upon deductive consistency of a particular mathematical physics, employed to create the relevant array of cumulative experimental material.

Just to restate that in a few sentences, to make the point absolutely clear.

We have, for example, the definition of the quark. The quark has no experimental existence. The quark, and associated features of that kind of theory, arise from the attempt to explain actual experimental evidence from the standpoint of consistency with standard theory. Thus, quark theory represents the creation of assumed ontological existences, purely on the basis of the requirement of establishing consistency with experimental evidence for a standard theory.

The problem here is that the standard theory is, we know, absurd. That is, any physics which is based on a deductive mathematics, is absurd, to the extent that the physics is dominated by deductive mathematics.

Let us put this another way.

Given a valid experiment, one which is reproducible by almost any standard, irrespective of, say, a deductive design of experiments:

First take an observation, which is agreed to be an anomaly, in which the event is not structured (the observation may not be structured by any experimental design, in which the experimental design might be contaminated by deductive assumptions).

So, in this case, we may use the deductive system to describe the phenomenon. It will not correspond to the phenomenon; that is, the mathematics will not correspond, except as a matter of approximation, in a sense of, shall we say, linear curve-fitting, as the famous four Archimedean propositions deal with that sort of thing.

The improper mathematics has described, in terms of approximation, an experimental result. Fine. The experimental result pertains to something which has ontological significance.

However, suppose we stretch the theory, the mathematical theory, to such a degree, that we attempt to account for the margin of error in curve-fitting, between the curve-fitting construction, i.e., the linear construction, and the actual phenomenon whose description is approximated.

Now, let us suppose that we say, that we must account for the existence of the phenomenon described in respect to the margin of error between itself and the curve-fitting involved. In that case, we would have created an entity, an apparent, but fictitious entity, which is the margin of error between the object and the approximation. This action, of course, would be subject to experimental verification. One could verify, repeatedly, under repeated experiments, that such a discrepancy exists. Therefore, one might leap, foolishly, to the assumption that the entity has an ontological existence, which it does not.

That is a very crude, simple, but I think effective, illustration of the point. It is the same point which Newton made (and probably he reasoned in a similar way in making it), in warning the reader, in his famous clock-winder treatment of the universe, that the universe was not running down (i.e., the Second Law of Thermodynamics does not exist), but that the appearance of this (that is, that the Second Law of Thermodynamics exists), is merely a product of the superimposition of a defective mathematics upon the process of description of the empirical evidence.

Now, how do we avoid this?

This takes us to the requirement of a different mathematical form of physics, something different than the standard classroom physics or standard classroom theory, for example.

Let us take the case of the alternative, which I have proposed: *a Riemann-Beltrami surface function*, as a general mode of describing all of these anomalous (otherwise called nonlinear) phenomena, which do not precisely fit neat standard physics. Also, look at some of the things in so-called standard physics, which might appear subtle to some, but which are not subtle, rather fallacious, because they involve the assumption of entities, where none exists: that is, pseudo-entities like the epicycles of Ptolemy, in order to make the system seem to work.

For example, I indicated the discrepancy between the attempted curve-fitting approximation, and the actual curvature. That margin of error and the epicycles, of course, come out very similarly. The existence of the epicycles is based on the margin of error introduced by a bad theory, a bad attempt at description.

In order to escape such bad attempts at description, let us take all the cases which are really wildly anomalous, obviously nonlinear; and let us take those which we should be looking at as anomalous, in which the entity, like the quark, comes into existence in our mind, solely as an attempt to reconcile a margin of error between the events actually observed and the error of approximation inherent in the method of description employed to represent that event.

So, the Riemann-Beltrami surface function is a very useful

way of subsuming the relationship among, and between, weak and strong nuclear forces.

In this case, when we bring that into play and deal with the relationship of electromagnetic and gravitational phenomena, for example, in these terms of reference, particularly on the nuclear scale, we get a completely different kind of result than we do with, say, the quark theory.

For example, there is a problem which arises in the published version of Wells's model for the solar system, in the sense that he is using a standard classroom theory-approach for describing something which was actually developed from a different standpoint. So, there is a discrepancy. I think that in that case, in Wells's construction, we'd have to go back, away from the standard theory which he is using for the IEEE publications, and so forth, and go back to the source to eliminate the "curve-smoothing errors" which arise from the use of linearity of standard theory, to represent the approximation of the process discussed.

Another part of this, which has to be emphasized, is that among Anglo-Americans, most emphatically (I keep away from the special problem of neo-Cartesianism among the French), there is absolutely lacking, in virtually every case, any understanding of what a strong rigor is. Not only do they show a lack of strong rigor in their work; but, in general, they do not even know what it is that they lack. They do not know what a strong rigor should be. None of them, for example, is trained profoundly in the Socratic method, which virtually all of the great classical discoverers in physics were, including Leibniz, or including all of the leaders of the work in developing the theories of elliptic functions and so forth, during the nineteenth century.

The greats approach these ontological and other questions and questions of axiomatics with an understanding of the Socratic method. The average Anglo-American, with terminal degrees of the highest qualifications, is educated *to avoid* any consideration of that sort of material, to avoid any conception of geometry which is inconsistent with that approach.

Thus, the typical American today (I'm talking about the Anglo-American scientist), by a margin of 99.9999 percent, is incapable of understanding the kind of rigor which is employed by the best scientists, the best continental scientists in particular, of the nineteenth century. This makes it doubly important to shift the emphasis away from standard theory and to compel some of these scientists, ones who are more viable, and perhaps a bit younger in some cases (if they can rebuild themselves), to take this Platonic approach. Because only on that basis can they become acquainted with a strong rigor. There is no sense in trying to educate people merely in constructive geometry per se. I suppose there is some sense in it, but you are not going to get the student to the kind of desired result from that. You must accomplish what must be done from the Platonic kind of approach, of which I have represented a reflection here.

For example, I would give examples of cases which are relevant, apart from Gauss, Riemann, Beltrami, and so forth. Look at the less profoundly rigorous figures, such as Felix Klein, Max Planck, and so forth. These people were much less rigorous about the turn of the century than their leading predecessors in the same institutions a half-century earlier. They'd gone down in terms of rigor. But still, the rigor of people such as this is overwhelming, astonishing, awesome, compared to the loose, almost gossipy character of standard theory today.

Max Planck, From This Standpoint

There are two Max Plancks. One is the Max Planck who derives the concept with which his name is associated; and there is the other Max Planck, the mythical Max Planck, who was created by Albert Einstein in 1917 approximately, with that terrible abomination that Einstein produced at that point on the subject, or as reified through the radical positivist version, which, coming out of Niels Bohr and company and others, seems to be hegemonic, more or less, today. So, we have this "multichotomy" among the so-called classical version (which is not classical at all), and the positivist version, of quantum mechanics, and the positivist version of relativity; these three kinds of things bobbing around, none of them really good physics. Everything has been misunderstood from the attempt to reconcile the irreconcilable among these three things, none of which should exist.

The original Planck and his derivation of the concept, is rich and exciting; at least it was for me, as opposed to the dull and arbitrary assumptions, not only of an Einstein, who was probably one of the better cases among the bunglers later on in this positivist tendency.

You will observe, going to Planck's own published account of his derivation of the concept,[1] that there is a precise affinity between my attack on Euler's attack on the *Monadology* and Planck's method. (See above, Chapter VI, and Appendix XI.) That is, rather than taking smallest of smallest of smallest or arbitrary division of line lengths, linearity, one must reduce the thing to action in the form of isoperimetric action and the question of division of rotation, as the division of an angle, and then the division of that angle. And that is exactly the way the Planck constant actually develops. So, looking at it in those terms, keep to Planck's original terms, in using the quantum relationships—that is, in this notion of rotation, this isoperimetric motion—and a lot of the nonsense which commonly arises, is averted. Then, put that back into the approach I have outlined to a Riemann-Beltrami surface function, and Planck's concept, as he describes his derivation of it in his autobiographical note on this, applies beautifully. It lends itself to comprehension and avoids this terrible, positivist, statistical mysticism, and convolutions which come along commonly in this connection.

Planck made a wonderful, great discovery, and he made

it in an extremely rigorous way. People seem to be deprived of the beauty of that rigorous discovery, and prefer the after-the-fact reification of that from a positivist standpoint; but the discussion of Planck should be situated, as I have recommended it be situated.

XVII | On the Subject of Unity

We have written and discussed earlier the subject of substance. Let us look at the same matter from a slightly different vantage-point. Let us take the issue of unity, the issue which I addressed glancingly in commenting on the *Theaetetus*, the Waterford translation, earlier.[1]

Let us take this maximum-minimum relationship. *One* is the individual. Why one? Well, because the individual is sovereign and that within the individual which is sovereign, that is, creative reason, is potentiality. It is not divided, in any way, beyond the individual. So, it is one.

Let us take the universe as a whole. It is sovereign. Its existence is not divisible. Therefore, its existence is also an indivisible one.

In both cases, the *One*[2] refers to substantiality, or to the quality of existence we associate with substantiality: efficient existence. (I'll make an observation on this efficient existence and problematic feature of literal interpretation later on.)

So, both are one. They're equal in that sense. Equal, why?

Equal because the ordering of the universe, for reasons we have given earlier, is coherent, consistent with creative reason as a potentiality, as the potentiality of the individual. And, the future order of the universe, in the sense of past, present, and future, is also equal to the present, in respect to the fact, that if we measure the *present* substantiality of the universe as *One,* with emphasis on the word *present,* as *potentiality,* it contains past and future, as well as present.

We can speak of the unity of the individual in respect to the potentiality of creative reason, in a somewhat similar vein, with certain qualifications. The individual is not really self-subsisting, the individual person, in this respect, *except* as the individual is in an efficient, unmediated relationship to the universe as a whole. But, in respect to the universe as a whole, the individual, in that relationship, does, in the present, reflect as potentiality, past and future, in the way we have indicated earlier.

Now, the interesting thing is the content of this *One.* And we shall see promptly why I'm doing what I'm doing right now.

What is the content of *One?* Creative reason. What does creative reason correspond to? Let us reference *In Defense of Common Sense.* In this case, we have the successive deductive theorem-lattices, *A, B, C, D, E,* and so forth. Creative reason occurs, or is reflected in, the *efficient* character of the apparent mathematical discontinuities, both separating *A* from *B,* and so forth, and also provoking, or prompting, the coming into being of *B* out of the catastrophe affecting *A.*

This representation, just identified, is not adequate. We have to go to a higher level, because we have to see this not really as a succession of independent discontinuities, or apparent discontinuities; we must see this as a *recurring function* of apparent discontinuity. And it is in that *function* that we begin to approximate creative reason.

We also then observe, that this function may be more or less efficient in the sense of being more or less dense. That is, we can have higher and lower rates of scientific progress, which, with the higher rate, would be measured in terms of a higher density of such discontinuities of the type we're referencing per lapse of time, or per unit of universal action (the same thing). This would mean that we would have different isochronic scales, in the following sense.

Let a function, which gives us a certain rate of scientific progress or scientific revolutions, as *A, B, C, D, E,* and so forth, represent a pathway of scientific revolution; let that be represented by an isochronic scaling. No problem.

Now, let us have a higher rate of scientific progress. That would be a slightly different isochronic scaling.

Just a note to bear in mind, as we think about these things: Make sure we're thinking rigorously about where we are at all times when we do these kinds of excursions; otherwise, we drift off into detours which become wild fantasies.

So, therefore, the notion of a variability in the rate of scientific progress, comes as close to the elaboration of creative reason as we can conceptualize it, from this approach. And the highest notion we can approach, is the notion of a unity of that kind of variability of function or functional variability.

So, that highest notion is that which corresponds, as an articulate notion, to the notion of efficient, existent, substantiality. This is true for the mind of the individual monad, the person; it is also true for the substantiality of the universe as a whole.

So, the number *One,* as a cardinal number, stands for that function.

To restate what we have just said: It is the accepted standard of classroom practice of mathematical physics, to start with the number *One* as a cardinal number (once we have defined it as a cardinal number), and to associate cardinal numbers with elementarities of physics, the smallest possible parts. And then to show how pair-wise relations and multiples of pair-wise relations, or multipliers of pair-wise relations, can be left to account for the universe as a whole. And thus, the search amid the flurry of quarks for colorful stories.

Obviously, that approach is absurd, because elementarity, in the terms of unity as we have just defined unity for the individual and the universe, and the relationship between the two, is the most complex of all number notations, or geometrical number notations.

So, we start with the most complex of all number notations, which defines the significance of simple counting numbers associated with things in the long run.

The idea of equality of one to one, and so forth, all depends upon the determination of the *One* by a function of the type we have just referenced. Therein lies a very great secret, so to speak, which should not be a secret. (We do not wish to spread any gnosticism around here.) It is not really a secret; it is only a secret from those who blind themselves. But that is the nature of the problem.

A Point of Clarification

There is one particular point which I wish to make very clear and which has two aspects.

The first is my reference to the distinction between the subjective and objective. It is clear, I think, that there is no strictly necessary distinction between subjective and objective knowledge, as in, for example, science. There is not an objective worldview which might be seen by some other being, as distinct from the scientific worldview of the physical universe which we are able to construct by virtue of the special features, including limitations, of our mental perceptual apparatus. Rather, on the level of creative reasoning, the representation of the laws of the universe in the language of creative reasoning and the actual laws of the physical universe are one and the same—both in fact and as to form.

Now, essentially, this bears directly upon the role of the monad: the fact that in the mental image of scientific knowledge of the universe, the monad is crucial. That is, the relationship between the universe as a whole, as a unity, and the individual creative reason as a unity, in direct, unmediated relationship to the universe as a whole, is the essence of scientific knowledge, is the essence of an *efficient* relationship between creative reason and the universe as a whole. For that reason, there could be no discrepancy as to form between the laws of the universe and a correctly devised representation in terms of creative reason's construction of a picture, shall we say, of the laws of the universe.

There can be a discrepancy only to the degree that there is imperfection in the application of reason.

So, the subjective element arises as a discrepancy *only* to the degree that this imperfection exists. There is no *inherent* discrepancy, but only the discrepancy of relative imperfection.

That is the essential point to be stressed. This bears upon the fact, which is the crucial fact of all physical geometry or all economic science (the two terms being really the same), that the increase in technology, which is the increase of the per capita power of existence of the human species or of a society, is caused by the generation of scientific progress by a purely subjective agency (apparently): creative reason.

Thus, the spiritual action, a creative-reason action of discovery, is the efficient cause of a physical result, the increase in productivity, for example, as one aspect of that physical result.

These two things permeate the entirety of Project A: the complementarity between this ostensibly anomalous relationship between the spiritual, i.e., creative reason and the physical result of creative reason as the cause, and the (in principle) exact correspondence between what we might think is the subjective view of science and objective reality, which we're representing by science.

The only time that we can speak of, significantly, a principled discrepancy between reason's picture of the universe and the actual universe, is in, for example, a deductive method or inductive method.

Amusingly and usefully, Newton points this out in stating that the imposition of his mathematics (in this case, a linear, i.e., deductive, mathematics) upon the physical evidence, leads apparently necessarily to an image of the universe which is in part false to fact, the running-down-clock image of the universe, the Second Law of Thermodynamics universe. In that case, there is a principled discrepancy between science and reality, such that we call science in this case the *subjective,* and the reality which it fails to represent, the *objective.*

In contrast, from the standpoint of creative reason, when that is employed rather than the deductive/inductive mode, then that discrepancy-in-principle vanishes, though a discrepancy may exist in terms of the margin of error. That is the point which permeates the Project A undertaking.

I thought I would restate it in this form, in case I do not make the point clear. Or, at least by contrasting what I say here with what is said in the text as delivered, so far, perhaps the comparison of the two will force to the reader's attention the nature of the issues involved. If the reader finds the thing a bit confusing at first glance, that is not exactly the reader's fault; this is a profound matter and the correct answer to the implicit questions goes far afield from what is generally considered, although wrongly, the accepted classroom view of the subject.

XVIII | On the Subject of Ontology, Again

As we have indicated so far, in reflections upon the material I have covered in this series and in other writings published earlier, the crucial issue of science, of knowledge in general, and of policy-shaping, therefore, is the issue of the notion of *ontology*, of being in the sense of substance: What is substantial?

In general, I have cautioned people that causality is the key to being. That which efficiently causes something to occur and which is the subject of causation in a reciprocal manner, is essentially what we should mean by *being*. As to how being elaborates itself, that is something for us to discover. But in starting out, we must reject simple perception, sense perception, as a definition of being and must have a more general notion of being which covers all cases, that is, which is of universal applicability.

I shall indicate some of that now and go through an exercise, essentially Socratic in its character, though not necessarily always Socratic in its form. I tend, in summation of the argument, more to the didactic, and leave the Socratic to the pedagogy of the classroom or similar circumstances.

Let us start with being.

All being is associated with *motion*. This motion occurs in two primary ways: Either we perceive the being, the entity in question, to move with respect to the physical space-time in which it is situated, or we see it not to move, but, that is, relatively to move with respect to the motion occurring about it. So in both cases, the notion of being is associated with motion.

It is associated more generally with *becoming,* with *change.* And change has two aspects: the linear aspect of change, or that which is representable in a linear way; and that which is not representable in a linear way, i.e., a *qualitative* change, we tend to say.

In this vein, on the simplest level, the preconditions for defining simple existence are, in order, first of all, motion, which signifies, generally, matter-motion, as a most common reading of that. And secondly, the motion of change of quality of motion, accompanying a simple matter-motion. This relationship of the two, as qualified in the second observation, is very important to bear in mind.

There is another consideration of universality which comes in in a different way here, negatively. Suppose we were to reject either of these two conditions or to qualify them. Then we would have a real problem, because our definition of substance, of being, implicitly, is that it is substantial in respect to all possible conditions of the universe.

Now how would we observe all possible conditions? What would we mean by "all possible conditions"? Or, reciprocally, what would we mean by failing to meet the standard of all possible conditions? In other words, all we would have to do, according to this line of argument, is to prove, that in one case the entity responded to the universe in a manner as if the universe did not exist.

For example, if you imagine a great explosion, a couple of kilotons or megatons of dynamite goes off next to a fellow who is walking. Everything around him is blasted, tattered, ruined, but he continues to walk through blithely, as if nothing had happened. We would say, well, this fellow cannot possibly exist. This must be a phantasm. It cannot be a real person.

Therefore, something that fails to respond appropriately

to action of the universe more generally, even in one case, puts upon itself a question mark as to its existence.

This may involve, in some exceptional cases, all kinds of subtleties, which might be explained away, as in the kind of case I just used for illustration. But, nonetheless, if we cannot explain it away in a consistent manner, then it does not meet the criteria of being.

Therefore, that is our crucial, negative test: It must be efficient in its action upon the universe, and the universe must be efficient in its action upon it. And that must be universal. A single exception tends to call that being into question. Therefore, universality of substance implies universality of response, as well as universality of its causal efficiency as an existence. It must respond as an efficient existence, in all possible motions and states, i.e., qualities of motion, in the universe. There may be, according to the rules, reasons why it should not appear to react in certain cases, though it actually must react in all cases, whether it appears to or not.

This sort of notion leads us to the question of *transfinite being*. Transfinite being, as a notion, starts out as a very simple kind of Socratic idea.

Let us take, for example, numbers. We have all kinds of numbers. Let us take the numbers in the proper fashion, not arbitrarily. Let us take them without fooling anybody; let us take them geometrically. Well, the number *one* has a very simple significance. And so does *zero*. *One* and *zero* have a very simple significance in geometry. Well, we make constructions. And as we make constructions, the simplest plane figure we can make is the triangle and so forth. We can make quadrilaterals, and so forth and so on, plane figures. Out of this we get notions of construction, which are generating plane areas and their roots by products of linear magnitudes. A very simple kind of case. One can try to generate the field of integers, so far, in that way, and other numbers that fill in between integers. We find out that we have rational numbers, which can be constructed that way. Then we have a number of irrational numbers. Then we have various orders above the irrational. We have the transcendental numbers and we have much higher orders than simple transcendental numbers, which can be generated in the manner which Gauss has indicated and as Cantor has indicated this problem.

We get into larger geometric numbers, as Gauss does. We get into the so-called imaginary and complex numbers, which are not really imaginary and which are quite clearly classes of geometric numbers. They tend to fill up the gaps in between, leftover in-betweennesses not filled in by all inferior sorts of numbers.

So, a general notion of number arises, not from particular experience, but by trying to approach universality by the method of successive transfinite orderings. So, hard proofs and strong proofs all involve universality. They involve universality positively, and they involve it negatively. We have referred to the negative above. We have referred to the single crucial experiment, which is a negative demonstration, tending to jeopardize the claims to being of something. And we have the more profound sort of negative inquiry, which may cause us either to abandon the definition of being for something, or to redefine it in a qualitatively new way.

In this process, as we have done in the foregoing sections, with intermezzi and affirmation, we have defined, that the change of quality of motion comes close to the proper definition of substance, that is, it covers universality. This must be the case, because any simple motion cannot be universal. There will be cases in which this particular motion does not exist or in which the universe is expressing itself in a different quality of motion, in which the universe is changing the quality of motion. So we cannot have a response, unless we fill up the gap of change of quality of motion. That leads us to a further consideration: the rate of change of change of quality of motion, or rate of change of rate of change of quality of motion. That begins to bring us to a kind of universality, in which the higher ordering of the functional notion of rate of change of rate of change, does pretty much, on the third level of change of quality of motion, everything we need to do in an ordinary way in representation.

Very simply, having come that far, let us look at our mathematics.

Simple, discrete matter does not exist, as in the sense of a perceptual discreteness, as an object of touch, as an object divorced from motion. That kind of substance does not exist. It cannot exist in our universe. Secondly, even simple motion cannot exist as something primary in our universe. It does not meet the qualifications of substance in any aspects of substantiality. It is not being, it is not substance. Nor is a rate of change in quality of motion adequate. We have to generalize the notion of a rate of change of rate of change of quality of motion and then we have, at least verbally, encompassed in a general way the kind of definition of being we require.

That being the case, let us do a very simple thing. Let us look at the domain of physics. Let us not be totally naive. Let us take into account the notion of curvature of physical space-time, which has been explored and pretty well refined, and which we have dealt with in various ways, in qualifying the implications of Kepler from a more advanced standpoint, say that of Gauss, Riemann, and Beltrami, and so forth. Into that space-time, let us introduce this notion of rate of change of rate of change of quality of motion, of matter-motion. And let us put that into any relativistic physics whose relativism is defined from the constructive geometric basis in terms of a curvature of physical space-time.

If that is the most primitive substance, look at what we have said earlier about the relationship between the individual monad and the universal. Let us suppose the monad is some-where in the order of a Planck distance. Suppose we squeeze it down in there someplace. We do not simply have a little black hole there; we have something that is very busy, with more lights than the thousand points of light that George Bush has been looking for lately. Very complicated, very active substance in there, nonlinear also. But from our standpoint, the substance in there, since it is cognate with the universal in particular, the substance of universality and the substance in that monad is of this nature: It is of the nature of a function describing a rate of change of a rate of change of the quality of motion. It is not only that: The function implies the ability, a method, for increasing that function; an increase which we can measure, in the first approximation, with a notion borrowed from Georg Cantor of an increase of the enumerable density of apparent mathematical discontinuities for interval of action. The interval of action being, say, this Planck distance. There is an arbitrary choice, consistent with Cantor's definition of an arbitrary choice, for that kind of comparison.

That becomes, then, simple matter. It is simple matter, of course, in the case of an individual human being endowed with sovereign, creative reason.

But we also referred earlier to the other kinds of little monads kicking around the universe, that do not have any intelligence, that do not have any creativity—little pieces of dirt, for example. We said that these things have to *react* to the universe, which is *characterized* by the relationship between that higher monad and the universe as a whole. Therefore, the lawfulness governing that little piece of dirt there, in its motion, is determined in reference to the higher degree of motion, that is, the motion of the mind of man, of reason, with respect to the universe as a whole. Thus, the laws that we adduce concerning the nature of substance, from the primary relationship, that of the individual human being to the universe as a whole, define the laws of the universe in which that little piece of dirt is functioning and having its relationships.

Thus, the simplest rigor of reason requires us to turn, so to speak, the entirety of physics on its head, in the sense that physics and simplistic physics, or accepted classroom versions of physics, attempt to reduce everything to derivation of the articulated from the simple, where in point of fact, the simple is determined by the increasingly self-articulated substance, in the sense we have defined. So, this defines another way of looking at the problem we have been discussing so far; a way which, of course, must be included in an all-sided treatment of the problem.

Finally, let us return our attention to the subject of creative reason as experienced by the human mind, as the map of physics and as the proper reflection, within itself, of the laws of the universe as a whole.

Let us look at this from a different standpoint, the standpoint of method, historically, and recognize that this is precisely the secret of what is called the Socratic dialectical method.

By recognizing that the individual creative reason, as a sovereign capacity of the person, was essentially in unmediated relationship to the universal, that is, directly, Socrates struck upon—whatever sources he used for this discovery—the essence of all science and all knowledge. We seek universality by eliminating those underlying assumptions which fail to be universal, and whose failure is demonstrated to us, or can be demonstrated to us, by the means internal to the sovereign faculty of creative reason within each person.

The limitations placed upon this are, of course, empirical. That is, the mind cannot know more than it knows as an interpretation, in a sense, of the falseness of perception. In order to understand the falseness of the misleading character of perception, we must have empirical perception, or we must have the absence of a perception where that perception is to be expected, according to some prevailing, accepted set of assumptions. That is really all there is to it.

The Socratic method rests, in fact as it does implicitly, by the very use of it, upon the evidence that the sovereign, creative reason, intrinsic to the individual human mind as potential, is in an unmediated direct relationship with the universal. And that, by exploring that, we have, in a sense, the perfect mathematical physics, given to us, as it were, *a priori,* but not in Kant's sense; not a specific physics, but we have the map of mathematical physics, which enables us to exclude all formulations which we attempt to force upon that map, which do not fit the map. Otherwise, it is as I have said, that the relationship between the monad, as a monad, which we are, and the universality, particularly the unmediated aspect of that relationship, which enables us to know and to prove, that the Socratic method is a true one and a uniquely true one.

That completes Project A.

The Geometers

Filippo Brunelleschi (1379–1446) architect of the Florence Cathedral dome

Gaspard Monge (1746–1818) father of France's Ecole Polytechnique

Jacob Steiner (1796–1863) Swiss mathematician

Peter Lejeune Dirichlet (1805–1859) German mathematician

The Classical Musicians

J.S. Bach (1683–1750)

Wolfgang Amadeus Mozart
(1756–1791)

Ludwig van Beethoven
(1770–1827)

Robert Schumann
(1810–1856)

Franz Schubert
(1797–1828)

Frederich Chopin
(1810–1849)

Johannes Brahms
(1833–1897)

Propagandists of British Liberalism

Thomas Hobbes (1558–1679)
author of *Leviathan*

David Hume (1711–1776)
asserted man's bestial nature

Adam Smith (1723–1790) author
of *The Wealth of Nations*

Jeremy Bentham (1748–1832)
agent of Lord Shelburne

Immanuel Kant (1724–1804)
German exponent of British
liberalism

Dark Voices of the Enlightenment

François Marie Arouet de Voltaire
(1694–1778)

Jean-Jacques Rousseau
(1712–1778)

John Locke (1632–1704)

The Deployment Against Science

Francis Bacon (1561–1626)
author of *New Atlantis*

Issac Newton (1642–1727)
plagiarized the calculus invented
by Leibniz

René Descartes
(1596–1650)
described the
universe as
mechanistic

James Clerk Maxwell
(1831–1879) Royal Society
hoaxster

John W. Rayleigh (1842–1919)
dabbled in pyschic research

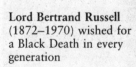

Lord Bertrand Russell
(1872–1970) wished for
a Black Death in every
generation

Landmarks of Political Economy

Alexander Hamilton (1757–1804) wrote *Report to the Congress on the Subject of Manufactures* 1791

Friedrich List (1789–1846) one of the founders of the American System of Political Economy

Henry Carey (1793–1879) advised Abraham Lincoln

Pope Leo XIII (1878–1903) author of the encyclical *Rerum Novarum* 1891

The Science of
Christian Economy

On the 100th Anniversary of *Rerum Novarum*

During the course of these next several pages, we shall come to the point at which we shall turn the attention of our ecumenical readership to numbered section 72, of the famous 1891 encyclical of Pope Leo XIII (1878–1903), *Rerum Novarum*.[1] We shall then focus upon the concluding sentence of that section, and also upon the passage from St. Thomas Aquinas's (1225 ca.–1274) *Summa Theologica* which the author of the encyclical has footnoted there.[2] The referenced sentence of the encyclical's text reads thus: "For laws are to be obeyed only insofar as they conform with right reason and thus with the eternal law of God."

The footnoted passage from St. Thomas Aquinas's *Summa Theologica* reads: "Human law is law only in virtue of its accordance with right reason: and, thus it is manifest that it flows from the eternal law. And insofar as it [manmade law—LHL] deviates from right reason it is called an unjust law; in such case it is not law at all, but rather a species of violence."

A hundred years ago, *Rerum Novarum* treated the remedying of the evil then being run by a "devouring usury," which, "although often condemned by the Church, but practiced nevertheless under another form by avaricious and grasping men, has increased the evil" effected by the handing over of workers, "each alone and defenseless, to the inhumanity of employers and the unbridled greed of competitors."[3]

At the time of the assassination of U.S. President John F. Kennedy at the end of 1963, approximately three-quarters of a century had passed. It appeared to most observers then, that the pleas for economic justice in *Rerum Novarum*, if not yet successful, were assuredly on the way to becoming so.

In the so-called "industrialized capitalist" sectors of this planet, the trade-union movement and other meliorist agencies had won, and were continuing to win, cumulatively invaluable and putatively permanent gains in human rights for most strata of the populations. Although a vicious form of neo-colonialism had been established at the end of the 1939–45 World War, the spirit of the United Nations Organization's First Development Decade Project, and the U.S. Kennedy administration's Alliance For Progress, suggested a commitment to global justice paralleling and perhaps echoing the rise of the civil rights movement inside the U.S.A. itself.

During the middle of the 1960s, that hopeful direction of development was reversed. During the recent quarter-century, social conditions in most parts of the world are far worse, on the average, than during the 1960s, and threaten to become soon far worse than 100 years ago.

The impulses for evil which have caused this recent calamity are not altogether new. A conspicuously leading cause of the greatly increased immiseration and endangerment of the human species, during the past quarter-century, has been the willful murderousness with which such forms of the old "devouring usury" as so-called "International Monetary Fund (IMF) conditionalities" have been so widely, so murderously, so shamelessly applied to the precalculable effect of rapid and large-scale increases of death rates by means of malnutrition, disease, and related mechanisms.

The most striking of the various included features of the new evil, is the dominant influence of the so-called "New

Age." This feature includes such presently pandemic expressions of this as the "rock-drug-sex counterculture," and increasingly irrationalist mass-murderous expressions of self-styled "ecologism," or "neo-Malthusianism."

The "New Age" is not itself an entirely new form of evil. It is as old an evil as the pagan roots of gnosticism. Prior to the 1963 launching of the "New Age" as a mass movement within the United States, this form of New Age satanism was an endemic cancer in such forms as the theosophical existentialism of the followers of the proto-Nazi Friedrich Nietzsche (1844–1900) and the pro-freemasonic satanists of Aleister Crowley's networks.

What is notable on these accounts is the increasingly emboldened way in which the two evils, the "New Age" and usury, have exhibited their natural affinities for one another, combining their forces in even the highest places of Anglo-American power, to demand, in the misused name of "freedom" and "ecology," the rapid extermination and global outlawing of every scientific and moral barrier which has hitherto existed as impediments to rampaging immiseration and dictatorial oppression of mankind.

Such are the leading characteristic distinctions between the problems immediately addressed 100 years ago and today.

The former hegemony of scientific and technological progress, upon whose continuation the existence of our populations depends, is being suppressed by both the loss of simple rationality in the education of the young, and by the spread of the paganist cults of anti-science, irrationalist "ecologism." Concomitant with such specific, catastrophic effects as this one, those European and American forces which are committed to calculated mass-murder of populations of all developing nations, and which are committed to the extermination of the Christian faith and conscience, have come plainly into the ascendancy in the policy-making processes of most of the governing international and national governmental institutions which have gained leadership and dominance over this planet today.[4]

The Ecumenical Standpoint

We propose that it is necessary, but not sufficient, to view the referenced state of affairs from a Christian standpoint; for practical reasons, it is essential that even the Christian standpoint itself be presented here from an ecumenical standpoint, as *ecumenical* is typified by Cardinal Nicolaus of Cusa's (1401–64) dialogue, *De Pace Fidei.*[5] On that account, we have considered it most important to reference the explicitly cited sentence and attached footnote from the encyclical. (Also see Appendix XIII.)

Different faiths, religious and/or secularist, can be brought to principled agreement only in two possible alternate ways of manifesting mutual good will. In the one case, they may agree on a common point of taught doctrine, such as the principle of monotheism, as in opposition to the pantheistic pluralism of pagan Babylon, Rome, or the Apollo Cult at Delphi. Or, otherwise, differing faiths may reach coincidence of principled views by the means indicated in the referenced features of the encyclical's section 72. It is the latter alternative upon which we concentrate attention here.

It is the obvious intent of the author of the encyclical that his own intention and that of the referenced passage from the *Summa Theologica,* respecting reason, should be received as identical. We adopt that intent here.

Faith may read those writings it deems sacred, or authoritative, commentaries on such writings. Or, faith may "read the bare book of universal nature," a book which plainly has been written directly by none other than the Creator Himself. It is certain to all men and women of ecumenical good will, that the two kinds of books—the written ones and the book of nature—cannot contradict one another, on condition that the written one be true, and that both the written and the natural one be read by means of the inner eye of true reason. (See Appendix XIV.)

So, where doctrinal writings differ, we may turn the eye of ecumenical reason to the common book of nature.

Let us argue the point in the following, twofold way. We emphasize, on the one side, the ecumenical notion of *intelligible representation* of a principle of knowledge of cause-effect in our universe, a means by which all men and women, despite differences in profession of monotheistic faith, may be brought by their own powers of reason to agreement upon a common principle of law. Second, we emphasize the importance of stressing *Christian* principles of Christian civilization as *Christian*, even within the framework of a monotheistic ecumenicism. (See Appendix XIII.)

Consider next this simple illustration.

The most ancient among known astronomies, that of the ancient Vedic peoples of Central Asia,[6] illustrates the obvious manner in which a so-called "primitive" people may construct a reliable solar astronomical calendar from scratch. Observe successively the position of the Sun, at dawn, mid-day, and sunset. Mark these observations each in stone. At night, observe the constellations and their stars, to which each of the respective three day-time observations point. After five years, we have thus the data on which to base a solar astronomical calendar of approximately 365 days per calendar year, measuring the year either from the winter solstice to winter solstice or from the vernal equinox to vernal equinox.

By the same method, the long millenarian equinoctial cycle is adduced. So, a system of solar astronomy, free of the whore-goddesses Shakti's and Ishtar's lunacies, is built up by aid of reason. So the book of nature may be read—God's book of nature.

In such successive revolutions and related ways, *reason* reveals to us that our universe has the apparent form of a unified cause-effect process of *Becoming,* a process of *Becom-*

ing which is subsumed by an indivisible Supreme Being, who embodies, among other qualities, what Plato (427–347 B.C.) admired as the *Good*.[7] Of such matters of principle, in such a manner, do the very stones cry out.[8]

Consequently, when we demonstrate, by access to reason, that a certain universal or approximately universal principle must be true, a monotheistic ecumenicism has gained a two-fold advantage. Since all of human knowledge is finally supplied by reason, there can be no valid teaching presented by any religion which contradicts true reason, as we define *reason* in the following chapters; there can be no valid objection to this principle which is to be tolerated on premise of secularist rejection of religious precept.

Physical Economy

By the nature of the case, there is no field of inquiry which unites all subjects of human reason—law, science, art—as directly, as immediately, as the science of Physical Economy, which was founded by Gottfried Wilhelm Leibniz (1646–1716). That is a special standpoint of the work we preface here.

As is to be seen in summary in the appended document, *Physical Economy* is the science of *successful change*, a study of the dependency of the continued existence of a society upon *successful* forms of successive generation, transmission, and efficient assimilation of fundamental scientific progress. The measure of that effective progress is an increase in what Physical Economy defines as the rate of increase of the potential population-density[9] of that society as a whole. That thus serves as an efficient empirical measurement of both the appropriateness of the society's way of changing its method of reasoning, and, therefore, the appropriateness of the principle of change adopted for that practice.

Any society which defies those considerations, is threatening its own continued existence, and is a society implicitly becoming an abomination in God's eye; a society which is not only losing the moral fitness to survive, but which, by God's clock, will not long survive in its present form.

Historically, to date, the closest approximation of a form of political economy consistent with Christian principles is the so-called *mercantilist* form growing out of *Colbertism* in France, and the far-reaching influence of Leibniz. This outgrowth came to be known by the name given to it officially by U.S. Treasury Secretary Alexander Hamilton (1757–1804), "the American System of political-economy."[10] This name came to be associated with the work of the U.S. economists Mathew Carey (1760–1839) and Henry Carey (1793–1879), and of Germany's Friedrich List.[11]

The deadly adversaries of the so-called "mercantilist," or "American" system, were the Anglo-French-Swiss, known in the early eighteenth century as the "Venetian Party."[12] This was the political faction allied against Leibniz and his friends, and allied with the first Duke of Marlborough (1650–1722), allied with the networks of Voltaire (1694–1778), with the physiocrats, and with so-called eighteenth-century "British liberalism" of Hugh Walpole, David Hume, Shelburne, Adam Smith, Jeremy Bentham, and Thomas Malthus (1766–1834) generally. These physiocrats and liberals were the chief guise for the pro-usury faction of that century.

That issue of the eighteenth century is more efficiently understood by emphasizing that the liberals and *illuminati* of Voltaire's eighteenth century were committed to a return to the model of a pagan Imperial Rome. Hence we call them "Romantics." These Romantics were dedicated to the overthrow of Christianity for the purpose of advancing their *Romantic imperial utopianism*. That is the root of the *structures of sin*[13] in Western European and North American civilization today. These were then, and are still today, both the pro-usury faction and the utopian cultural form from which the present-day satanic "New Age" utopianisms have sprung.

The 'American System' Model

We do not uphold the Leibniz-Hamilton-List form of "American System" to be a perfect model. We do not propose that the American leading stratum of 1776–89 was a pure embodiment of Christian principles.

We make two modest claims for that system. First, it was, in the domain of political economy, the only significant resistance at the time to the evils of eighteenth-century British imperialism, and for as long as it did resist that evil thereafter. Second, that, relative to the British liberal and communist systems, the Leibniz-Hamilton-List form of American System is the only historically notable form of modern political economy which is a proven successful alternative to the twin, catastrophic moral failures of British liberalism and communism. Thus, historically, this American System is the only significant approximation of a modern agro-industrial system which tends to afford the means to satisfy the requirements of *Rerum Novarum*. In contrast, British liberalism, intrinsically, implicitly fosters even in the worst degree all of the principal evils addressed by that encyclical.

In the relatively shorter or even the medium term, sweeping changes in general practice can be successful only if much of the population can be induced to regard innovations as bearing the historical authority of a successful precedent.

So, in the United States of America, for example, nearly every person over 40 years of age today has a vivid recollection of the moment and circumstances each first heard the news of the assassination of President John F. Kennedy. So, it is relatively easy to recall the happier economic policy trends of the Kennedy administration, relative to the comparably depressing trends of the adjacent Eisenhower and Lyndon Johnson administrations. So, the idea of reviving anti-recession policies referencing successful precedents from the 1961–63 period, is one which must tend to enjoy support under the rudest economic circumstances of the United States today.

Similarly, it requires only a slightly longer reach of the American or European mind to recall the happier "mercantilist" policies of the American System, Friedrich List, Charles de Gaulle, Konrad Adenauer, or Italy's Enrico Mattei.

So, those of us looking at today's global conditions from the standpoint of an ecumenical reading of *Rerum Novarum*, are compelled to take a practical historical view of available meliorative measures, whose employment represents a philosophically *unobjectionable* tactic for furthering the cause of principles. Thus, we are obliged to inquire, formally and historically, why the American System of Hamilton, List et al. is consistent with Christian principles, when British liberalism is adversary to those principles. We are not thus adopting the American System as a point of Christian or ecumenical doctrine.

Nonetheless, although we are obliged to recommend such attention to historically proven methods, that required work does not allow us to descend into the moral mediocrity of mere pragmatism. It does not free us from the duty of setting forth principles which are fully consistent with the eternal laws which reason may make accessible to our knowledge. So, if we recommend the American System as a historically proven precedent for modeling short-term and medium-term remedial policies today, we must also set forth the lawful principles which must guide us through the medium term into the long term, which may be different than those of the American System precedent.

On the Subject of The Science of Christian Economy

Today, three years after the great financial crash of 1987 and two years after my October 1988 Berlin address on impending German reunification,[1] it is increasingly clear that the two formerly reigning economic dogmas of this planet, those of Adam Smith (1723–1790) and Karl Marx(1818-83), are being buried, perhaps forever, under an avalanche of post-industrial rubble and usury. Unless the specifically appropriate replacement for these two failed dogmas of yesterday is adopted soon, every part of this planet is to be judged now as already plunging into a New Dark Age, worse than that which crushed Europe, with nearly apocalyptic force, during the middle of the fourteenth century.

Had we been confronted so immediately by such awesome truths at the beginning of our century, when a significant minority among economists and historians were still literate in their professions, the latter would have responded to the preceding paragraph here, with words to the following effect: "You are proposing an immediate return to the original 'American System of political-economy.' " They would signify so, those (anti-British) principles of economy and banking associated traditionally with such prominent names as U.S. President George Washington, Treasury Secretary Alexander Hamilton,[2] the two Careys,[3] Speaker of the House Henry

Clay,[4] and the founder of the German economy's nineteenth-century economic supremacy, Friedrich List.[5]

The now freshly discredited such liberal dogmas were the leading evils of the eighteenth century, against which the U.S. War of Independence was then so justly fought by all patriotic Americans.[6] The American System, on whose behalf those patriots fought against their British oppressor then, is imbedded implicitly in the 1776 Declaration of Independence and in the Preamble and in Article I of the 1787–89 federal Constitution draft.

Since 1787, whenever the U.S. government has applied the policies of that American System, the nation has prospered, to the net effect that it once became for a period of decades the leading economy of this world. Whenever that same government committed the great folly of imitating the lunatic ideas of Adam Smith, as it did under Presidents Jefferson, Madison, and Jackson, for example, the United States has been plunged into economic ruin, as it is now being ruined by the accumulated follies of the past six Presidents following the assassination of President John F. Kennedy.

It would be a very good thing if the United States today would overturn by a single law, immediately, every change in U.S. economic, financial, and monetary policy which has been introduced since the assassination of President Kennedy. That would be a good thing; but more must be done. We must reaffirm the American System of political-economy, upon which all of the United States's economic successes to date have been premised implicitly; but, even that is not quite enough.

For reasons to be shown here, we must not admire the proven superiority of the American System so much, that we overlook the fact that the American System is merely a successful approximation, for purposes of application, of something much deeper, of something less imperfect, something truly fundamental. The American System was chiefly a reflection of the combined direct and indirect influence of the founder of economic science, Gottfried Wilhelm Leibniz, upon certain leading thinkers of the English colonies in the Americas.[7] Yet, there is something still more profound at issue here.

Let us focus for a moment upon the issue of the several wars which Britain either led or orchestrated against the United States during the interval 1775–1865.[8] The central issue of those wars was Britain's refusal to tolerate those economic policies identified by President George Washington's treasury secretary, Alexander Hamilton, as "the American System of political-economy."

The more general significance of that economic conflict with Britain, is that the American System of political-economy is broadly in agreement with the principles upon which Christian civilization is premised. The original and continuing enemy of that American System is an anti-Christian dogma, an explicitly immoral dogma, which was conceived, originally, by Adam Smith and other agents of the eighteenth-century British East India Company, as an emulation of the model of ancient, pagan Imperial Rome.

In the following pages, we summarize the deeper issues between Christianity and British neo-paganism, underlying London liberalism's continuing efforts to exterminate even the memory of that "American System." We address and defend thus, the "axiomatically" Christian features of Gottfried Leibniz's founding of economic science. We defend the "American System" in its implicit aspect as a reflection of Leibniz's influence among leading eighteenth-century American patriots.[9] We indicate in this way, the crucial importance of those Christian principles and related matters for defining efficiently the strategic crisis of the 1990s.

I | The Strategic Setting

The underlying, principal strategic conflict dominating the planet today, is expressed as the varied threat of generalized warfare, famine, and epidemic disease, the which is caused, ultimately, and that almost entirely, by the several efforts to employ the model of pagan Imperial Rome, and pagan Roman, or more barbaric ethics, for the purpose of establishing "one-world government," through the United Nations Organization, or some alternate instrument. Today, as in the earlier Roman times of the evil Tiberius, Nero, Caligula, and Diocletian, the chief impediment to consolidating such an evil "New World Order" is the force of Christianity.

This has been a persisting threat throughout modern European history, and a particular such threat since the time of the evil first Duke of Marlborough and Marlborough's "Venetian Party" cronies, at the beginning of the eighteenth century.[1] The displacement of Christian morality and that morality's replacement by the amoral ethics of such sundry paganists as the British liberals and Romanticists generally, was a characteristic feature of such would-be revivals of pagan Roman world empire as Pax Britannica, Bonapartism, or Dostoevsky's and Hitler's sordid dreams of a Third Rome.

Inevitably, the campaigns for a pagan form of one-world imperialism appeared frequently in the guise of attempts to

219

eradicate Christianity, and always appeared in the guise of preferring some form of secularist, pagan ethics to Christian morality. This paganist impulse is typified by the work of such figures as Francis Bacon, Thomas Hobbes, John Locke, David Hume, Voltaire, and Adam Smith. Another name for this paganist one-world cult is the "New Age." Strictly speaking, the paganist Francis Bacon was already a New Ager, as were Marlborough's cronies and followers among the British eighteenth-century liberals, and the cronies and followers of Montesquieu, Voltaire, and so forth, among French-speaking Romantics. Usually, today, "New Age" signifies those beginning with Oxford's John Ruskin, for example, who espouse the astrological slogan of self-proposed Anti-Christs Friedrich Nietzsche and Aleister Crowley: to end the Age of Reason (Socrates, Christ) and bring on the Age of Aquarius (Dionysios-Apollo, Lucifer-Lucis, Satan).

II | The Genesis of Economic Science

The science of political-economy is premised upon the conclusive, empirical evidence of a fundamental difference which sets the human species absolutely apart from, and above, all of the animal species, as Moses specifies in Genesis 1:26.

This crucial difference is mankind's power to increase the potential population-density of the human species as a whole by means of the voluntary generation, transmission, and efficient assimilation of scientific and technological progress. Mankind is capable of increasing, intentionally, the maximum size of the human population which could be self-sustained by its own labor, per average square kilometer of land area, while also raising the average physical standard of living.

No animal species can accomplish this. The range of successful adaptation of an animal species is delimited, as if by genetic determination; mankind incurs no such limitation upon our population, nor the development of the individual members of that population.

This increase in man's physical productivity is properly measured in both per capita and per square kilometer terms: the rate of useful physical output per person and per square kilometer.

We are obliged to measure that output not merely as simple quantities of objects. Since we are referencing the rate

221

of production in terms of the self-reproduction of the human species, we must measure both the inputs and outputs in reproductive-actuarial terms. We must take into account, as functionally causal variables, a system of measurement of inputs and outputs premised upon the family household as the social unit of both the quantitative and qualitative reproduction of not only entire societies, but mankind as a whole.

Therefore, we are obliged to measure the characteristics of the individual members of the family household in such terms of differentiation as *generations, age intervals, health, mental development, life expectancies,* and *fecundity.* We must measure not individual objects, but the average market basket of consumption required, per person and per household, in terms of the corresponding cause-effect relations.

We are obliged to measure the productive relationship in terms of mankind's productive changes in nature: that is, man's increase of the present and future fecundity of land for human reproduction. This is as we read the message of Moses in Genesis 1:28–30.

In order that such a process might be continued according to the instruction of Genesis 1:28, mankind must effect willfully those successively more truthful scientific discoveries by means of which increase in the well-being and productive powers of labor is accomplished. This willful progress can occur only under condition that there is *a knowable principle of ordering* governing the progress from inferior to higher levels of knowledge for practice.

Since the better ideas so discovered, must correspond to a superior mastery of nature as a whole, the knowable laws which govern progress in fundamental scientific discoveries must be the laws which govern the universe as a whole. If this were not possible, then the human species population would never have risen above 10 million persons living upon our planet at any one time: a fair estimate of the potential population-density for a "primitive hunting and gathering society."

Man might employ successfully a tool whose design he

did not understand correctly; however, the principle which orders successfully successive, fundamental improvements in tool design, must be in at least approximate agreement with the principle ordering the underlying lawful ordering of our universe.

Indeed, the name of *science* is properly reserved to designating the discovery of those underlying principles which can be shown empirically to order a successful ordering of successive scientific revolutions.

In that view of *science* as activity, man's conscious knowledge and mental activity is approximating the principles of organization of all creation. In that respect and degree, the mind of the individual scientist is mirroring the creative will of God the Creator. Thus, as we read in Genesis 1:27, man is created in God's own image.

This much which we have just outlined is all demonstrable to human reason by means of crucial, incontestable empirical evidence. More is similarly demonstrable. What has been outlined thus far bears chiefly upon mankind's interaction with the universe, and that only in a general, if nonetheless conclusive form. Examine next, somewhat more deeply, man's living likeness to the image of God the Creator.

III | Imago Viva Dei

Although the development of the creative mental powers of the human individual occurs within a social process, the creative processes by means of which each individual may generate, transmit, or assimilate practically valid discoveries, are processes of concept-generation which are, demonstrably, wholly internal to each individual person. Therefore, those creative powers of the individual are *sovereign* powers of each individual in which that *divine spark* of potential for creative reasoning is developed. (See Appendix XIII.)

It is not only the existence of the creative powers which defines man as in the image of the Creator; it is the fact that this creative power is in each instance a sovereign capability of the person, a sovereign essence of that individual, which defines the human individual as individually in the living image of the Creator or, in Latin, *imago viva Dei*.

Therefore, all human life is sacred. If a human life may be taken in the heat of morally justified warfare, or other mortal combat, no Christian may ever terminate a human life at leisure, when the individual is helplessly in our power to sustain or kill. Otherwise, we sin directly against God.

The sacredness of human life is perhaps better understood, if we take into account the practical importance, to all of mankind, of each individual person who adds in the slightest degree to our fundamental scientific knowledge.

Put more simply, every improvement in mankind's store of fundamental scientific knowledge adds implicitly to the potential productive power and moral development of every person, present and future, of society as a whole. The rate of human progress thus tends to increase as we increase the total number of living persons whose mental powers are developed to generate, transmit, and assimilate the fruits of fundamental scientific progress. This is a simply demonstrated fact, which involves the deepest principles of the science of economic practice.

The Malthusians argue, that the rate at which a society produces is the rate at which the entire society is depleting raw materials and other natural preconditions for human life. It should be obvious, that if the rate of scientific progress is great enough, no depletion will occur through expanded scale of production and consumption. Thus, on the condition that we develop and employ the creative potential of each new individual, a higher birth rate increases the relative scale of natural resources—a result directly opposite to the Malthusians' well-known, but anti-scientific assertions.

Indeed, those create who might merely rear children in such a fashion as to nurture the creative potential of those children.

Any society which persists in what is commonly called today "a zero technological growth" policy of practice indefinitely, must first stagnate and then collapse ultimately into ruin. Archeology is occupied chiefly, although not exclusively, with the pitiful remains of such failed, inferior cultures.

The more obvious among the contributing causes for such a wretched failure, is the inflationary and other ruinous effects of depletion of raw materials and analogous "environmental resources." There are also deeper, underlying causes for the ruinous outcome, requiring deeper insight. We treat the case for raw materials first.

Generally, the relative quality of "ore" is defined in terms of the processes required, successively, to find it, to get it, and to refine it into the desired form of semi-finished or so-called "intermediate" commodity. The predominant consideration, in nearly all of the cases, is the ratio of labor required to bring a per capita market-basket's ration of consumption of that intermediate commodity to the appropriate place, in the appropriate state of refinement. What we term "energy" has a leading bearing upon this determination of relative cost.

The case for metallic ores illustrates the principle.

The feasibility of reducing an ore to produce good-quality ingot, involves the relative temperature (energy-flux density) delivered to each relevant molecule. For example, to go as directly as possible to the working point, if we can put a reduction-process into suitable magnetic confinement ("magnetic bottles"), and raise the operating temperature inside there to the critical level of temperature-equivalent (energy-flux density) at which tungsten exists only in the plasma state, every kind of rock or solid or liquid waste in the universe becomes a more or less economical form of ore.

If we have available sufficient energy, at a sufficiently high energy-flux density, if we can handle that energy-flux density in production processes, and if the labor-cost of that energy and its productive application is a sufficiently small ratio of the average amount of productive labor employed by that society, there is virtually no limit to the supply of commercial grades of ore. On the condition, that the *increasing* quantity of energy, the *rising* level of energy densities, and advances in employed technologies are proceeding in a properly coordinated way, at adequate rates, there is no "limit to growth" on the horizon of mankind today.

To show the fallacy of the obvious objections to what has just been stated, note the following.

Once we have achieved what is usually termed a "second-generation" fusion-energy device, in the range of a terawatt unit-output, mankind implicitly has escaped the bounds of planet Earth, to as far distant as the asteroid belt. Similarly, after that, the next energy sources, controlled matter/anti-matter reactions, should be achieved by approximately the close of the coming century—on condition we are determined to bring this about. That upward step takes us to the outer limits of our solar system and into technologies carrying us far beyond that.[1]

That destroys the implied objection to our observation respecting the limitlessness of prospects for growth.

Limits to growth appear and close in upon us, only if our society is a foolish one. If a society is foolish enough to suppress the increasing of per capita consumption of energy, the society will be crushed by its own stupidity. If a society is suicidal enough to call a halt to capital-intensive, energy-intensive investment in scientific and technological progress, or, even worse, to substitute labor-intensive "services" for capital-intensive, energy-intensive manufacturing, that society is implicitly dooming itself to collapse.

The possibility of a successful society depends upon two conditions. First, the society must generate scientific and technological progress; to do this, the society must have developed in its members the disposition and capacity for scientific progress. Second, the society must adopt policies which cause (the physical equivalent of) productive investment in scientific and technological progress to prevail over opposition to such policies.

Thus, with certain qualifications, we must speak now of "man the creator." The mental-creative powers, which mankind demonstrates through the use of scientific revolutions, to increase qualitatively the potential population-density of our species, is the generality referenced. This generality shows mankind to mirror the Creator. Thus, man is designed to become the "little creator," the small mirror-image of the universal Creator. The former, the "little creator," we call the "Minimum"; the universal, the Creator, we call "the Maximum."

Not only is this creative power uniquely characteristic of mankind, among all species; this creative power is located within the individual human personality, as a *sovereign* potential contained within that individual personality. Thus, it is the individual person who, by virtue of representing this *sovereign* power, is, individually, in the *living image of the Creator* (*imago viva Dei*).

In the frequent case, that we may think that particular persons fail to express this living image of God in their conduct, those persons were born with the potential for creative reason, even though they may have abused or rejected that divine spark of potential within themselves. Thus, all individual human life is sacred.

IV | The Sovereign Personality

So far, we have indicated some of the leading facts which show all intelligent men and women, that scientific and technological progress is the essential characteristic which distinguishes the economy of the successful society from the relatively inferior, failing culture.

In the author's short book, *In Defense of Common Sense,* the reader will find the required elementary definition of the term *creative.* That book makes clear the difference between deductive argument—the lower order of rational thought—and creative mental activity.

The discovery of a brand new, valid scientific conception is the expression of a process which, by its very nature, occurs entirely and uniquely within the mind of an individual person. No matter how numerous the external, social influences participating in developing that person's creative-mental potentials, the generation of a new concept is a process which occurs exclusively within the mind of that thinking person. As we have already emphasized, the process of generation of that conception is therefore a *sovereign* process.

These two conceptions, *the role of scientific and technological progress,* and the fact that *each creative mental act is a sovereign process of an individual personality,* are the essence of all economic science. Such an economic science is in a unique form of agreement with Christian principles. Moreover, economic science was developed, in fact, by Christianity; furthermore, the evidence is that perhaps economic science could not have been developed except by Christianity. The essence of this connection is expressed by the *Filioque* of the Latin Creed; only Christianity, through the view of Jesus Christ reflected in this feature of that creed, organizes society implicitly according to the principle of the sovereignty of the human individual, defined in the way we have defined it here.

When we hear ourselves speaking solemnly words and phrases such as *survival, national interest, individual rights, human rights, equality, freedom,* and so forth, what do we really mean?

Given the foregoing outline of the matter, it should be clear, that the essential self-interest of the individual person is the self-interest implicitly associated with this notion of "sovereign creative process" of the individual personality. We now explore, summarily, step by step, the way in which such essential self-interest is adduced.

Firstly, since we are each mortal and thus must die, our highest self-interest is associated with the best of our life's productions, which we leave after us. This donation which we make to our posterity presumes that there will be a posterity to receive the gift. These reflections guide us toward the understanding which we should be seeking here and now.

Think of the productions we might so bequeath. Begin with the most obvious of the implied queries.

Can this production be an object?

Suppose a man and his wife take a poor piece of wild or depleted land; suppose that this pair is raising a family there and develop that poor patch of soil into a fecund farm. Suppose an architect designs a city, better than most in utility and aesthetical merits, which may endure to mankind's admiring advantage for several thousand years to come. Are these or other worthy *objects, in and of themselves as objects,* the kinds of production we wish to bequeath to our posterity?

It is good to provide our immediate posterity with useful objects; but no object could embody, merely as an object, the quality of almost timeless, virtually inexhaustible durability respecting its benefits to future generations. We ought to desire, that our brief, mortal existence might contribute something of virtually timeless benefit to future generations.

This matter is examined rigorously in *In Defense of Common Sense* to the following effect.

Any *object* we might fashion may crumble or become relatively useless by virtue of technological attrition.[1] In contrast, no valid scientific discovery of today can ever be rendered as having been historically unnecessary. All valid scientific discoveries will be superseded by more valid ones; but, nonetheless, each is the necessary foundation for each and all of its successors; in the latter fashion it enjoys a splendid immortality in the whole of human existence.

In this sense, valid scientific discovery of a more truthful comprehension of natural law, typifies the immortal fruit of a mortal life. In this sense, to contribute or even merely to service such a discovery, typifies, by reflection, what is truly the essential self-interest of any person. It is only a reflection; it is not yet an adequate representation of the true, deeper self-interest; but, this reflection points our thinking along the right pathway.

So far, here, we have said implicitly, that a person is expressing his or her self-interest as an individual human personality, only as he or she is engaged in activity which employs the same, sovereign, creative process of powers of reason which we associate most readily with the generation, transmission, and efficient assimilation of valid forms of fundamental scientific discovery.

That argument implies, in its turn, that the only true self-interest of the human personality is to express, and also defend, one's own human nature. Since mankind is set apart from and above the beasts, solely by the person's sovereign

potential for creative reason, only the individual's expression
and defense of the supremacy of such creative reason is a truly
self-interested action by a member of the human species.

For pedagogical and kindred reasons, we have considered
here so far only one among the expressions of creative reason,
those kinds of valid, fundamental, physical scientific discoveries often termed both "crucial" and "revolutionary." We do
not intend to exclude or disregard other expressions of creative reason. We are implying that, whatever is true for the
case of scientific discovery, is also true, to kindred effect, for
each and all other individuals' expressions of creative reason.
For that reason, it is permitted to present our case for economic science as we do here initially, limiting our attention
to the implications associated with valid forms of fundamental
scientific discovery.

The achievement of a valid revolutionary discovery in
physical science yields implicitly an array of useful objects.
These objects may be judged "useful," only in the degree that,
by class, they elevate significantly the productive powers of
our species, and thus tend to increase our species's self-reproductive power. That defines the notion of "usefulness" of the
object generated, as a by-product of creative reason's action.

Consider an outstanding example. The first comprehensive mathematical physics is that of Johannes Kepler (1571–
1630). (See Appendix V.) In connection with Kepler's work
of founding a comprehensive mathematical physics, sundry
instruments were generated as by-products of his creative
reasoning. This included the first mechanical computer.

The weight of Kepler's influence lies in the success of the
method by means of which he founded the first comprehensive
mathematical physics. This is said in the sense, that we might
overlook all of the useful objects generated then as by-products of this work, without diminishing thereby the enduring
historical importance, and continuing usefulness of the discoveries.

To the present date, although improvements in the secondary features of Kepler's solar astrophysics are necessary, the underlying conception of Kepler's design remains essentially competent, whereas the Newtonian and other proposed alternatives of the past are all discredited by means of crucial-experimental evidence. (See Appendix V).

Look at the pre-history of the digital computer.

Kepler designed, built, and used the first mechanical calculator. The same principle was central to the later design and construction by Blaise Pascal. Kepler and Pascal were directly forerunners of Gottfried Wilhelm Leibniz's development of the mechanical calculator. The essential features of the modern electronic digital computer are nothing better than an application of Leibniz's principles for mechanical devices.

It is also illustrative of the same argument, that two of Kepler's most durable contributions to scientific progress were problems he posed for solutions by his successors: the development of the differential calculus, and the solution of elliptic functions. Pascal contributed to establishing a Keplerian differential calculus, followed by Leibniz, whose first successful discovery of such a calculus was completed by 1676.[2] The mastery of elliptic functions was effected by Carl Friedrich Gauss (1777–1855) et al., during the first half of the nineteenth century, more than 200 years after Kepler had posed this challenge.

Today, more than 350 years since Kepler's death, Kepler's method for astrophysics has been proven also crucial for correction of the common failings of quantum mechanics respecting the atomic nucleus.[3]

Any valid "revolutionary" discovery in natural philosophy does much more than correct deep errors of contemporary popular textbook opinion. Each valid such discovery increases the rigor and creative power of the method available for effecting new, greater "revolutionary" discoveries. This point is made clear by imagining a proper form of secondary-school and university curricula, from which that abomination known as the course textbook is outlawed.

In physical science, as in geometry, too, the student masters the comprehension of the subject by reliving, as nearly as possible, the mental experience of the original discovery by the original discoverers. In that approach, a collection of original sources replaces the course textbook. The original crucial experiments are relived by the student; and improved, better experimental versions of the same crucial hypotheses are also scrutinized.[4]

Most important, physics discoveries are to be accomplished by aid of recognizing a faulty assumption imbedded historically in the supposed proof of a hallowed truism of contemporary professional certainties. A grounding in crucial historical source materials, is obviously the virtually indispensable foundation for scientific rigor.

So, the creative physicist will be forever, periodically, reexamining the work of Kepler and Kepler's predecessors, again and again; in this and kindred Socratic enterprises, the foundations of coming scientific revolutions are being established, reaching so the indefinitely distant horizons of the future.

Thus, the essence of the scientist's true self-interest is that which he contributes, as sovereign creative activity, to furthering the endlessly continuing process of fundamental scientific progress. To restate this same point: The most essential contribution which the scientific discoverer may make, is less a particular scientific discovery, than an improvement of the known principles by means of which subsequent generations effect entire new generations of valid, fundamental scientific discoveries. In this way, the mortal sovereign person becomes the necessary individual mortal existence, who has enriched the power of the human species as a whole, for all time to come.

The way in which such a mortal life benefits present and future generations should be more readily obvious. To consider, next, the benefit to the past, touches the subject of our inquiry more profoundly.

Let us return our attention to the two cited challenges

which Kepler left to his successors: the development of a differential calculus, as accomplished by Leibniz; and the general solution of elliptic functions, solved essentially by Gauss et al. Did not Leibniz and Gauss benefit Kepler in a readily intelligible way? Does my work die with me, or is it reinvigorated to continued, efficient life, by the work of my successors? Kepler clearly sought a Leibniz, a Gauss: In good time, each responded to Kepler.

If and when relations of individuals across time, in the future and into the past, are seen in these terms, mortality is cheated of its fearfulness. For this author, for example, some such scientific figures as the fifteenth-century Nicolaus of Cusa and turn-of-the-eighteenth-century Leibniz are, in many ways, efficiently as if living contemporaries, as are unknown figures from the distant future to whom this author is also morally accountable.

Science, thus, gives an *isochronic* quality to the linking of the work of diverse persons across even great expanses of past and future time. The same is true in matters of classical forms of art and in all other matters truly important, by their nature, to the human species as a whole.

How shall we define here the purpose of this development to which the sovereign creative powers of the mortal individual contribute so *transfinitely?* The answer can be summed up on two successive levels.

On the first level, it is a physical advantage. The continued existence of the human species depends upon technological progress. We have already considered an illustration of that point. Technological progress increases the per capita productive powers of society; at the same time, technological progress transforms and improves nature. That improvement is essential, or else human depletion of fixed varieties of so-called "natural resources" would doom us.

On the second, higher level, *it is a spiritual advantage.* It is the development of the quality of man by means of which the twofold, subsumed, physical gain is effected.

The net effect of a valid fundamental sort of scientific

discovery, is to increase the sovereign creative power of virtually everyone who assimilates that discovery. Thus, through fostering the development and expression of individual sovereign creative powers, the net result is the self-increase of the sovereign creative powers of the members of the human species as a whole.

Let us, next, re-examine what we have said thus far, introducing a slight, but crucial change in our choice of standpoint.

V | Agapē

Let us turn our attention to Paul's first Epistle to the Corinthians (I Corinthians 13). Let us now examine from this vantage point what we have argued thus far.

For the benefit of the layman, before proceeding to our point at hand, we interpolate here a preliminary, introductory definition of *agapē*.

Most literate adult persons have encountered, somewhere in their readings or related encounters, a reference to a distinction between "sacred" and "profane" *love*. As a preliminary step, assume that the *love (agapē, caritas, charity)* referenced in I Corinthians 13 and other New Testament locations, is defined approximately as equivalent to "sacred love," the latter as opposed to "profane love."

"Sacred love" is exemplified by "love of God." We signify otherwise *love of truth, love of (classical forms of) beauty,*[1] and *love for mankind*. We signify the love of the parents and the grandparents for the development of the human potentials of the child.

We signify this set of *isochronic* relations, encompassing past, present, and future, defined always in terms of the individual creative reason's generation, transmission, and effective assimilation of valid, fundamental scientific discoveries. This is usually restated in terms of our preceding section's discussion: These isochronic relations, encompassing past, present, and future, are defined always in terms of individual creative reason's perfection of individual creative reason, *by means of* the generation, transmission, and efficient assimilation of valid, fundamental scientific discoveries. This latter can be restated usefully in all the broader terms implied, respecting classical art forms, creative acts of love for mankind, and so on.

What is emphasized at this immediate juncture, is the agreement, the coextensive congruence of *agapē* and of universal acts of creative reason. The reaching out to the universality of mankind's past, present, and future, *for the love of God,* is *agapē* expressed practically, as a creative act directed toward perfection of the creative powers of mankind.

Without such *agapē,* there is no creative power, no creative act. It is by means of creative acts, as we have defined "creative" here and in other published locations,[2] that the emotional state associated with *agapē* is expressed and communicated.

To act is, implicitly and efficiently, to prefer, to choose one way, among others, of using a portion of that pitiably finite resource which is the entirety of the allowed mortal existence of an individual person. It is the highest, true self-interest of that individual person, to prefer, to choose an act which is of the relatively greatest and most far-reaching benefit to future generations of all mankind. Such a choice is implicitly an act of *sacred love* toward mankind, on the condition that the chosen act is of appropriate quality and is motivated by such a specific intention.

We have considered already some of the reasons the quality of that chosen act must express efficiently the individual's sovereign power of creative reason. We have included among such universal acts, the development of the potential of moral character and creative reason in the individual child. However, we have also indicated, it is the higher development of the creative power of future generations, which, in itself, expresses the essential form of the true good, which expresses, less imperfectly, the choice of act which is consistent with one's true, deeper self-interest.

In his *Of Learned Ignorance (De Docta Ignorantia)* and other locations, Cardinal Nicolaus of Cusa develops a conception termed "Maximum-Minimum." (See Appendix II.) This conception has several subsumed significances for physical science; it has a more generally inclusive significance which we stress in connection with *agapē*.

We have defined the human individual engaged in use of developed powers of creative reason as a sovereign entity. This use of "sovereign" signifies, among related notions, that the process of constructing a single conception, a conception of the form of a valid, revolutionary, scientific discovery, or germ of a great artistic composition by Raphael or Beethoven, is a process which occurs entirely within the mind of the individual person. That also signifies, that whoever relives that mental act of discovery—for example, as a student—in transmitting and assimilating that integral idea, is also acting in the same sovereign capacity.

In the case of such revolutionary discoveries in physical science, it is implicit that this sovereign creative process of the individual is engaged practically with the laws governing the universe as a whole. The latter represents the efficient will of the Creator. The latter is the Maximum; the individual creative mind is the Minimum, which echoes the Maximum. To the degree this mirroring is in the process of being perfected, the Minimum is in *the living image of the Maximum (imago viva Dei)*.

The relationship so described between Maximum and Minimum is *agapic,* if it fulfills what we have pointed toward as requirements of both quality and intent. Also, this sort of relationship among two or more *Minima* (Leibniz's *monads*), participates in the higher ordering, the higher, mirroring relationship to the Maximum. Thus, does *agapē (caritas, charity, sacred love)* permeate all; nothing is a true good unless it is so permeated with *agapē*.

The purpose of human existence, the truest self-interest of both every individual person and every society, respectively and as one, is to order the proliferation and perfection of human existence in our universe, in this *agapic* way.

So, I Corinthians 13 may be read for the occasion of reflection upon this topic.

VI | Reproduction of Man

The most crucial among the *specifically economic* facts which set a Christian economic practice apart from any other, is the manner in which both Leibniz's *science of Physical Economy* and Christianity define the *sovereignty* of the individual personality. As we have stressed that fact thus far, this *individual sovereignty* is strictly defined, uniquely, by that *universal historical fact* which sets the human species perfectly, and absolutely, apart from any and all beasts: *the divine spark* of developable potential for *creative reason,* as this is defined in *In Defense of Common Sense* and *Project A.* (Also see Appendix XIII.)

The crucial historical fact referenced by economic science, is the combined increase in per capita productive powers, and in potential productive fruitfulness of land-area, resulting from society's progress from lower to higher levels of culture. This progressive change, in both elements, is incorporated within and subsumed by the notion of *a rate of increase of potential population-density.* In other words, this is a notion of *an implicitly continuous function,* expressed in terms of *a functionally variable such rate of increase.* That function expresses the effect of increase of the level of development of the *potential mental-creative powers of the individual* in the society.

At first inspection, these considerations may appear al-

241

most self-evident. After a more thoughtful, more rigorous second glance, we find that we are among the most troublesome axiomatic problems of any mathematical physical science. There are principally two such issues posed by any rigorous reflection upon the reproduction of the successive generations of a society.

First, we discover that, although we do employ the ordinary counting-numbers to effect a raw measurement of actual population-densities, we may not employ simple arithmetic methods in defining the number of individuals represented by an increase (or decrease) of potential population-density.

Second, the sovereignty of the person's creative-mental processes defines the "economic individual" of per capita calculations as both *formally* and *ontologically* a member of the higher of the species of Gottfried Wilhelm Leibniz's *monads*.[1]

A student of mathematical physics who had considered only carelessly what we have summarized thus far, would probably fall into a blunder of the following description. Imagine a teacher-student exchange including these elements.

We have identified the precondition for durable survival of a social culture, as a continuing process of successive increases in the average *physical* productive powers of labor, effected through the generation, transmission, and *efficient* assimilation of valid, revolutionary, scientific (and related) discoveries, transforming generally social practice. We have summed up the interdependent benefits of such continuing scientific and technological progress, respecting persons, society, and non-human nature, in terms of *rate of increase* of potential population-density.

How shall we measure *potential population-density*?

To come directly to the crucial point: Shall we, or shall we not, use simple counting-numbers to identify *actual* or *potential* population-density per unit area of land-use? Essentially, it would be a sophomoric blunder to use the simple counting-numbers to measure their respective population-densities, for simple comparisons. The nub of the issue, stated most simply, is, as we noted in an earlier section, that the typical individual in one set is an individual of a different *quality* than that of another set.

Let us examine one aspect of this difference in *quality*. Let us view this first in the most rudimentary and one-sided way: in terms of social costs of producing a unit per capita value of a family household's basket of consumption requirements.[2]

As we raise the level of technology in productive and related practice, we increase the required quality of educational and correlated development of the *creative potential* and moral character of the young.[3] Therefore, we increase the per capita consumption and constructive-leisure requirements of the family household which produces these young. Nonetheless, we increase the per capita productivity potential of the society by a greater amount than the increased costs of the family household's basket of necessary consumption.[4]

Also, as society's condition is elaborated in these ways, the demographic characteristics of the society are altered.

Yet, important as these two kinds of changes are, they reflect something more profound, more spiritual, if you please.

In counting potential population-density, we do not count the number of individuals as one; sometimes, we may make an approximation by counting the number of persons as individuals, in terms of countable numbers, for an estimation of *actual* population-density statistically, but that does not define *potential* population-density.

The obvious reason is that the use of the counting-numbers, particularly for the purposes of measuring density functions of any kind of counting magnitude, usually assumes that there is an equal magnitude, in some sense an equal magnitude, associated with each counting-number.

But that is not the case with human individuals. The quality in the human individual, relative to the density function, changes, as the *potential* population-density function increases. That is, both the activity, in the first approximation, and the consumption per person increases.

Furthermore, these functions are associated with a primary function, which is creativity: the higher order of creativity.

Thus, we have this nonlinear magnitude, the higher order of creativity, with its predicated features of required consumption levels, and potential productivity, in one sense or the other, occurring within a sovereign individual.

So, when we're counting individuals, we're counting sovereign entities, whose internal magnitude is determined in the indicated, nonlinear way.

Thus, when we compare populations in two periods of time, assuming a process of development, we are comparing *non-comparable* magnitudes. What we're comparing, is simply the absolute number of equivalent of sovereign individuals who might satisfy that potential population-density function. We are not assuming that you are simply increasing the number of persons per se.

Suppose we increase our population-density over a significant interval between points A and B in historical time, the same society. The higher potential population-density at B does not mean a higher density of the *same* individuals as at A; but rather, a higher population of *transformed* individuals, or individuals of a transformed quality.

So, that's the point we're at in terms of reproduction. And it's on this point of the *quality* of the individual, that the crux of the Christian aspect, that is, the *Filioque* Christian aspect of the science of Christian economy, pivots.

The Ontological Paradox of Social Production

If we attempt to represent rising population-density in an appropriate mathematical function, as an effect of a capital-intensive, energy-intensive mode of social investment in scien-

tific and technological progress, we should recognize immediately why we cannot compare the individual persons undergo-
ing such a transformation in this productive potential as
simply-countable individuals. Over the duration of a continuing function, each person represents a changing magnitude
from the standpoint of this function.

The same difficulty confronts the student in the case
where a more generalized form of the same mathematical
function is employed, describing the condition of declining
potential population-density.

In a third case, the apparent steady-state, constant political population-density must be determined as a representation
by the same generalized function employed to represent both
increased and decreased potential population-density.

However, the idea of a constant apparent potential value
for potential population-density, that is, the notion of society
maintaining such an apparent "steady state" over an extended
period, presents the student with some provocative and very
relevant results.

Remember, that even the simply-continued existence of
the society, at *a constant level of employed technology,* must
determine a declining potential population-density. This fact
is illustrated by the rise in marginal social costs and lowered
average productivity, caused by a marginal depletion of quality of resources. Only technological progress can offset that
factor of marginal depletion. So, to achieve a net-zero-growth,
"steady-state" condition for an economy (as measured in potential population-density), it is indispensable to maintain
a corresponding level of capital-intensive, energy-intensive
investment in scientific and technological progress.

Imagine a three-dimensional graph, in which the *x-axis*
is *time,* the *y-axis technological progress,* and the *z-axis potential population-density.* So, for a constant value of potential
population-density, in terms of a function of z, there must be
an associated rate of increase in the function of y. This function of y is implicitly not a linear function.

246 *The Science of Christian Economy*

This simplified imagery illustrates the point that, in order even simply to maintain a "steady state" of potential population-density, there must be constantly a significant increase in the level of capital-intensive, energy-intensive investment in scientific and technological progress. This defines a "world-line," as the locus of a steady value for potential population-density.

In this sort of function, the individual person corresponds to a *changing* function of activity. Thus, even in our hypothetical steady-state society, the average individuals in successive intervals are not equivalent individuals. Not only are they different magnitudes, but over the longer term, at least, the differences are reflections of a nonlinear function.

Although each person's biological individuality is simply countable, human and animal population functions are not comparable; we cannot count humans competently as merely biological individuals, as we are permitted to count animals with fair approximation. The functional significance of the differences in human individuality is not *merely* the biological individuality; rather, the biological individuality is, essentially, merely an indispensable *vehicle* for a different kind of individuality. That latter, different kind of individuality, is the sovereignty characteristic of the individual's developed mental creative powers.

In shorthand, even in a steady-state society, the average individual is a *nonlinearly* changing quality.

Land and People

We must never lose sight of the fact, that it is not sufficient to improve nonlinearly the quality of the average individual person. We must also transform the wilderness into a fertile place, a place where fertility is defined in terms of both the existing and emerging levels of productive and related technology.

In the earlier civilized or quasi-civilized cultures, until

modern times, more than nine-tenths of households were occupied in agricultural and related production of physical wealth. The introduction of steam-powered machinery along lines pioneered by Gottfried Wilhelm Leibniz, brought a drastic change over the course of the nineteenth century, a change which, when successful, has followed the outlines of cooperative relationships between township and countryside, presented by U.S. Treasury Secretary Alexander Hamilton's 1791 *Report to the U.S. Congress On the Subject of Manufactures*.

From this modern vantage-point, we are able to offer certain securely proven generalizations respecting the mathematical forms of representation of physical-economic history up to this time.

To this effect, we define the required improvement of land-area and waters in terms of physical-economic categories of *land-usage*. To this effect, we associate each among the family households of which the total population is comprised, with the primary physical-economic (e.g., household-income-related) activity of the adult labor force members of that household.

In the statistical practice of physical economy, the functional relationship between per capita potential for *physical productivity,* and fecundity of improved or other land- and water-area, is treated in first approximation in terms of *categories of land use*. Seven such rough categories are employed: 1) basic economic infrastructure; 2) agriculture and related; 3) mining and closely related; 4) manufacturing; 5) residential; 6) commercial and administration; 7) other (including unused reserve land, wasteland, etc.).

Basic economic infrastructure includes the development, maintenance, and operation of water management, general transportation, generation and distribution of power, general sanitation, and general communications. It also includes general education and related cultural support for the population as a whole, and general medical and related health-care delivery. These represent the categories of essential capital improvements in the society's total environment, required to sustain a population in its production at a given level of range of both technology in use and potential population-density.

In sane, civilized nations, the development and maintenance of basic economic infrastructure is the economic responsibility and customary function of government. In successful modern economies, such as the pre-1964 U.S.A. (prior to the mid-1960s "post-industrial," "neo-Malthusian," "rock-drug-sex counterculture" "cultural paradigm-shift"), basic economic infrastructure was supplied either as an economic activity of government or as the function of a government-regulated, if privately owned public utility. If the state does not adopt and maintain efficiently its moral and economic obligation to provide adequate basic economic infrastructure, private enterprise generally will fail and misery abound.

In the matter of determining the upper bounds attainable by production in general, three aspects of basic economic infrastructure are most conspicuous: water, power, and transportation. Examine each briefly.

Water. In the history of Physical Economy, the emergence of civilization is the history of oceans, shores, rivers, lakes, and of water-supplies generally, as food for the hunger of the land. Water is the source of food from the sea, lakes, and streams, chiefly animal protein. Water is transportation, historically the superhighway along which every civilization has advanced.

Look to the development of Europe from the time of Charlemagne (742–814). The history of the development of Europe's nations, cities, and population-densities, is the history of Europe's seas, coastlines, navigable rivers, and canals. Even today, water-borne freight has inherently the lowest cost per ton-mile, and, as cheaper, bulk cargo, also the lowest cost per ton-mile-hour or ton-kilometer-hour, better than the next-best competitor, rail. Coasts and rivers are the oldest arteries of civilized life.

The productivity of land, for agriculture, mining and

refining, manufacturing, and residence, is expressed in correlation with cubic meters of water.

Similarly for *power*. For purposes of first approximation, we may measure power in linear quantities per square kilometer and also in the simplest qualitative terms, *energy-flux density*, e.g., watts per square centimeter of a cross-section area of direct application.

Transportation. In order of increasing cost per ton-kilometer, we have transport of freight by navigable water, rail, highway, and air.

If cost per ton-kilometer were the only factor of cost in determining the per capita physical-economic productivity, then water would predominate, such that long-haul highways and air freight transport would virtually not exist. Two considerations require heavy emphasis on rails, upon relatively short-haul highway freightways, and air freight.

First, there is the matter of *density*. The possibility for economical development of navigable waterways is limited, such that rails and highways are indispensable additions to our freight-transport repertoire. Otherwise, much of the world's land area could not be significantly productive.

Second, there is the cost factor of *time*. We are required to supplement measurement of costs in terms of ton-kilometers, by cost in terms of ton-kilometer-hours. *Spoilage* is an obvious consideration, a point which should require no further elaboration in this location. Here, we emphasize the great increase required in a nation's work-in-progress inventory, if the average time to move goods from production point A to production or distribution point B is significantly increased. This consideration is taken into account in fair degree or first approximation, by replacing ton-kilometers with ton-dollars or ton-ECUs per kilometer-hour.[5]

Among the cost variables to be considered in such computational systems of linear inequalities' approximations, are the capital and maintenance energy-costs of the respective modes of freight transport.

The physical costs of these cited, various forms of basic

economic infrastructure vary significantly in two degrees. They vary in correlation with the normal level of productivity, expressing, in turn, a degree of development of employed technology. They vary also according to category of land-use.

For example: At any level of quality of existence, the average member of a family household requires a minimum to maximum range of daily potable water consumption. This is reflected chiefly in the statistician's *residential* category of land-use. If we apportion the per hectare output of the *agricultural* land-use by average per capita requirements, we must associate these per hectare and per capita rations of output with a corresponding fresh water requirement. We have a similar case for *mining, manufacturing,* and so forth, land-usages.

So, in this vein, the rising physical, plus absolute educational and medical costs, per capita consumption market-basket, required by a rising level of applied technology, is associated with infrastructural requirements expressed in such *combined* per capita and per hectare terms as *liters of potable* (and other qualities) *water, watts per hectare,* and *watts per square-centimeter cross-section, ton-dollars per kilometer-hour,* and so forth.

So, our physical-economic universe is examined statistically in such units of infrastructure measurement per capita unit of land-use.

As to other categories of land-use, other than physical infrastructure, it is not necessary to elaborate these in any significant degree in a paper dedicated to the specific purpose of this one. A few further remarks should be sufficient.

Capital-Intensity

The technological progress of society is reflected in the form of a changing composition of the division of (physical) labor within and among the family households of which a healthy society is composed predominantly.

As longevity is increased, as physical productivity per capita increases, the possible required modal "school-leaving age" converges asymptotically upon the mid-20s age of biological maturity. The ratio of the total labor force employed in agriculture and other rural occupations, declines toward what is apparently an asymptotic lower limit of perhaps approximately one percent. Within urban centers, the ratio employed in production of producers' goods, increases relatively to the shrinking urban ratio employed in production of household goods—although the absolute physical content of the per capita household ration of consumption increases.

These shifts correlate with an increase in the power employed by society per capita. This is a *humanizing* of production, away from muscular, to mental means of controlling willfully the minutiae and end results of the productive process. This is expressed, in great part, in a rise of capital-intensity, a capital-intensity reflected in the ratio of productive employment, between producers' and households' goods, of the urban labor force.

The Invisible Hand

There is no reasonable doubt, but that David Hume and his follower, Adam Smith, were dedicated adversaries of both Christianity and Western civilization. It ought to be clear also, that to the degree that private entrepreneurship in high-technology family farms and manufacturing is essential to the superiority of modern European forms of economy, the superiority owes nothing to Smith's famous "invisible hand" dogma, but rather to the Christian fostering of the value of the human individual as *imago viva Dei*.

Smith is associated with a single conception so crucial to all his work, that everything else he asserts, or that which is asserted by his modern devotees, stands or falls absolutely upon that one point. It appears as the centerpiece of argument in Smith's two principal published pieces, his 1759 *Theory of the Moral Sentiments,* and as the "invisible hand" dogma, in his more famous, plagiaristic, physiocratic piece, the 1776 *The Wealth of Nations.*

Smith, following in the proto-positivist school of British philosophical irrationalism of such predecessors as Franc(i/e)s Bacon (1561–1626), Thomas Hobbes (1558–1679), John Locke (1632–1704), and David Hume (1711–76), asserts in the 1759 book, that man is incapable of foreknowledge of the larger consequences of his actions and is advised therefore, to be indifferent to the ultimate effects of his acts of commission and omission. In *The Wealth of Nations,* Smith's advocacy of pagan (epicurean) immoralism assumes the form of the dogma of the "invisible hand."

In fact, the net result of every generalized application of the British model of liberal political-economy, has been the contraction or even collapse of the region of the world subject to the rule of that so-called "free trade" system. In the case of the U.S.A., whose federal constitutional government was founded successfully upon rejection of Smith's British liberalism, we see that each time the "free trade" dogma was introduced to the government, as under Jefferson, Madison, or Jackson, for example, the direct result was national economic catastrophe. The accelerating of the collapse of the home physical economies of Britain and the U.S.A., during the past 25 years, is a fresh example of this.

London's prosperity during the nineteenth century is no exception to this rule. It was Britain's looting of the population and natural resources of its empire, the looting of much of the rest of the world through the London market's dominant part in the bloodsucking practices of international usury, which was the source of Britain's economic power during that century.

Similarly, today, it is the Anglo-Americans' looting of most of the planet, through aid of the mass-murderously bloodsucking practices of the International Monetary Fund's "conditionalities," which produced hundreds of billions of dollars of loot annually into the collapsing U.S. economy of the mid-1980s.

Nonetheless, throwing Adam Smith and his dupes to one side, there is something of great practical importance to be said in behalf of private entrepreneurship. The history of England, even prior to Adam Smith, provides an important clue.

Private Entrepreneurship

The efforts to promote scientific and technological progress during the fifteenth century, for example, met powerful resistance in the form of craft guilds' stubborn enmity against technologies offering economy of labor. Against this backwardness, stand out such notable cases as Brunelleschi's (1379–1446) solution to the challenge of constructing the cupola of the Cathedral of Florence. One of the practical solutions found for the backwardness of the guilds, was the use of the power of government to create limited corporate monopolies, of fixed duration, to promote the production and sales of useful inventions.

In the emergence of such patents in sixteenth-century England, for example, must be seen reflected the work of such as Nicolaus of Cusa and others, during the fifteenth century, in stipulating from the vantage-point of natural law, the right of nations to the benefits of scientific and technological progress.

Those and related aspects of modern economic history serve here to illuminate a deeper principle. It is only through the direction of human society's behavior by means of the faculty of creative reason, that the human species exists, is able to survive as human, is in *imago viva Dei*. This is the connection to the true basis for promoting private entrepreneurship.

Scientific and technological progress is a characteristic reflection of that divine spark of potential for creative reason which defines man as *imago viva Dei.* As we have illustrated by reference to *classical forms of art,* and to the function of *agapē* more broadly, scientific and technological progress, as narrowly defined, is not the exclusive expression of the divine spark, but is the only form in which that spark is reflected as an approximate form of physical-economic practice of entire societies. Most simply, any contrary policy of practice for economy would negate man as *imago viva Dei.*

The qualification implicit in the foregoing paragraph taken into account, the object of society is to produce individuals who are *imago viva Dei,* individuals expressing a lessening of the imperfection, the realization of their true nature as in the living image of the Creator. Society and its individual members must thus live and work in a manner consistent with the purpose of human existence.

Insofar as entrepreneurship in physical-economy is a means for placing economic processes under the *lawful domination* of the creative principle, reflected as fundamental scientific progress, either entrepreneurship or something equivalent to this specific effect, ought to be considered in principles of economic science.

This point is directly relevant to the causes and failure of the now-collapsing Soviet economy; but, it is equally cause for opposing and despising those "free trade" dogmas which have brought about the presently ongoing collapse of the Anglo-American economies. In particular, what is called "Thatcherism" by some—notably the policy of "deregulation" and "privatization," which former Prime Minister Margaret Thatcher adopted as her principal stock of professed political wisdom—deserves to be put into the bottom of the rubbish bin of history as quickly and as permanently as possible.

We must demand, on the one side, that the hierarchy of governments' parts—national, regional, and local—each assumes, in a mutually coherent way, appropriate respective responsibilities for maintaining effective monopolies of regulatory power to supply an adequate development and maintenance of basic economic infrastructure. Yet, we must also insist upon preference for quasi-sovereign entrepreneurship, as the form of ownership and direction in agriculture and industrial production and distribution of physical goods. That is no contradiction in our policy; rather, the differences in our political treatment of basic economic infrastructure and private owner-management of production, flows coherently from a single principle.

Our principle is the nonlinear process of increase of society's potential population-density through the direction of society's physical-economy in a capital-intensive, energy-intensive mode of investment and production, under the rule of a form of scientific and technological progress consistent with the creative principle of *imago viva Dei*.

Wherever possible, we rely on social-economic forms which are consistent with the sovereignty of the relevant creative processes: hence, the principle of entrepreneurship. Yet, wherever the continued advancement of scientific and technological progress demands it, the government must be responsible for establishing and maintaining the necessary preconditions, both as basic economic infrastructure and as regulation of the market.

For example, it is insane and immoral in the extreme, to foster a market in which the prices paid to farmers are below the average cost of production of an adequate supply of food. Today, when governments intervene to prevent those usurers called food-cartel monopolies from drawing down the farmer's price so, the "free traders" howl insane moral indignation against "subsidies," and demand a practice of "free trade," which, in fact, will be globally more mass-murderous in effect than Hitler's Schachtian slave-labor system.

Similarly, society's political economy must provide, through government, an adequate development of basic economic infrastructure. That governmental provision is not an *exception* to private entrepreneurship in agriculture and industry; it is, rather, an indispensable precondition for a successful form of entrepreneurial economy.

The moral object of production is the reproduction of mankind according to the individual person in the living image of the Creator. On that account, we are each morally responsible for all mankind, past as much as present and future. We are responsible to our forebears, to fulfill in the richest degree possible, the potential of good they've contributed in their time. We are each responsible to the limit of what we may develop our creative powers to become.

VII | 'The One and the Many'

We have considered already the fact, that without those changes both in nature and in human social practice, which we associate with "scientific and technological progress," any culture is on the pathway toward, sooner or later, self-induced "entropic" collapse. Thus, the continued existence of the human species as a whole, depends upon the relatively hegemonic influence of those cultures which uplift humanity as a whole, evermore, to higher plateaus of scientific and technological practice.

Thus, we have stressed, human existence taken as a whole requires, inclusively, but nonetheless absolutely, the generation, transmission, and efficient assimilation of scientific progress. This must occur to the specific effect of causing an increase in the human species's per capita power over nature. This principled policy defines the realm of Gottfried Wilhelm Leibniz's *science of Physical Economy*.[1]

It should be recognized readily, that the empirics of such a *science of Physical Economy* are two-faceted. On the one side, *effects,* we study the increase in mankind's physical power over nature. At the same time, the *source* of this increase in power is scientific progress. So, at first glance, Physical Economy not only measures changes in man's per capita power over nature, but studies these changes as *the material effect of a spiritual (mental) cause.*

257

That descriptive definition of Physical Economy brings us, now, directly to the deepest among the common underpinnings of all classical European philosophy, and all physical science. We address directly the deepest of the quasi-axiomatic principles upon which the bare concept of the verb "to know" must be based.

Physical Economy, as established in an exemplary way by Gottfried Wilhelm Leibniz, is the aspect of physical science as a whole, which addresses *most directly* this concept of principle. Inasmuch as Physical Economy is the science of social reproduction of mankind, it is the science of the way in which human survival is accomplished by the indispensable aid of scientific and technological progress. It is survival so effected, which is the test of the process of generating improved human knowledge, and thus, it defines the crucial experiments which must be referenced for a proper definition of the verb "to know."

As indicated earlier, this emphasis upon physical science has a twofold import. First, there is no principled distinction between the qualities of creative thought associated with valid fundamental discovery in physical science and in classical humanist art forms. The widely popularized contrary arguments, such as those by Immanuel Kant (1724–1804), Friedrich Carl von Savigny, and others, are essentially absurd in fact. Thus, what is said of physical scientific activity is implicitly also true of all expressions of *agapic creative work*. Second, in Physical Economy, it is fundamental scientific discovery which is the most prominent causal feature of increase of the productive powers of labor.

In all of the literature of European culture's classical philosophy, the most succinct statement of the related deeply underlying principle of human knowledge is *implicitly* uttered by Plato's *Parmenides* dialogue. That dialogue's *implicit* statement occurs not explicitly within the dialogue, but, rather, as a unique, required solution to the ontological paradox which the dialogue as a whole describes. Once that required solution is recognized, this implied definition of the verb "to know" is realized, without reasonable doubt. Physical Economy is the empirical domain in which this solution to the Parmenides Paradox is most readily demonstrated.

The argument underlying the solution to that Parmenides Paradox is summarized as follows.

Any formal (deductive) system of argument is implicitly reducible to a deductive theorem-lattice derived from an original, integral (indivisible) set of (inseparable) axioms and postulates. (See Appendix IV.) Any idealized form of deductive mathematics is one such case; a deductive formalist's ideal mathematical physics is another such case. Deductive formalism prevails in academic practice, in most mathematical physics today; therefore, in present, customary professional practice, actual mathematical physics is treated, formally, as an imperfected approximation of a mathematical deductive theorem-lattice.[2]

The characteristic feature of such a theorem-lattice is what is often termed today a "hereditary principle." This principle can be described most easily in terms of two corollaries.

A. No theorem of a deductive lattice may claim any essential quality or predicate of existence, which is not already implicitly claimed by the underlying, integral set of "axioms and postulates," from which the theorem-lattice as a whole takes, putatively with perfect consistency, its origins.

B. Any theorem which is required by nature, which is not perfectly consistent with the underlying, integral set of "axioms and postulates," disproves each and all other theorems, hereditarily, and requires implicitly, a new, integral set of underlying "axioms and postulates" consistent with that theorem.

In actual scientific practice today, these two corollaries are at the center.[3] The term "fundamental scientific research" is usefully circumscribed in use, to signify a special, higher class of experiments and crucial observations. This higher

260 The Science of Christian Economy

class of empirical studies is associated typically with the expression "crucial experiments." A "crucial experiment" is intended to test the kind of hypothesis (theorem) which has been rigorously defined to affirm or overthrow the theorem-lattice's underlying, integral set of "axioms and postulates."[4]

So, from the modern formalist's standpoint, the internal history of successful advances in fundamentals of physical science is defined implicitly by a succession of successful such, revolutionary "crucial experiments." Repeatedly, the heretofore-established scientific worldview is overthrown, replaced by a new one. The association of such revolutionary ("crucial") transformations is the formal representation of fundamental physical principles, with resulting increased social-reproductive power of mankind, both per capita and per hectare; and is the key to the formal proof and empirical definition of "scientific and technological progress."

In the case of successive scientific revolutions satisfying that practical requirement, we are confronted implicitly by three levels of scientific principle integral to a formalist (deductive) view of this process of successive revolutionary progress taken as a whole. Each of the levels of principle is represented by conceptions which are each *one* and *indivisible*.

1) On the lowest level, we have each of the one and indivisible "hereditary principles" associated, respectively, with each of the successive, each relatively more successful theorem-lattices, A, B, C, D, E. . . .[5]

2) The fact that the empirically proven theorem-lattices are each and all ordered according to increasing per capita and per hectare powers of society (A less than B, less than C, less than D, less than E, less than. . .), defines the hereditary principle of each and all of A, B, C, D, E, . . . as ordered commonly by a uniquely subsuming principle, which, in turn, is one and indivisible.

3) The existence of alternative orderings, on level 2, implies that the choices among each such are also ordered, as a

set, according to a subsuming, ordered principle, which is *one* and *indivisible*.

Plato's *Parmenides* dialogue confronts us with a threefold paradox. The elaboration of the dialogue confronts us with a problem in *knowledge,* a paradox defined in formal, deductive terms of reference. The dialogue taken as a whole defines an *ontological paradox*. Both facts taken together define a single paradox, subsuming both others: a paradox respecting man's possibility of effecting a truthful, intelligible representation of the elementary nature of universal and subsumed (manifold) states of *being*.

Plato includes in that dialogue only one explicit clue to the required solution: that the formal argument has ignored the fact that *change* is an elementary condition subsumed by *being*.[6] The significance of that reference to change, as a crucial feature of the dialogue as a whole, is not comprehended by most commentators. From the standpoint of Physical Economy, for example, the solution is derived more or less directly.

The connection may be represented as follows.

As in *In Defense of Common Sense*, A, B, C, D, E . . . is a series, in which each term represents a linear (deductive) description (approximation) of an indivisible *theorem-lattice* in encompassing the current technology of productive and related practice. Each successive term represents the twofold effect of a scientific revolution: 1) the replacement of one theorem-lattice by a second which has no formal consistency with the predecessor; 2) the resulting increase in per capita and per hectare (physical) productive powers of labor (i.e., increase of *potential population-density*).

The combination of the two aspects, the ordered character of the formal change in theorem-lattice and the increase of potential population-density, indicates that each member of the series, A, B, C, D, E . . ., when taken as a whole, is a member of a well-ordered series. That ordering principle is thus an *indivisible unity*, a *transfinite one*, with a qualification that this is the lowest of the readily defined levels of transfinite ordering. (See Appendix VII.)

For most readers' benefit, the following qualifying observations are required here.

This strict usage of the term "transfinite," and earlier references to the "sovereignty of the individual's creative process," represent two coherent expressions of the same conception.

The easiest argument defining the indivisible unity of any truly *transfinite* conception, is the deductive case. For example, the "hereditary principle" specific to any deductive theorem-lattice is related to the associated, integral set of (inseparable) variously stated and implied axioms and postulates, and also to each and all subsumed theorems. Yet, relative to these two "Manys," the "hereditary principle" is *relatively transfinite,* and corresponds directly to a unitary notion which is indivisible; it is indivisible in the sense that it vanishes instantly the moment we might attempt to represent the principle itself as *composite* from the vantage point of either the axioms or subsumed theorems.[7]

In the illustrative case, referenced from *In Defense of Common Sense,* A, B, C, D, E . . ., the ordering-principle defining the series as a species of series, is the *change* determining each and all of the successors to each and every term of the array. This feature of our illustrative case is exemplary of the solution to the ontological paradox of Plato's *Parmenides* dialogue.

What "lies between" A and B, for example, is a "mathematical discontinuity" from a formal (deductive) standpoint. It is distinct and efficient, yet cannot be expressed by a theorem of any possible deductive theorem-lattice.

Its *ontological,* and *formal* character is, implicitly, "change." In the case of a well-ordered series, A, B, C, D, E . . ., the "change" exists *ontologically* in the transfinite principle subsuming the ordering of the series as a whole.

If such an ordering does indeed define as subsumed a

series corresponding to the *Becoming,* the increase of potential population-density, the transfinite converges upon the principle which expresses the lawful ordering of the universe. That latter principle is also represented by the transfinite which subsumes the *Becoming* of all series which converge so. (See Appendix VII.)

Thus, the proper definition of the term "science" ought to be limited to describing *efficient consciousness of those forms of consciousness which generate the kind of efficient transfinite convergence indicated here.* In other words, consciousness not of discrete objects, but rather consciousness of revolutionary *change,* as the proximate cause of increase of potential population-density.

Clearly, from this standpoint, we have already proven conclusively that no proposition consistent with any formalist theorem-lattice can state a *law of nature* explicitly, directly. It may reflect a law of nature, but it never states a law of nature.

Speaking most strictly, there are not *many* "laws of nature," but, rather, what may appear as laws, are, ultimately, only reflections of a single *One,* highest-ordered, transfinite conception, which expresses a single, indivisible, universal law of nature. Leibniz's principle of least action points in the direction of the kind of physics conceptions *subsumed by* the relevant single, indivisible, transfinite notion.

The Definition of Science

Science is a matter of man's increasing mastery of the universe, a mastery expressed in such forms as increase of potential population-density, and accomplished through man's conscious ordering of his willful ordering of revolutions in scientific consciousness governing increase of the power of human practice.

There is no non-anthropocentric science, no so-called "objective science" in the positivist's sense of the latter term.

What we know, is *not* that which we have experienced with our senses. What we know, are those principles for generating successive, successful, revolutionary advances in our mastery of fundamental laws of our universe; these are the principles that are efficient voluntary action upon the universe, by means of which actions we maintain and may increase the potential population-density of present and future generations of mankind as a whole.

By implication, that definition of *science* defines the following paradox. Is the primary importance of creative scientific (and related artistic) discovery, that it is the indispensable means for fulfilling man's obligation to satisfy Genesis 1:28? Or, is it, that by satisfying that obligation, by this means, man is forced to recognize himself as in the living image of the Creator? Is it not, rather, the case, that the two are inseparable, an indivisible oneness? The two are thus portrayed as the mirror-image of the relationship between Becoming and the Good.

We do not know truth in the form of sense-perception. We know scientific truth only by means of a Socratic form of successively successful criticisms of our problem-solving interpretation of those aspects of our sense-experience which bear upon increasing the potential population-density of the human species as an indivisible whole. *Truthfulness* is expressed practically, only in a correct view of the reciprocal relationship between mankind as a whole and the universe as a whole. The essence of truthfulness, is situated in the eternal practical contribution which a sovereign individual creative-mental process may add to the potential population-density of the present, future, and past generations of mankind as a whole.

This locates scientific consciousness in the activity of our own critical consciousness of our own critical consciousness: We employ the term *self-consciousness* in this restricted sense. Insofar as such critical consciousness is occupied with what is termed here an *intelligible representation of* an ordered succession of successful revolutionary advances in the implicit integral set of deductive axioms and postulates of physical science, *self-consciousness* treats the indivisible ordering-principle of this series as an indivisible conception, an *object* of *self-consciousness*. This latter is a true *transfinite*.

It is the comparison of such transfinite objects of *self-consciousness* with the appropriate quality of crucial-experimental evidence, which serves as the center of focus for scientific truthfulness.

It is in that perspective that truthful concepts of universal concepts of universal physical law are situated. This rejects, obviously, the view, that "laws of physics" pertain to observation of what appear to be repeatable pair-wise interactions among phenomena. This situation demands what is associated with Gottfried Wilhelm Leibniz as the notion of *necessary and sufficient reason*. It is not conclusive that pair-wise interaction repeats; it is required that there be necessary and sufficient reason that it repeats or not.

The "scientific law" is situated in the proven transfinite ordering principle subsuming successive, successful scientific revolutions. This is not a perfect representation of "God's law"; it is therefore not *perfected truth*, but is, rather, *scientific truthfulness*.

There exists, clearly, a higher ordering which subsumes a positive ordering of successively less imperfect scientific truthfulnesses. Approach the concept of the Good and Becoming in these terms of reference. To this immediate end, define a few essential terms of distinction.

Begin at the relatively low end of the scale, with a *crucial-*experimental *hypothesis,* a hypothesis of the quality associated immediately with a single successful scientific revolution. Consider what Bernhard Riemann (1826–66) has termed Lejeune Dirichlet's (1805–59) discovery of "Dirichlet's Principle" of topology (see Appendix IX), as an example of such a hypothesis, or Kepler's hypothesis of Golden-Section-harmonic ordering of solar planetary orbits. (See Appendix VI.)

The transfinite which subsumes an orderable series of such crucial hypotheses is then termed a *higher hypothesis*.

The fact that the latter is demonstrably subject to imperfection, obliges us to hypothesize (crucially) upon the subject of perfection of the higher hypothesis.

That, in turn, demonstrates to us the *ontological negativity* of the relatively best higher hypotheses; in this way, scientific truthfulness, by recognizing that ontological negativity, knows the certainty of the Good as the changeless cause of the changes represented by the necessity for perfection of the higher hypothesis.

This relationship between Becoming (higher hypothesis) and the indivisible being of the Good, is also the conception of the *One and the Many*. The *One* is the cause of the necessity of the Many.

This adoption of scientific truthfulness were impossible, except as the individual is consciously self-defined for practice as a servant of the Good for the work of the perfection of human existence as a whole.

The function of society may be represented as the duty of society as a whole to develop sovereign individual creative minds, who each develop society as a whole in this way. That, in the last analysis, is the mission and the true definition of the science of Physical Economy. The truthful notion of *economic value* cannot be different than this moral one.

Physics, Briefly

We conclude this penultimate chapter with some necessary observations on, first, the physical notions of cause-effect implicitly imbedded in that which we have just previously developed, and second, some historically illustrated implications of those physical principles for statecraft in general.

To put the physics matter as simply as seems possible, we have argued earlier to the following effect: Given three relatively nearby, discrete bodies in space, we are implying that these bodies do *not* react with one another in terms susceptible of a simple pair-wise analysis: Rather, we have appeared to imply, that each discrete body might be a *monad*, which reacts *indirectly* to its neighbors, by interacting immediately with the universe as a whole. In other words, we defend the inference, that each body acts primarily by interacting with the universe as an indivisible entirety, and that it is through these interactions of each body with a whole, that the bodies act indirectly upon one another.

Is that as wild a depiction of the situation as some critics might hasten to argue? Not really, not if a handful of elementary facts in the modern internal history of physical science is taken adequately into account.

A. Kepler versus Newton on Gravitation

It is readily shown[8], that Newton's famous formulation for universal gravitation is simply a consistent algebraic manipulation of Johannes Kepler's Third Law, a manipulation which references not only centrifugal impulses but also the then-well-established inverse square law for electromagnetic radiation. It is notable that Kepler, rather than Galileo (1564–1642) or Newton, was, at least relatively, the original discoverer of universal gravitation. There is no deductive inconsistency in the derivation of Newton's formulation from Kepler's Third Law; however, in this simple algebraic derivation, there is a relevant problem. (Also see Appendix V.)

In Newton's case, we incurred the notorious, insoluble paradox of the "three-body problem"; in Kepler's physics, this paradox does not arise. Given the demonstration, that Newton's algebra is consistently derived from Kepler's Third Law, how is it to be explained, that this paradox occurs in the copy and not in the deductively consistent original? This takes us to the next, related point to be considered.

B. The Orbital Characteristics of the Asteroids

Carl Friedrich Gauss's successful demonstration, that the orbital characteristics of the asteroids Ceres and Pallas conform to Kepler's orbital calculation for the missing, exploded planet between Mars and Jupiter, proves that the axiomatic assumptions underlying the physical space-time of Descartes (1596–1650), Newton et al., are absurd relative to Kepler's physics.[9] The crucial-experimental point to make, is that Kepler's physics *as a whole* requires this planetary orbit, whereas the opposing, empiricist physics does not.

These two anomalies, the three-body paradox and the asteroid orbits, cited thus far, combine to the following effect. Kepler's physics, as an entirety, depends, axiomatically, in a crucial, pervasive way, upon the preceding work of Leonardo da Vinci et al., on the subject of the Golden Section and related physics implications of the Platonic solids: notably, that on the ordinary scale, the physical geometry of living processes is ordered in harmonic congruence with the Golden Section; whereas, on that same scale, non-living processes are not. This crucial, empirical fact determines the attributable, axiomatic structure of Kepler's mathematical-physical method in its entirety. This is the feature of Kepler's physics most emphatically rejected by Galileo, Descartes, Newton et al. The origin of the Newtonian three-body paradox lies in this axiomatic difference.

In Kepler, the available orbits and their mutual harmonic ordering, are determined by what we term, since Georg Cantor's (1845–1918) work, a *transfinite* principle. (See Appendix VII.) Since the universe efficiently contains living processes, *the which are negentropically ordered,* the universe as a whole (a transfinite process of Becoming) must become, according to Cantor's definition and proof, characteristically transfinitely ordered (negentropically). This notion of *transfinite negentropy* is expressed within the work of Leonardo da Vinci, Kepler et al., as *harmonic* (least action) *ordering* cohering with a determination of the Platonic solids.

Thus, in Kepler, bodies are situated in those Planck quantum-like orbits determined by the kind of least-action principle cohering with the constructive determination of the Platonic solids. In the Newtonian case, the availability of orbits is indeterminate.

In the Kepler-Gauss configuration and also in a synthetic, electromagnetic "history" of our solar system's derivation from application of Kepler's laws to slowing of the rotation of our aboriginal Sun, the mass of (polarized-fusion)-determined plasma shed as the slowing rotation of the Sun, is distributed among available solar planetary orbits, according to the relevant Keplerian principle of harmonic ordering. Thus, implicitly, in this model, the Gaussian toroidal distribution of the mass of material along the elliptical torus of the orbit, forms a planetary mass as a singularity generated within the continuing action within that orbit as a whole.[10]

The contrast which the two referenced axiomatic schemas illustrate as existing between the Keplerian and Newtonian schemas, can be received as confronting us with the notion of a Keplerian curvature of physical space-time, as opposed to the linear matter, space, and time, of the Descartes/Newton schema.

C. Non-Algebraic Functions

In both of the instances just cited, the crucial issue is some physical evidence which affects rigorous scientific thinking in two ways. First, there is the physical evidence which forces us to construct a *hypothesis* upsetting established opinion; second, there is the crucial evidence supporting either that hypothesis or a modified version of it. The same principle applies to the additional examples now to be considered.

The great battle within mathematical physics during the seventeenth and eighteenth centuries, was between the neo-Aristotelian gnostics, including the Cartesians and the Newtonians, the so-called analytical faction on the one side, and, on the opposing side, the current of geometricians, followers of Nicolaus of Cusa, Leonardo da Vinci, and Kepler, through Christiaan Huygens, Gottfried Leibniz, and Jean and Jakob Bernoulli. By the end of the seventeenth century, the characteristic essential feature of this factional affray within mathematical physics, was the Leibnizians' emphasis upon the so-called non-algebraic geometrical functions, and the analytical school's rejection of this non-algebraic systematic view. (See Appendix VIII.)

The systematic study of these "non-algebraic" geometrical functions, was most intimately associated with three classes of physical phenomena. First, from Leonardo da Vinci (1452–1519) into the nineteenth century, the study of the phenomenon of electromagnetic radiation (e.g., light). Second, the manifestation of the crucial isochronic processes in nature corresponding to the cycloids and related non-algebraic functions (e.g., the tautochrone, brachistochrone, optics, and the relevant evolutes and involutes). The union of these matters of light and isochronism is found in the general (Leibniz) physical principles of *least-action*.

The geometrical school of Leibniz et al. continues, by way of the circles of Gaspard Monge (1746–1818) and Carl Friedrich Gauss, through the nineteenth-century work of such exemplary figures as Bernhard Riemann, Eugenio Beltrami (negative curvature), and Georg Cantor. (See Appendix VIII.) There, fundamental (e.g., axiomatic) progress in mathematical physics comes not to an end, but a zone of rapidly attenuating rates of progress respecting the axiomatic issues. During the twentieth century to date, there has been significant progress in experimental work, but very little progress in established scientific doctrine respecting deeper axiomatic issues. Indeed, the very mention of those deeper issues, formerly central topics of all serious scientific discovery, is virtually banned under the rubric of "philosophizing."

The feature of this on which our attention is focused at the moment, at the point where a principle of *least-action* presupposes a definite, universal, nonlinear *curvature* of physical space-time. This is already clear, even if only the relative validity of Kepler's physics—relative to the Cartesian and Newtonian—is taken into account. The deeper implications of non-algebraic isochronism confront us more directly with the evidence of the pervasively ruling efficiency of that principle of curvature.

When we situate our working definition of the term "curvature of physical space-time" so, in respect to such notions as tautochrone and brachistochrone, we ought to begin to see the argument here more readily, more clearly, with a geometrical-experimental insight. The very term "curvature of physical space-time," were either gibberish, or, at best, mere license of romantic poetizing, if it meant anything other than the startling statement of principle we have cited above: *Reactions among bodies are determined as secondary features of each body's primary immediate interaction with the universe as a whole.* The refraction of a composite beam of light, composed of several distinct frequencies, is an illustrative image of this statement's import for the novice. The (Leibniz) principle of least-action exemplifies implicitly the content of the statement.

It is by means of the sovereignly individual agency of individual creative reason, a *spiritual* force emanating from the person as *monad*, that the universe as a whole is altered isochronically, by means of that generation, communication, and assimilation of valid, fundamental discoveries, which, individually and cumulatively, increase the power of the entire human species over the universe.

This "entire human species" is not a finite, nor a "potentially infinite" collection. It is represented, in first approximation, as a transfinite magnitude. It is isochronic, in the specific sense that the relevant forms of present action affect the past as efficiently as the present and future. For, valid fundamental scientific discovery, for example, acts principally upon the outcome of a generation's activity, and thus the present efficiently changes the past by altering its relevant quality of *outcome.*

The true self-interest of the individual person, and of the society, are so made known to law-giving reason.

The individual person is a sovereign individual person by virtue of that *divine spark,* which we recognize in such form as the capacity for super-logical, creative reason, to generate, to communicate, and to assimilate efficiently valid, revolutionary transformations in science and technology. It is only from that standpoint, that a person or nation has the quality and rights of sovereignty.

The relationship of the sovereign personality and of the properly sovereign state, to the universe as a whole, is ostensibly ambiguous; at least this seems to be the case at first impression. For, from the standpoint of potential population-density, we can show that the individual sovereignty, as *One* of *Many,* locates its superior *One* practically in the relevant, subsuming scientific-historical process of *Becoming* potentially a society of increased population-density. This *Becoming* is a true *transfinite,* as we elaborate the proof of this in *In Defense of Common Sense.* It is therefore *not* the Good. Thus, the apparent ambiguity of connection to the *Becoming,* and also to the *Good.* We must clear up this ambiguity.

It is sufficient, for purposes of this aspect of statecraft, to note the following. The process of change defining a transfinite *Becoming,* defines perfection in terms of lessening imperfection and thus identifies, as unchanging, an ordering principle of lessening imperfection. (This ordering principle is equally congruent with the difference among any three, arbitrarily chosen, successive elements of the series, and is therefore a true transfinite.) The Good were such a transfinite, which is everywhere equal to every aspect of itself considered as part of a continuing process.

Thus, our reason grasps the meaning of the Good less unclearly than otherwise from the standpoint of the *Becoming;* but we do not perceive the Good directly. We perceive and know its efficient reflections in the *Becoming.* We know it as that which is reflected by the Becoming. Thus, from day to day and place to place, as we express our true self-interests in our work, the Becoming, as typified by the increase of potential population-density, has for us the character of the pathway which the Good and the biblical book of Genesis, oblige us to follow.

Adam Smith and Karl Marx

Among other uses, the immediately preceding set of tightly interconnected sub-topics permits us to address, with more devastating force and greater relevance, the principal, twin economic pestilences of this century: British liberalism's ruinous cult-dogma of "free trade," and also Adam Smith's terrible grandchild, the economic doctrine of Karl Marx. The examination of that connection, from the standpoint of the immediately preceding topics, demonstrates a broader principle of statecraft to which we attribute great importance for reference here.

To begin, consider the superficial history of the connection between the "free trade" dogma of Adam Smith, and Karl Marx.

Smith (1723–90) was a follower of the British secret intelligence service's David Hume (1711–76) in the teaching of an immoralist concoction perversely named "moral philosophy."[11] When, about 1763, Smith came directly into the employment of the second Earl of Shelburne's British East India Company,[12] Shelburne et al. provided Smith access to Hume's old physiocrat and Rousseauvian cronies in France and Geneva. Out of this latter apprenticeship in the French physiocrats' dogma, came Smith's famous 1776 apology for the established anti-French and anti-American policies of his narcotics-trafficking employer, the British East India Company: *The Wealth of Nations*, 1776.

It is relevant to the consideration of the Smith-Marx connection here, that Smith's *The Wealth of Nations* has been viewed as, substantially, a parody of Turgot's published work of that period.[13]

Karl Marx, recruited to the Mazzinian freemasonry's "Young Europe" association by no later than the early 1840s, ended up, beginning a few years later, in London under the protection and virtual control of the same Lord Palmerston who maintained British secret intelligence service's connections to Mazzini's continental "Young Europe," through such channels as the British Museum's David Urquhart. Urquhart was noted by Marx as among those who steered Marx into indoctrination in the fraudulent myth of Britain's supremacy in scientific progress, British East India Company economics included.[14]

Thereafter, Marx always professed his intellectual debt in economic thinking to Smith, Smith's follower David Ricardo, and to Smith's instructors among the French physiocrats, Dr. Quesnay most notably. Marx, in addition to Friedrich Engels, was always thereafter vile in his praise of the British East India Company's usury-based political-economy, especially in his attacks on the American System of Leibniz,[15] Hamilton, the Careys, and Friedrich List. In summary, he attributed the origin of "scientific economics" to the physiocrats and to the East India Company's usurers; and, the relevant British academics, especially the Fabians, have accepted this view by Marx without remarkable quibbling on the point.

Now, re-examine the same historical connection from the vantage-point of our earlier discussion of curvatures of physical space-time. Let us recognize thus, that the approximate simultaneity of the collapse of the Anglo-American and

Muscovite economic systems shows the convergence of effects of the two systems sharing in common certain among the most flawed axiomatic assumptions implicit in each.

The history of European civilization, including the post-1492 Americas, is essentially, as Friedrich Schiller (1759–1805) portrays this,[16] the struggle of *republicanism* (such as Solon's [638–558 B.C.] reforms in Athens) against the barbaric heritage of ancient Mesopotamia, usury-ridden oligarchism. British philosophical liberalism, the root of the "moral philosophy" of Hobbes, Locke, Hume, Bentham, Mill, and Smith, and of eighteenth-century British (Haileybury) political-economy, is in all essential features, a utopian *pantheistic* ethical dogma, modeled chiefly upon pagan Imperial Rome, but also upon the ancient, pantheistic Delphic cult of Gaia, Python-Dionysios, and Apollo. The principal forerunners of Delphic oligarchism in Ancient Greece are the so-called "Babylonian" model of ancient Mesopotamia and Canaan.[17]

Common Axioms

To adduce the relevant common axioms implicit in the collapses of the British and communist political-economies, it is perhaps sufficient to compare British and communist dogmas of national and super-national economic practice with the following reference-points in ancient and Renaissance history. We begin with the succession of usury-caused collapses of the ancient Mesopotamian "bow-tenure" system of agriculture. We include reference to the circumstances of Solon's anti-usury reforms in Athens. We examine the crucial, related issues associated, successively, with the Flaminian and Gracchi reforms of pre-Imperial Rome. We include the process of collapse inhering in the axiomatic features of the pagan Imperial Rome of Augustus, Tiberius, Nero, and Diocletian.

We examine the reasons for the upsurge of economic power generated by the Golden Renaissance, and view this

276 The Science of Christian Economy

with a reflection upon the great enterprises set into motion by
Charlemagne earlier.

Two opposing features of these cases are emphasized:
the role of usury and the issue of increase of the per capita
productive powers of labor, scientific, and technological
progress.

The forms in which usury's systematic taking of unearned
income (i.e., "theft") occurs may be reduced to three general
sub-classifications. First, there is *simple usury:* payment taken
on account of debt, whether the original principal amount of
that nominal debt may have been created in payment of money
or real value advanced, or simply imposed upon the debtor
either by fiat, or kindred means. Second, there is the role of
monopolies (e.g., the international grain cartel), in exacting
usuriously (un)earned income from both producers and con-
sumers of some essential commodity. Third, there are the
sundry guises of *ground-rent usury.* We include among these
the evolution in the modern British model of central banking
and related forms of public indebtedness, from roots in ancient
(e.g., Mesopotamian) tax-farming.

We counterpose *earned profit* of *physically productive*
enterprise to the merely nominally "*earned*" *profit* and *interest*
of usurious activities. We explain the necessity and functional
basis for this distinction.

For various reasons, including the durability of clay cune-
iform tablets, the earliest good accounting record of physical
economies is our knowledge of ancient lower Mesopotamia.[18]

What we know of the economic history of the region
shows that the critical physical factor causing the depopulat-
ing collapse of society after society successively in this region,
was the effect of usury. This pattern extends from the most
ancient cases known, through the usury-caused collapse of
the post-Abbasid Baghdad.

Without the anti-usury, referenced reform of the type
introduced by Solon, Athens as a center for classical Greek
culture would never have come into existence.

The monstrous failure of the pagan Imperial Rome of

Augustus, Tiberius, Nero, and Diocletian, serves us as probably the best available case-in-point for classroom and related
uses.

Whatever else may be said of the Gracchi and their proposed reforms, something akin to their leading proposal was
the mandatory alternative to the nightmare which gripped
Italy for centuries following the Gracchian faction's defeat.

Essentially, insofar as the returning farmer, from the
ranks of Rome's legions, was settled productively on his farm,
the aggregate families of Italy produced significantly more
than Italy consumed. This, relatively speaking, represents a
state of prosperity.

If that same returning legionnaire is denied his farm and
is relegated, instead, to the company of a proletariat enflaming
the Roman piazzas, internal physical-economic bankruptcy
grips ancient Roman Italy. As the engorged mass of a parasitical Roman idle rich gobbles up the formerly productive farm-
lands, to establish and to enlarge a sybarite's unproductive
slave plantations, ancient Italy depends increasingly upon
looting by force from the subjugated foreign colonies and
satrapies, which provided, as tribute, the needed grain and
other prime necessities.

This was the circumstance of the breakdown of the Roman republic, the condition of the civil wars and imperial
designs of Julius Caesar and Augustus Caesar. Thus, like the
bankrupt United States and Britain of 1990–91, the Romans
of the first century B.C., rather than remedying the evil policies
ruining the nation at home, prolonged their reign at home
by foreign adventures. A parasitical mask of global imperial
grandeur adorned that portion of the imperial capital abutting
a sea of slum-helotry. Thus, then, as the Washington and
London of 1990–91, did the rise and persistence of an Imperial
Rome, rotting at its core, accomplish the general ruin not
only of Italy, but of those colonies upon whose looting depended the Roman helotry's TV-like diversions at such places
as the Circus Maximus.

So, the U.S.A. of 1991 is gripped, like the rotting Victorian British Empire before it. By permitting the internally rotting imperial British Empire of Castlereagh (1739–1821), Palmerston, Russell, Mackinder, and Milner to drag European civilization into the ruinous new Thirty Years' War of 1912–45, a war which London orchestrated against the Eurasian development perspective of France's Gabriel Hanotaux (1853–1944), the complicit governments of France, Germany, Austro-Hungary, and the United States (among others) did bring a monstrous ruin upon themselves and upon the planet as a whole.

As a self-bankrupted ancient Rome sought an empire based upon naked force, thus to postpone the inevitable time of social-economic collapse at home, so the Anglo-American Liberal Establishment, the fanatical proponents of Smith's lunatic "free trade" dogma, have based their policy upon establishing a worldwide "Pax Americana" (directed from London), a "World Federalist," "one world," "new world order," a utopian's parody of the imperial pagan Rome of Augustus, Tiberius, Nero, and Diocletian.

To understand adequately the controlling impulses of the relevant neo-imperialist, Anglo-American liberals, we are obliged to study the rise, especially in eighteenth-century England and France, of the corrupting influence of so-called "Enlightenment Liberalism."

In Britain, this included most prominently the doctrinaires Francis Bacon, Thomas Hobbes, John Locke, David Hume, Adam Smith, the second Earl of Shelburne, Jeremy Bentham, and Thomas Malthus. In France, this must include René Descartes's neo-Aristotelian formalism as well as such shamelessly romantic figures as Montesquieu, Voltaire, Rousseau (1712–1778), the reactionary, oligarchical physiocrats of the Jacobins' salon of free-trader Jacques Necker and his notorious daughter, the Madame de Staël. Considering such exemplary cases, we trace the manner in which the axiomatic features of pagan Roman imperialism were imbedded in the "mind-set" of the modern liberal, fascist, and communist.

From that latter vantage-point, we may see more clearly, not only the true nature of the genetic links of Adam Smith to Karl Marx, but we see also why this connection is of such importance for understanding the common roots of the almost simultaneous collapse of the Muscovite and Anglo-American political-economic systems.

Roots in Ancient Pantheism

The pagan Roman imperial model of "new world order" adopted by the Thatcher-Bush circles of 1990, is, like most things of pagan Rome, a parody of someone else's earlier designs. Two precedents are of outstanding relevance. The nearest was the Cult of Apollo at Delphi and Delos in Greece. Ultimately, all significant European oligarchism and pantheism are either ancient Mesopotamian or Dravidian in traceable origins. By examining Mesopotamian and Delphic pantheism as the relevant models for modern liberalism and anti-republican oligarchism generally, we show how the archetypical oligarchical religious idea, ancient *pantheism*, is the axiomatic root of such modern phenomena as *gnosticism* and *Satan-worship*, in religion; *liberalism, fascism,* and *communism* in social philosophy; *positivism* in legal philosophy of practice; and *imperialism* in statecraft. Also, most directly to the point, oligarchism is impelled axiomatically not only to the parasitical practice of usury, but is impelled, as if by instinct, to destroy any society which bases itself upon the fostering of investment in scientific progress for increase of the productive powers of labor.

The known root of ancient pantheism is the cult of an Earth-Mother/fertility goddess, known variously as *Shakti* (Dravidian "Harappan"), *Ishtar* (Chaldean), *Athtar* (Sheba-Ethiopia), *Astarte* (Canaan), *Isis* (Hellenistic Egypt), *Cybele* (Phrygia), or *Gaia* (Delphi). She is associated with a *satanic* male deity, a phallus-serpent deity known as, for example, *Siva* ("Harappan"), *Python* (Delphi), *Dionysios* (Phrygia), or

Osiris (Hellenistic Egypt). Probably, the earliest of the known origins of the Earth-Mother/phallus cult was "Harappan," spreading through "Harappan" maritime colonies of the "black-headed people," such as Sumer and Sheba, to appear as such Semitic and Hellenistic utterances of "Shakti," as "Ishtar," "Astarte," and "Isis."

It is relevant in several ways, that these ancient Satan-(serpent/phallus-) worshipping cults are associated with the Moon-goddess and the lunar calendar, rather than the solar astronomical calendars of the earlier, Central Asian Indo-European cultures. The gods of the pagan pantheistic cults are, like Imperial Roman emperors, apotheoses of the most degraded forms of lustful, existentialist, cupidity-ridden irrationalism. These pantheisms' ministry is the cult of fear, not agapic love; their so-called "law" is but the *ukase* of power's capricious passing whim.

The endemic political impulse of pantheism is seen in the instances the gods of the new vassal are induced to submit to the gods of the conqueror. So, Anglo-American imperialism today demands submission to the Supreme Architect of Usury and to the cult-dogma of universal "free trade."

The syncretic fusion of many pagan deities into an "Olympian" ethical pantheon, prohibits any reasonable distinction between truth and falsehood, right and wrong. Morality is prohibited. And its place is fully occupied by a mere ethics, as Aristotle's *Ethics* and *Politics* show this.

Here is the pantheistic root of British liberalism and its political-economy; here is that axiomatically pantheistic[19] feature of liberalism which leads consistently toward its self-expression in such forms of manifestation as British neo-Roman imperialism, fascism, and communism.

Two Historical Illustrations

Two historical illustrations of this point are supplied here now. The first case is the apparent anomaly, that certain leading U.S. and British financier circles, including those associated with Morgan, Harriman, and Theodore Roosevelt, should have been not only actually or nearly "card-carrying Bolsheviks" during the course of the years soon following the so-called October Revolution of 1917, but also, later, supporters of the Mussolini and Hitler fascist regimes. The second case is the common characteristics of Karl Marx's political-economy and the "Malthusian" socialistic decrees of the Roman Emperor Diocletian.

During the period 1917–27, certain among the most powerful financiers and related political circles, chiefly in Britain and the U.S.A., were not only partners of the young Soviet government, but also "co-owners" of those sections of the Communist International (and its communist spy service) later known variously as the "left" and "right" Comintern "oppositions," including circles associated with the U.S.A. "neo-conservative" extremists of today. In those terms, 120 Broadway in New York City's Lower Manhattan was a leading center of the obvious capitalist-Bolshevist entente.

Later, the same Harrimanite circles who were once associated so with Trotsky, Stalin, and so forth, moved to provide very significant support for both Benito Mussolni's and Adolf Hitler's neo-Roman cult of fascism.[20] The latter included the Harriman circle's public support for the Nazi Party's "racial purification" dogma,[21] and the key role of Harriman's company, as bankers, in moving funds to aid Hitler's "legal coup d'état" in Germany in 1932–33.

The second historical reference, is the connection between the characteristic feature of Roman Emperor Diocletian's repressive, "Malthusian" socialist decrees, and the most crucial among the attributable axioms imbedded in the formal side of Karl Marx's political-economic doctrine.

Although the establishment of Romantic liberalism in Britain began with the evil circles of Francis Bacon and Thomas Hobbes, and although liberalism was formally established as the state philosophy of the United Kingdom with the accession of the first Duke of Marlborough's King George I, for all practical immediate purposes here, we begin with

Castlereagh's role in establishing London's institutionalized control over the internal affairs of continental Europe, by means of the 1815 Treaty of Vienna and the Holy Alliance.

Although it had been the German friends of anti-oligarchic Friedrich Schiller, who had led in bringing about the downfall of an entrapped Emperor Napoleon Bonaparte, it was London and Hapsburg carrion crows of usurious oligarchism who carved up the spoiled peace at the 1815 sessions of the Congress of Vienna. The infamous Holy Alliance's instincts were an echo of the Malthusianism of Gian Maria Ortes and of the odious "socialist" edicts of the Roman Emperor Diocletian. The repressive "Karlsbad Decrees" expressed the essence of the matter.

The 1832 Hambach event signaled a de facto reversal of the Karlsbad Decrees and the approaching erosion of the Holy Alliance itself. Against the renewed threat of a nationalist effort for an anti-Jacobin republican renaissance in the domains of science and economy, the challenged oligarchy unleashed its own neo-Jacobin, Dionysiac forces of chaos, forces which soon came to be centered around the Mazzini freemasonry of "Young Europe."

U.S. President Abraham Lincoln's qualified defeat of London's Confederacy, was a crucial turn in the post-Holy Alliance world order. Not only had republican, agro-industrial, mercantilist progress triumphed over London-backed chattel slavery and usury. Defeat of London in this affair and the doom of that bloated British special constable, Napoleon III of France, had been ensured by the intervention of Russia's Czar Alexander II, threatening the use of Russia's naval and land forces against London and Paris, should those capitals carry through their intention, not only to invade and loot Mexico, but to intervene more directly in military assistance to London's otherwise doomed puppet, the Confederacy.

At the same time, Czar Alexander continued to reverse the barbaric devolution of Russia's social and economic life, under his two predecessors, and to reform Russia along the lines of the reforms which Peter the Great had launched on the prompting of Gottfried Wilhelm Leibniz.

Russia's action for European neutrality in the U.S. Civil War of the 1860s, was one of the three major developments of the second half of the nineteenth century which impelled London to unleash the "new Thirty Years' War" of 1912–45 in Europe, and which prompted also the strange 1920s cohabitation between the Harrimans and the Bolsheviks. The other two developments were, first, the friendship between St. Petersburg and Bismarck's Germany, and second, the efforts of France's great statesman Gabriel Hanotaux, to establish an anti-British bloc of North Eurasian cooperation for economic development.

The Fabians' Britain of Cecil Rhodes, Milner, and Mackinder,[22] caused World Wars I and II for the same, deeper, geopolitical motives, which, more recently, prompted the Britain of Margaret Thatcher, British agent-of-influence Henry A. Kissinger, Nicholas Ridley, and Conor Cruise O'Brien to launch what might very well become the plunge into World War III.

Britain then responded to the cooperation between St. Petersburg and Bismarck's Berlin, by seeking war between Russia and Germany, more or less as Prime Minister Margaret Thatcher's circles reacted to German reunification in 1990. The convergence of Hanotaux's Paris upon continental European general cooperation with Sergei Count Witte's (1849–1915) and, later, Stolypin's (1863–1911) Russia, brought from London a more profound reaction: World War I.

The Western oligarchical interests used traditional special channels into the old, anti-Petrine boyars generally, and into such institutions as the Third Section and Okhrana in particular, to unleash *raskolniki* forms of madness and terrorism against those institutions of Russia, from the czar on down, responsible for the friendship with Bismarck's Berlin and Hanotaux's France. This is the key to the anomalies of Averell Harriman's Broadway.[23]

London's perception of this 1880–1900 threat of Eurasian continental economic development and cooperation, impelled the circles associated with Milner and Mackinder not only to corrupt France successfully with the *Entente Cordiale,* but to use the "messianized" *raskolniki* followers of Fyodor Dostoevsky and Nikolai Bakunin to destroy from within a czarist Russia, which might seek cooperation with both Germany and France against Britain's control of the continental "balance of power." The greatest threat currently to civilization as a whole is, that the pattern of Mrs. Margaret Thatcher's apparent "Svengali-like" control over the United States's George "Trilby" Bush might persist, even under a change of the specific personalities occupying those official positions. If so, then, as the pre-history of Britain's authorship of World War I was re-enacted during 1990, against the implications of Germany's reunification, so the danger of a "new Thirty Years' War" threatens the planet with a plunge into a "New Dark Age" by the beginning of this coming century.

Before turning to the second of the two historical examples, let us underscore a crucial lesson demonstrated by the case just outlined.

The popularized, ignorant opinion of history presumes axiomatically the misanthropic Hobbes-Locke-Hume-Smith notions of peoples as hedonistic, instinct-ridden little *homunculi,* each born—bestial instincts apart—a *tabula rasa.* For such poor dolts, real history never existed, but, rather, nothing more than a "Zeno's Paradox" sort of "Achilles and the Tortoise" kind of separated, short intervals of current events. History for them is a succession of kinescopic still photographs, within which "current events" are determined with little or no regard for the cultural heritages of the preceding kinescopic frames.

The simplest, empirically-based disproof of the cited British man's Hobbes-Locke view of history, emphasizes two interrelated sets of facts. The first fact is the known, millennial, philologist's history of those languages in which all contemporary conscious behavior is molded. The second fact is the relationship between a language-family's classical poetry, and the singing of that poetry under the influence of harmonic principles of vocalization genetically intrinsic to all healthy specimens of the human species, regardless of race or national origin.

No person ever existed as a *tabula rasa*. What is transmitted to each new member of society from preceding generations, includes not only that which is transmitted by the memory-medium of language, but those ideas which are characteristic of the developed grammatical and other structure of the spoken language itself. It is shown that a language's organic structure is itself a kind of physical geometry, which reflects actual orderings of social relations, as well as inorganic ones generally. A language, also, as the biophysical harmonic laws of, for example, *bel canto* vocalization illustrate the point, is formed to conform more or less wittingly to the biophysical requirements of transmitting and receiving "profound and impassioned conceptions respecting man and nature" (Shelley) among the processes of *thinking to utter, uttering, hearing, thinking, and thinking what is heard.* (See Appendix X.)

In real history, like the most important, millennia-spanning example referenced, the underlying structure determining today's crucial events reaches back across centuries. In the more adequate view of historical processes, it is the transmission of embattled ideas, over successive generations, which determines the course and outcome of each moment of history taken in the shorter term. The free will of the individual is not a matter of indifference in the process, but the individual free will is historically efficient, only to the degree that its action, wittingly or not, alters the quality and interaction of those ideas which pour, like a mighty avalanche of political-cultural traditions, out of the long past, into each momentary present.

Schiller argues, that European history in totality can be

understood as essentially a millennia-spanning, continuing conflict between but two historical-cultural traditions: *oligarchism* as typified by Lycurgus's Sparta, and *republicanism* as typified by Solon of Athens. Our example references directly the span 1812–1990; implicitly, we trace the same conflict to the end of the sixteenth century (Francis Bacon); we could have traced it to the times of Socrates (469–401 B.C.) and Solon, or, with increasing fuzziness of vision, to more remotely ancient times.

Given the fact, that the free will of individuals does alter the course of what is in the long term a culturally determined history, history is not determined by simple ideas as such. Determinism exists in the long-span cultural determination of history, not in terms of simply fixed kinds of ideas, but in terms of the transfinite principle, which expresses the continuing characteristic of a cultural-factional body of thought throughout the multi-century span of the many changes introduced to it by action of sovereign, individual free will.

Thus, it is types of cultural ideas, as, for example, oligarchism or classical-humanist republicanism, which act efficiently upon history. The sovereign, individual free will acts upon such cultural ideas, that is, upon the efficient, transfinite characteristic of such ideas, to the effect of rendering the power of such ideas relatively greater or less.

Marx, Smith, and Diocletian

This brings us to the case of the second historical example, the common axiomatics of the economic doctrines of Adam Smith, Karl Marx, and the Emperor Diocletian.

If we consider Marx's four-volume *Capital*, and his related writings, only in their narrower aspect, as a system of political-economic analysis as such, Marxian economics can be reduced essentially to a set of simultaneous linear inequalities, purporting to represent a linear mathematical model of

what Marx terms "extended reproduction." Two explicitly
adopted assumptions of a formally axiomatic-deductive quality are then shown to be direct points of equivalency between
Marx, on the one side, and, on the other side, Adam Smith
and his physiocratic teachers.

First, Marx not only accepts, and defends fanatically,
Adam Smith's "free trade" model of competition; he adopts it
as a method of linear statistical determination of the marginal
distribution of what he terms "exchange value." He is viru-
lently anti-mercantilist and a faithful apostle of Adam Smith
and the physiocrats on this account. He is, otherwise, an
avowed, pro-British adversary of the American System of
political-economy, as he underlines this in connection with
his vile defamation of Friedrich List (1789–1846) and Henry
C. Carey.

Second, Marx observes, accurately, that in his constructing linear inequalities intended to describe "extended
reproduction," he has ignored both technological progress
and what Henry C. Carey describes as the "economy of labor"
determined by technological progress. (Although Marx does
seek to bring in technology as a depreciator of price, after the
fact, in *Capital*, Volume III.) In respect to principles of Physical
Economy, it could be said that Karl Marx is a "knuckle-
headed" populist, and a physiocratic one at that.

Among the relevant other absurdities in *Capital* which
cohere with these two axiomatic assumptions, are Marx's—in
fact, pro-usury—distinctions among *profit*, *rent*, and *interest*,
and his fool's-errand quest for the "primeval hoard" of
money.

The better approach to recognizing the disastrous folly of
the two axiomatic assumptions, is to reflect upon the practical
significance of what they require implicitly be excluded from
policy-shaping considerations. The summation of this line of
argument is partly repetition of points made earlier in this
text, but usefully so: We see, one hopes, more clearly, how

the issue of "the One and the Many" bears directly upon the determining connection between Physical Economy and statecraft in general.

Since the practice of statecraft must be concerned with the *durable survival*[24] of the society and its included most essential social institutions,[25] there can be no competent statecraft whose practice fails to address efficiently the requirements of a science of Physical Economy. Diocletian's decrees and the physiocrats Adam Smith and Karl Marx demand, on common included ground, that political-economy evade those conditions which are indispensable for the *durable survival* of a society.

We have indicated, that a transfinite, positive ordering of increase of an entire society's potential population-density is the general precondition for durable survival. We know, by definition, that the capital-intensive, energy-intensive investment in scientific and technological progress, in both basic economic infrastructure and production of basic physical goods of producers' and households' consumption, is required policy and practice.

We can show, either on the basis of the physical geometry of those seemingly simple principles, or by reference to appropriate, crucial empirical evidence, or both, that the following set of general inequalities must be satisfied.

1) The per capita leisure and physical consumption of the family household must be improved, but under the condition, that the per capita and per hectare physical output of the whole society increase more rapidly than the per capita household goods consumption.

2) That, with technological progress, the school-leaving age must increase asymptotically toward an average upper limit. This requires corresponding increases in health and longevity.

3) That, with technological progress, the ratio of required employment in physical production and manufacturing and basic economic infrastructure must increase relative to agri-

culture and related activity, up to a lower asymptotic limit for the latter.

4) With technological progress, physical-productive employment in producers' goods must increase relative to that in physical production of household goods.

5) That, with technological progress, we must increase not only the quantity of energy available per capita area of potential population-density; we must increase the effective intensity of the applied energy.

That is enough detail for our immediate purposes. It is clear that successful growth must take into account the ratio of the labor force trained for employment in each category, and must establish correlated priorities for fostering credit and capital to admit realization of such goals. This requires corresponding forms of "mercantilist dirigism" in the economy, otherwise no rational result will occur. Indeed, no "free trade" economy has continued to prosper at home, except by looting both an "under-class" of actual and quasi-helots at home, and looting foreign populations most generously.

While the "free-traders" howl loudly of their freedom to steal, they deny real freedom, the freedom to create, and to obtain the conditions of family and general social life needed to foster the creative potential of the individual and his expression. This true freedom is exemplified by a truth-seeking commitment to valid, fundamental scientific progress, to related creative work in classical-humanist art forms, and so forth.

Without doubt, the Soviet political system denied true human freedom. Without doubt, as long as Moscow was perceived as a credible strategic adversary, there was still much greater political freedom in the West, than in Soviet society.

We should add, relevantly, that the best Soviet scientific workers lived and worked at the rim of a chasm of political dissidence on this account. Part of the depth of the crisis in the U.S.S.R. today, is that the pre-1917 stock of Russian intellectual capital has been almost used up, in science as in the eradication of those productive farmers who used to be named "kulaks."

One is reminded thus from Soviet history, of the "Malthusian" socialist decrees of the Emperor Diocletian.

The situation in the "free trade" West is not generally much better. Only by exception, such as following the leadership of a de Gaulle, Adenauer, Mattei, or John F. Kennedy, some genuine long-term growth was promoted during the postwar period. Otherwise, as in the case of occupied and quasi-occupied postwar Germany, the net relative productive potential has been declining from the high point reached, about 1944, during the course of World War II.[26] In general, apparent short-term growth has been realized by resort to what Marx, Rosa Luxemburg, and Yevgeny Preobrazhensky termed "primitive accumulation."

Look at the picture of the world from Japan circa 1983–84. Japan, which had made good use of purchase of otherwise idle U.S. patents, was faced with the collapse of U.S. expenditure in research and development, together with a catastrophic decline in quality of top-ranking U.S. science graduates. Thus, if Japan at that time were to maintain its rate of growth in "economies of labor," it had to increase rapidly and substantially its percentage of national employment in research and development.

Without "dirigist" decisions of that sort, in R&D, in basic economic infrastructure, and in education and employment generally, there can be no true opportunities for exercise of human freedom in the society.

Where cartels are permitted to loot agriculture, by the dropping of government parity-price protection for farmers, free agriculture vanishes, and, sooner or later, hunger enters. Where ultra-competition under conditions of reckless deregulation prevails, small industries, the bulwark of economic freedom, fail, and the margin available for freedom—technological progress—drops to below zero percent of the cost of sales.

Without real growth in potential population-density for the society taken as a whole, there is a net real decline, perhaps temporarily concealed by primitive accumulation, which means a disaster in the longer term.

For the most efficient route to uncovering the common, principal axiomatical characteristics of Adam Smith, Karl Marx, and the Diocletian decrees, turn attention next to the common physiocratic features of each. Following that, place emphasis upon the explicitly immoral, populist form of irrationalism, which is professed repeatedly, with shameless openness, by Smith, and which is his sole premise for his esoteric "invisible hand" dogma of "free trade."

Aristotelian Follies

As a matter of principle, the doctrine of "free trade" begins in history as the lunatic's[27] worship of the whore- or Earth-Mother goddess, Shakti, Ishtar, Gaia et al. As for the Diocletian decrees, also for the pagan physiocrats and Adam Smith, the source of *profit, interest,* and *rent*[28] is the mysterious "bounty of nature." The neo-Aristotelian René Descartes, assists us perversely in decoding this esoteric pagan dogma, implicitly placing the "bounty of nature" among matters under the more general heading of *deus ex machina.*

Read Adam Smith on this:

The administration of the great system of the universe . . . [and] the care of the universal happiness of all rational and sensible beings, is the business of God and not of man. To man is alloted a much humbler department, but one much more suitable to the weakness of his powers and to the narrowness of his comprehension: the care of his own happiness, of that of his family, his friends, his country. . . . But though we are . . . endowed with a very strong desire of those ends, it has been entrusted to the slow and uncertain determinations of our reason to find out the proper means of bringing them about. Nature has directed us to the greater part of these by original and immediate instincts. Hunger, thirst, the passion which unites the two sexes, the love of pleasure, and the dread of pain, prompt us to apply those means for their own sakes, and without any consideration of their tendency to those beneficent ends which the great Director of nature intended to produce them.[29]

We find a relevant observation in the work of Sir Isaac Newton, the apotheosized god of science among Britain's pagan imperialists. Newton confirmed that his formal physics contained the patent absurdity, of portraying the universe, mathematically, as running down in the fashion of a mechanical timepiece. He observed, that this faulty mathematical construction gave the appearance, that for the universe to continue to exist, God must rewind it periodically. This is Newton's fair representation of Descartes's dogma of *deus ex machina*. Newton qualified his argument, by confessing that his faulty choice of mathematics was the only one he found acceptable.[30]

Notably, Gottfried Wilhelm Leibniz referenced this "clock-winder" matter in the Newton-Clarke-Leibniz correspondence. The Cartesians' and Newtonians' fanatical refusal to accept the reality of *non-algebraic functions*, and to refuse to consider, therefore, a competent calculus, shows that Newton was indeed aware, that a mathematics schematically different from his own, was an available choice.[31]

Newton's fictional "clock-winder god," is Descartes's *deus ex machina*, and the omniscient, but impotent post-Creator of Aristotle's schema.[32] This pagan deity of Aristotle (384–22 B.C.) and Descartes is also the mechanistic Enlightenment's freemasonic concoction of Robespierre's Jacobin Supreme Being cult. The quarrel of Leibniz with Kepler's adversaries among the Cartesians and Newtonians, shows the crucial point at issue in the readily most intelligible and historically actual form.

Over the decades, from the late seventeenth century through to the beginning of the twentieth, this issue is embodied in the mathematical guises of the geometric magnitudes, the "non-algebraic," "transcendental," and "transfinite"; each, successively, represents but progress in comprehension of the same matter already addressed by the preceding usages. For reasons which we have already considered above, the possibility that an intelligible representation of the lawful ordering of both the *Becoming* within physical space-time and human knowledge of that *Becoming,* lies within the scope of combined notions of an underlying, harmonically ordered curvature of physical space-time, and the employment of those geometrical forms of mathematics associated successively with "non-algebraic," "transcendental," and "transfinite," to represent the ordering of events within that curvature.

Thus, the mechanistic axiomatics of Cartesian and Newtonian conceits, are the persisting source, not only of the cited, Newtonian "clock-winder" delusion, that our universe is entropically ordered. The same, Aristotelian folly, termed the deductive/inductive method, is the sole rationalist form of operation responsible for the belief in a *deus ex machina.* (See Appendix IV.) It is also, in the same way, the rationalist sophistry employed to support the physiocrats' version of the gnostic,[33] "fundamentalist," populist faith in the mysterious "bounty" of the Earth-Mother goddess Gaia, otherwise named "Mother Nature."

Like the ancient Aristotelian *organon,* the modern deductive/inductive method permits no consistent schema, but a universe of constantly linear physical space-time curvature, a universe of linear pair-wise interactions among bodies in linear space and linear time. The corollary of this, in a linear system situated within a constantly linear physical space-time, can be supplied a consistent representation within the terms of the deductive/inductive method.

That Aristotelian or "neo-Aristotelian" method could not represent the lawfulness of our real universe in general, or, most emphatically, a living or living-thinking process.

This is a corollary of the fact that the inevitably failed effort to understand the real universe, or living processes, or human thought from an Aristotelian, or neo-Aristotelian standpoint, must lead ultimately to something like a Cartesian or a Newtonian gnostic's occult phantasm, the *deus ex machina*. The Cartesian's and Newtonian's rejection of the non-algebraic form of a *valid* calculus,[34] illustrates the manner the gnostic's occultism slithers through the cracks inhering axiomatically in the deductive mind-set.

Most simply, the *non-algebraic* domain has two ostensibly equivalent modes of existence: a geometric one, and a physical one. At this moment, it is sufficient, to prove the case, to limit attention to the vicinity of the cycloid.

Formally, there is the demonstrable geometric existence of the cycloid and the system of related evolutes and involutes. These define non-algebraic functions which live within the cracks of the algebraic ones. Hence, even on formal grounds, the argument which suggests the existence of something functionally efficient outside the scope of algebraic functions, obliges that the non-algebraic functions be preferred to an irrationalist, occultism-ridden *deus ex machina*.

Physically, we have already indicated some of the crucial-experimental features of physical space-time which correspond to the cycloid and nearby forms of function. Since such conclusive crucial evidence is beyond doubt, to propose to maintain a representation of physics in stubborn defiance of this evidence, is a gnostic's arbitrary, occult irrationalism.

So, we cannot exclude these functions, even on formal grounds alone, since they came into formal geometrical existence by methods of construction, whose authority could not be denied without throwing out all of geometry. Not only do these functions possess a formal existence within mathematics; they are functions with a unique mathematical correspondence to elementary physical principles; so, no truly rational physics could exist without taking this non-algebraic domain into a leading position of authority. Finally, these functions address directly the location within which the gnostic Cartesians and Newtonians insist, that the "Maxwell demons" of the *deus ex machina* lurk; to ignore the transfinite in that case, is not a mistake but a willful fraud.

Not only is the transfinite conception of a "Becoming" indispensable to a rational representation of even a hypothetical inorganic physical space-time. There can be no rational comprehension of living or of thinking processes without it. This was already proven implicitly by the work of Leonardo da Vinci et al., as the harmonic ordering of living processes, in morphology of geometrically self-similar growth and function, in congruence with the Golden Section. As we have noted earlier here, this was made, successfully, the basis for determining the curvature of physical space-time by Kepler. The morphology of successful economic growth, measured so in terms of increase of potential population-density, has the same morphological harmonic characteristics. Thus, the creative mental processes, which cause such growth, define a function of the same general harmonic characteristics.

The highest form of functional activity which is known to exist within a process is, therefore, the minimal level of form of activity characteristic of that inclusive process as an entirety. Thus, if the universe includes lawfully living, e.g., *negentropic* processes, the *minimal* characteristic of the lawful ordering of the universe as a whole, is that that universe is negentropically ordered as a whole. Thus, since efficient creative reason among sovereign individuals, is a characteristic of a successful society's relationship between man and nature, the minimal characteristic of the universe is represented by a transfinite ordering of *Becoming* congruent with the definition of individual human creative reason.

That latter consideration, then, in turn, becomes the basis for assessing the conjecturable, axiomatic congruence among sub-phases of the universe generally, or human behaviorisms in particular. So, finally, the coincidence among the Diocletian decrees, the physiocrats, Adam Smith's "free trade" dogma, and Karl Marx is to be adjudged.

Some Common Immoralities

That forerunner of President George Bush's long-standing, Malthusian policies on population control, the Emperor Diocletian's decrees, prescribed a de facto ban on technological progress and a de facto fixing of a ceiling upon population totally, by limiting its growth locally. Yet, the population of the same region was far greater, far denser, longer-lived, and less impoverished 1600-odd years later. Indeed, it was the blocks on population growth and against fostering technological progress, which characterized the collapse of the Roman Empire in the West and also later in Byzantium.

Similarly, beginning approximately in the year of the assassination of President John F. Kennedy, the Dionysiac "New Age" of the neo-Malthusian rock-drug-sex-irrationalism counterculture was launched into a mass-recruiting mode, from Aleister Crowley's Britain, into the United States of America. Under President Kennedy's successors, the U.S. economy and its population's intelligence level were systematically, intentionally destroyed by successive policy changes in the direction of a "pro-ecology," "anti-technology," "anti-nuclear family," "post-industrial" society.

Hence, from approximately 1967–68, there has been a visible, overall accelerating decline in useful physical output per capita and per hectare. This began, during the 1966–68 interval, as an average decline in the rate of growth of physical productivity, and became an absolute decline by approximately 1968–70.

Official U.S. opinion chose to see this in contrary terms. Since the old economic yardsticks of the 1950s showed decline, the U.S. government and other institutions adopted newly designed yardsticks, designed to measure decline in accord with a new definition of growth. These changes in yardsticks were of the form of the change of behavior, which managed to replace the statement, "My family is going hungry," by "We have succeeded in eating less."

One of the first general indicators of a secular decline toward a new U.S. depression, was a net decline in per capita quality of maintained basic economic infrastructure, from 1970–71 onward. The precedent was the manner speculators earlier turned the New Haven Railroad into junk: Nearly all maintenance of roadway and rolling stock was terminated. In the short run, this curtailing of New Haven maintenance expenditures was employed to increase disbursements for New Haven stockholders' dividends. The price of New Haven stock shot up on markets. Later, the controlling New Haven speculators sold out the stockholding at a vast nominal profit, leaving the railway company resembling a bankrupt scrapheap. This is what has been done to the U.S. physical economy as a whole over the interval 1965–90.

The same principle of "business management," exemplified by the looting of the New Haven Railroad, can be applied to populations. Close down what appear to be relatively less profitable production facilities, thus presumably increasing the average nominal profitability of the surviving enterprises. Then, increase per capita incomes, by mass extermination of the unemployed and the poor: true "social Darwinism" at its most consistent. We have seen such "social Darwinism" practiced during 1990 against the new states of the Federal Republic of Germany, at the insistence of the co-thinkers of Margaret Thatcher and her anti-Germany trade and industry minister, Nicholas Ridley.

In the less extreme case, there is the national economy, in which the appearance is, that some are wealthy and some are poor. In the extreme case, some social Darwinists might propose to increase the per capita level of wealth, by eliminating a requisite ration from among the poor; that was done, in effect, to the eastern new states of the Federal Republic of Germany during the second half of 1990. In another extreme case, a nation, as the United States of America did over the 1970s and 1980s, most notably, shut down more and more employing industries and farms, in order to use cheap labor abroad to increase the profitability of sales of goods to pur- chasers within U.S. markets! The insanity of policy of practice illustrated by each of these examples ought to be apparent immediately.

Let us adduce the common principle characterizing such insane and immoral economic policies of practice.

The prosperity of a nation is, at first inspection, a matter of self-generated physical well-being per capita and per hec- tare. This requires not merely an average level of development of the productive powers of labor; it requires that, in each and all cases, the productive powers of labor not fall below a certain minimum. This latter requires that the standard of the nuclear family not fall below some minimum, from case to case. Thus, the general prosperity of a nation is not just a function of its average productive potential, but must also take into account as a determining variable term, each of both the minimum and also the maximum upper and lower ranges, inclusive of the proverbial 99 percent of the nuclear-family households.

In summary of this point, it is the relationship of each hectare to all hectares, and the productive potential of each individual to that of all individuals. It is such a relationship between the individual and the economy taken as a whole.

It may be the passionately asserted opinion of some, that what we're saying is either not true or is virtually irrelevant even if true. So much the worse for their mistaken opinions. We are addressing here adducible principles of Physical Economy.

A man may leap from a precipice to demonstrate his contempt for the principle of gravitation; the principle of gravitation responds to this by ignoring that man's opinion, with the relevant resulting consequences. So it is with the issues immediately at hand here. It is the inevitable consequence of the Diocletian decrees, "free trade," and Marxian doctrine, which exposes most conclusively the common axiomatic flaws of each and all of the opinions we've attacked here.

Thus, reality responds to those common axiomatical features of all three follies with consistent kinds of effects, as we see in the near-simultaneous collapse of the communist and Anglo-American economic systems today.

The greatest economic crime is to devalue that which sets sovereign individual human personality absolutely apart from and superior to each and all beasts. This crime includes the devaluation of the society's and each individual's own duty to foster the development of that divine spark of potential for creative reason which makes man absolutely superior to all beasts. That required development is not limited to the powers within the individual, but includes the appropriate environmental conditions for fruitful expression of that productive potential.

The benefit occurs for society, not as the arithmetic sum-total of individuals case-by-case; it occurs as the individual participation in the benefit to the society as a whole.

The reflection of the "One and the Many" into this phase-relationship, presents itself at first in a twofold way. In the one way, it is the society as an indivisible sovereign unity, which stands as analogous to a *One* in relationship to the *Many* individuals of which the society in this contrary aspect, as a mere aggregation, is apparently composed. In the second way, the sovereign creative processes of the person are the *transfinite* surrogate for the *One*, and the society is treated in its aspect as the *Many*. Finally, the two views are subsumed by the corresponding higher vantage-point.

In other words, we have, at first encounter, both the development of the individual, as of the *Many*, by the society, as *One* again, and also the transfinite corresponding to the ordering of the continuity of the *Many* successive phases of the development of the society, as determined by the action of the developing sovereign, creative potential of the individual personality. This twofold, reciprocal relationship implicitly defines that which integrates the two views, by subsuming both under one.

The true wealth of a society is only ephemerally its eroding, depleted current, static form of wealth, in the process of vanishing through consumption or other attrition. The true wealth of society is the rate, per capita and per hectare, at which the potential population-density of the future society is being generated.

To augment a part, at the expense of some other part, is what Luxemburg and Preobrazhensky recognized as primitive accumulation. It is what the U.S. Liberal Establishment and its relatively moronic, Friedman-like "conservative" fellow-travelers regard as their market-oriented "social Darwinism," their own Marxian "primitive accumulation," which has thus ruined the Anglo-American imperial economies at home, as the Roman Caesars and their forebears used similar means to collapse, first, the internal economies of Italy, and, then, the colonies which they looted to supply temporarily a ruined economy of Italy.

The *science of political-economy* is not less than overthrowing, by means of absolutely crucial proof, the *materialist* delusions of the British and French-speaking occultists' Enlightenment of René Descartes, John Locke, Isaac Newton, Adam Smith, and Bertrand Russell (1872–1970). *The continued existence of the human species has depended upon the efficient spiritual cause fostered by aid of the material effects of that cause.* The efficient relationship, the essential, efficient, ontological consubstantiality of the true material and the true spiritual, is the crucial evidence provided by historical human existence to date. In no other way and certainly by no contrary way, could our species continue to exist.

VIII | A World under the Rule of Law

I. The Principles of Modern Statecraft: A Summary

Let us now use illustrative references to some among the currently leading global issues of today's practice of statecraft, to summarize the practical import of the chapters preceding this one. Let us begin by identifying some ostensibly axiomatic features of our implicitly proposed general policy:

1) The essence of good modern statecraft is the fostering of societies, such as sovereign nation-state republics, the which, in turn, ensure the increase of the potential population-densities per capita of present and future generations of mankind as a whole, and which societies promote this result by the included indispensable, inseparable means of emphasis upon promoting the development and fruitful self-expression of that *divine spark* which is the sovereign individual's power of creative reason.

Here, as elsewhere, the definition of *sovereign power of creative reason* is exemplified by, but not limited to, indispensable, successively successful, valid, revolutionary scientific progress in advancing per capita and per hectare potential population-density, by means of increasing capital-intensive, power-intensive investment of productive resources in scientific and technological progress.

302 The Science of Christian Economy

2) The *anti-oligarchical* form of *sovereign nation-state republic,* itself based upon the nation's self-rule through the deliberative medium of a literate form of common language, is the most appropriate medium for the development of society.

By "literate form of common language," is signified not only the written and spoken verbal language, but also a rigorous constructive geometry, and a classical form of musical-poetic language. This combined notion of "literate language," should be understood to signify, in the words of Percy B. Shelley, a language corresponding to the power of "imparting and receiving the most profound and impassioned conceptions respecting man and nature."[1]

3) We emphasize that such anti-oligarchical, sovereign nation-state republics are *almost* perfectly sovereign. This sovereignty is to be subordinated to nothing but the universal role of what Christian humanists, such as St. Augustine, Nicolaus of Cusa, and Gottfried Wilhelm Leibniz, have defined as that natural law fully intelligible to all who share a developed commitment to the faculty of creative reason.[2]

4) As the statesman Charles de Gaulle, for one, has argued for this point, a truly sovereign nation-state republic finds a sense of national identity for each of its citizens, in a general spirit of commitment to the special mission which that republic fulfills on behalf of civilization as a whole.[3]

What we must establish soon upon this planet, is not a utopia, but a *Concordantia Catholica,*[4] a family of sovereign nation-state republics, each and all tolerating only one supra-national authority, *natural law,* as the classical Christian humanists recognized it. Yet, it is not sufficient that each, as a sovereign republic, be subject passively to natural law. A right reading of that natural law reveals our obligation to co-sponsor certain regional and global cooperative ventures, in addition to our national affairs.

The division of humanity's self-government among respectively sovereign nation-state-republics, is not a partition of the world's real estate, but a most preferable arrangement, by means of which all of humanity governs itself as a whole.

A. Literate Language and the Sovereign Republic

This last point of argument is illustrated by aid of a preliminary examination of the functions of a literate form of language in Dante Alighieri's (1264–1321) sense of such a popular literate language. By "language" we should understand the spoken form of communication of ideas; but we must also include a coherent constructive geometry, as "the language of vision," and also the development of the well-tempered polyphonic form of *bel canto* musical communication, the language of hearing.[5] (See Appendix X.)

We have witnessed, in the preceding chapter emphatically, that elementary forms of existence are necessarily not simple, and their relations are not intrinsically reducible to aggregations of linear, pair-wise ones. Therefore, just as a competent mathematical physics requires a suitably developed rigorous language, so do all important matters bearing upon the policy of nations. Without mastery of a language of such quality of literacy, no person is qualified to participate in shaping directly the policies of a nation. Without a common proficiency in a literate form of common language, a people lacks the competence in power of communication to govern itself. So, without a common literacy in geometry and music, in addition to the spoken language, a people is intellectually and morally crippled in its potential qualifications for *effective* self-government.

The political issue of literacy, as a qualification for full citizenship, faces strong, usually hypocritical, often more or less racialist, sometimes even violent objections. Those objections come partly from among populist fanatics. They come also from influential bodies of so-called "professional opin-

304 The Science of Christian Economy

ion." The most fanatical, and most relevant among the latter professionals, are academic and like-minded representatives of those radical positivist, inductive pseudo-sciences, which first mushroomed in Auguste Comte's and Emile Durkheim's France, during the sordid heydays of the Holy Alliance and Napoleon III. (See Appendix XIV.)

Respecting the positivists' objections, one need not rely upon conjecture; the Anglo-French nineteenth- and twentieth-century positivists and their spiritual brethren of Theodor Adorno's and Hannah Arendt's "Frankfurt School," have made their objections against the introduction of the issue of *truthfulness* in matters of statecraft a central feature of the entire history, and leading pre-history of positivism's existence as a sociological phenomenon.

The most obvious of the subsuming issues posed by the positivist's objections, is whether the well-being, or even perhaps the very survival of a form of society might be determined by that society's success in discovering and adopting policies consistent with laws of nature. (Let us begin with the simplest facets of the issue.) If that theorem is true, we demolish the positivist's objection with the observation, that it is urgent that the policy-shaping processes of society be weighted (vertically) in favor of those agencies and persons which have developed a capacity adequate to distinguish between scientific truth and any contrary assertion of a more strongly held majority opinion.

The classical illustration of the evil inherent in a populist's political dogma of "majority," is the 2,400-year-past trial of Socrates.

The immediate victim of that politically motivated judicial murder, was, of course, the innocent Socrates. The putative victors, if only for the short term, were the chief prosecutor Meletys and Meletys's Democratic Party, the latter then, for the moment, the ruling political party of Athens.

This ancient Athens Democratic Party was a concoction whose self-adulating conception would drown the hall at a

Thomas Jefferson-Andrew Jackson dinner, with reverent tears from the assembled multitudes. That Athens party's political show-trial charge against Socrates, embodies implicitly the kernel of the radical populist's and positivist's enmity against our observations on natural law and literate popular language.

Yet, the corrupt Democratic Party's prosecutor, Meletys, was himself later justly condemned by an Athens court for his party's capital crime against Socrates. The corpse of that Democratic Party itself soon found a permanent resting-place in history: obloquy. Athens itself, for allowing earlier the death sentence on Socrates, soon found itself conquered by those very forces against which Socrates had sought to defend it.

Turn the eye back to the time of Aeschylos (525–456 B.C.) and Aeschylos's surviving fragment of his *Prometheus* drama. The Delphic pantheon of Gaia, Python-Dionysios, and the rest of the would-be immortals of the Olympian oligarchy, reigned in smug, hubristic delusion, that no true God, no natural law existed to punish or to check the oligarchy's capricious pranks against poor human beings. For that, the Olympian pantheon was inevitably brought down, by the action of natural law; and those Greeks foolishly corrupted into adoring such false gods, suffered the conquest and en-slavement which their cowardly insolence, in serving such gods, had brought upon themselves and their posterity.

We, as human, may lack the direct access to perfection in our mortal selves, by means of which we might know the unblemished *truth* in a manner and form as if at an instant. Yet, we are equipped by the potential lodged within the *divine spark of reason* in each individual person, to walk the upward path of *truthfulness*. This transfinite pathway of truthfulness is efficient in respect to natural law, to such effect, that a society which prefers truthfulness efficiently benefits, and a society of contrary impulses must suffer.

A literate form of popular language has the formal merit, that it is a constructive geometry of an open-ended type, which permits the rigorous use of the hypothesis-forming capacity associated with the proper use of the subjunctive.

As for well-tempered polyphony cohering with what is termed today *bel canto* vocalization, how could Plato and Leonardo da Vinci et al., have led Johannes Kepler to establish the first valid form of a general mathematical physics without a *bel canto*-based polyphony? (See Appendix X.) Read the *Republic* and *Timaeus,* for example. Read the relevant work of Leonardo da Vinci. Read Kepler. See the failure ("the Newtonian three-body paradox") which punishes us (according to natural law) when we abandon the rigorous notion of a *bel canto*-based polyphony!

What is *bel canto,* but the result obtained when qualified teachers and their attentive pupils see the joy of singing *naturally,* as the normal genetic endowment of every human being endows virtually all with but one choice of developable least-action mode of singing? On what is this all based? Leonardo and Kepler are emphatic; on the scale of ordinary observation, all healthy living processes' morphology of growth and movement is harmonically congruent with the Golden Section; non-living processes are not—except, at both the maximum and minimum extremes of scale.

How does that bear directly upon a literate form of *musically spoken constructive geometry?*

The fact that living processes are harmonically ordered morphologically, negentropically, in congruence with the Golden Section, proves implicitly, and conclusively, that the universe as a whole is characterized thermodynamically by a *negentropic* ordering of itself as a whole. That is plainly *antipantheism,* although the actually or potentially *gnostic* deductive formalist will insist sophistically that it is pantheistic. This has also been shown experimentally for the microphysical domain. Thus on to *bel canto*-defined (i.e., well-tempered) polyphony.

The *bel canto*-ordered, well-tempered polyphony is also a reflection of (e.g., *negentropic*) harmonic congruence with the Golden Section. So, the combining of such polyphony with constructive geometry, as Plato's referenced locations illustrate this,[6] forces the issue of a non-algebraically (*transcendental*) ordered mathematical physics upon a bare physical geometry.

The common use of the term "music" is too narrow for our purposes here. All natural language must tend, as a Renaissance-revived healthy Italian language does, toward a natural, *bel canto* vocalization. This vocalization, as we might compare a literate form of *bel canto* Italian with Vedic hymns, for similarities, determines the musical structure of a literate form of language.

We state our theorem on literate popular language in this light.

The kernel of the issue of *literacy* in language, is central in the development and employment of the individual person's *divine spark* of creative reason for the functions of generating, communicating, and assimilating efficiently, conceptions equivalent to valid, fundamental, revolutionary advances in a (practiced) science and technology. There is no available medium for extending this process from one sovereign person to another, except the medium of literate language as we have defined it implicitly here.

In order that we may receive and impart "the most profound and impassioned conceptions respecting man and nature," the creative thought, sovereignly generated within the indivisible unity of our creative mental processes, must be *communicable.* If we are careless, and disposed to rush too quickly to a plausible conclusion, we might say, *mistakenly,* that to communicate a conception, we must express it as an image in the material of communicable language: Not so. Something far more interesting and useful must be said instead.

How do we teach, for example, secondary-level mathematical physics, effectively? Look closely and the textbook is ejected from your classrooms, to be replaced by both original sources and modern-language restatements of the content of those classical sources. What is it that the effective teacher does, which the textbook teacher usually does not do?

Look at such classical sources. Imagine presenting this to a class of secondary students. What ought to be your objective in this matter? Do you wish the pupil to swallow the text, word for word? You do not; you see our point, perhaps. We wish to have each pupil work through, not the text, but the process whose identifiable steps are *indicated by* the text.

What we should seek to communicate by use of such a source, is chiefly two results. First, one mind (essentially), the author of the source-text, issues a set of instructions to the mind of his audience (to you, and to the pupils), to relive the mental experiment outlined. Second, a similar mode is employed, to direct the mind of the individual audience-member to conceptualize an identified conclusion obtained from the experience. (That is enough said of that for our immediate purposes here).

The point so illustrated, is that the idea is not contained within the explicit communication. Rather, the communication is a more or less reliable guide, as a key to a locked compartment, to the secret of the message. The receiving mind does not "decode" the message. Rather, the receiving mind relives—"unlocks," in a sense—the sequence of mental actions prescribed as the explicit message (geometric construction is an example of this). It is the interior of the creative processes of mind, in response to the stimulus represented by the message, which regenerates more or less faithfully the concept which prompted the sender to compose the selected set of instructions, which are aggregately the relevant working-content of the message itself.

To oversimplify, without doubt, the relevant features of the process of communication are aggregately devised, by the sender, to set up the receiver's state of mind in such-and-such a combination of ways. Thus, respecting the essential idea to be regenerated in the mind of the receiver, *the message is not the medium.*

The study of topology, originally from the standpoint of Gottfried Wilhelm Leibniz's mind respecting *analysis situs,* past Riemann surfaces, through Georg Cantor, indicates to us, in significant part, the existence of general, transfinite principles of cardinal ordering of non-algebraic constructions, which are to a valid physics, in general, as the form of mathematical-physics-like aspect of language-communication is to the substance of the creative thinking on physics matters. (See Appendix VII.)

When we examine more intimately the role of a non-algebraic constructive geometry and also of well-tempered *bel canto* polyphony, in defining the morphological and physics qualities of a literate form of language, we see the matter in less inadequate terms of reference.

We ought to become thus more sensitive to the fact that, although language does not and could not "contain" important classes of ideas, the function of language in the social radiation of creative conceptions generated within an indivisibly sovereign individual mind, demands a kind of rigorous maintenance of the language-media (spoken, geometry, music), in its truer form and in its true form as a unified whole. This maintenance and development, which is the proper referent for the term "literacy," puts relatively upper limits on the yet-developed capacities of virtually all persons sharing the use of the commonly used form of this language and its various, subsumed phases.

Thus, the possibility that a society is able to achieve that *truthfulness* requisite for policy-shaping leading toward *durable survival,*[7] depends upon the level of literacy developed and maintained, especially, by those in the society in power to exert substantial influence upon policy-shaping. Indeed, in the extreme case, it were in the vital interests of those not so qualified, that they be disenfranchised, rather than put the entire nation in jeopardy because of their illiterate incompetence.

Howls of righteous indignation! "Elitism!"

We must respond. No, no, you asses! The issue here, is the modern republic's vital self-interest in fulfilling its implicit moral obligation, to have provided an adequate quality of education to all graduates of a virtually universal, compulsory secondary schooling. The term "adequate quality of education" must not be construed to mean other than or less than a twenty-first-century equivalent of a nineteenth-century Schiller-Humboldt program for development of both the individual moral character and, in the fullest possible, broadest intellectual potential *of each and all pupils.*

That requirement must not be construed to signify what, for example, numerous, themselves miseducated, "conservative" U.S.A. parents have been misled to support as a proposed educational form: a *Brotgelehrte*[8] quality of public education, "tracking" the student narrowly to receive shallow indoctrination in the "three R's," with no more breadth or depth of subject-matters than might not exceed the intellectual requirement of the student's projected future levels of employment and income.[9]

Every pupil must have experienced, by means of exemplary instances, a reliving, as by reliving the experience reflected in a crucial source-document, the successive development of those conceptions upon which the successful outcome of the past thousands of years of known history of development of civilization had been based.

The core of education in European and closely associated history, should be presented under such a descriptive heading as: *"The Republican Idea: the continuing struggle for individual human freedom, against the common enemy-forces of pantheism, usury, oligarchism, and imperialism."*

The idea of history to be presented is the history of ideas. Therefore, the idea of history itself is presented empirically upon the basis of a classical philology, which recognizes the language of generation, communication, and efficient assimilation of valid innovations and ideas as including the spoken, constructive-geometric, and musical facets. This is not a history of the mere contemplation of ideas, but of the advancement of the social-reproductive power, coordinately, of the sovereign individual person and of mankind as a whole. In this overview, that advancement of the individual in mankind, is both the general mission of human labor, and also the crucial-experimental domain in which the nature of the success and failures of customary and proposed ideas is rendered *intelligible,* by means of a literate language, to the human mind.

Thus, it is the paradox of individual mortality addressed implicitly. Here, in this connection, we confront education's task respecting the development of the moral character of the republic's prospective new citizen.

The positivist apologist may often seek to allege, that we propose to disenfranchise the relatively illiterate. On the contrary, the person who is denied that quality of compulsory education needed to attain literacy, is already disenfranchised, and those who disenfranchise him of that quality of education are the morally guilty parties. Contrary to our critics among "conservatives" and liberals, he who has denied the right to compulsory literacy, is the party who has injured the rights of the persons allowed to remain illiterate.

In pedagogy generally, we observe three general types. The populist liberal attempts to drag the subject-matter down to the level of illiteracy which he assumes the pupil to bring into the classroom; or, alternately, to his own level of illiteracy. The successful teacher works, in the image of a Swiss mountain-climbing guide, to bring the pupil up, step by step, to the level of literacy (proficiency) which competence in the subject-matter demands. The third recites litany, which artful, if uncomprehending pupils regurgitate successfully in examination papers. The practical issue confronted by the thoughtful teacher of the second persuasion, is what, concretely, defines the "level of literacy" at which competence in even the most rudimentary features of the subject-matter is possible.

To illustrate the point, consider as a subject-matter one of the most essential Christian subject-matters, *consubstantiality*. In known literature, the first effort to supply a rigorously intelligible representation of this conception is found as we approach the conclusion of Plato's *Timaeus* dialogue. To master the *Timaeus* to such effect, one must master the deductionist's ontological paradox, as delineated in Plato's earlier *Parmenides* dialogue.

Compare this with another illustration. The most distinguished, late Prof. Winston Bostick, has shown, out of a life's work in high-energy plasma physics, that all of the so-called "elementary particles," from photons on up, are not only far from "simple" in their composition, but are highly complex processes. Professor Bostick referred to these as "L'chaim" entities, signifying what we term their manifest *negentropic* characteristics. This is the same negentropy which Leonardo da Vinci showed in the Golden Section congruence of the characteristic harmonic ordering of living processes. Professor Bostick's work to this effect has the quality of "crucial-experimental"; it requires a revolution in the mathematical form of mathematical physics, before the generality of professionals will all begin to grasp efficiently the sweeping implications of these crucial-experimental discoveries in plasma physics.

In both of these illustrative cases, it is impossible to construct anything better than babbling gibberish on either of these topics, at the level of literacy from which the college-educated populist expresses his opinionation. Similarly, on matters of national economic policy bearing upon physical economy, most of today's prestigious business-school graduates babble gibberish. On other important matters of statecraft it is relatively the same.

Consider a third illustration, the ridding of the mathematics curriculum of a grounding in classical geometry. This was begun, at the close of the 1960s, with the fostering of the so-called "New Math," and was accelerated by the influence of the avowedly white-racialist neo-Malthusian, Dr. Alexander King,[10] in the 1963 education policy utterance from the Paris OECD office.[11]

The simple empirical evidence is, that today's university graduates are markedly inferior in quality to those of 25–15 years ago. The lack of a grounding in classical geometry[12] is an outstanding correlative of this decadence.

It is implicitly a straightforward matter, to show how all mathematical orders are derived from a synthetic constructive geometry. This includes, of course, the role of the "non-algebraic" (transcendental) geometric constructions to represent a nonlinear "curvature" of elementary physical space-time. (See Appendix VIII.) These qualities of a generalized synthetic geometry, are indispensable for full *transparency (intelligible representation)* of a coherent mathematical physics. Lacking that discipline, as a consequence of "overdose of the New Math," or kindred afflictions, the very notion of anything more advanced than the very simplest ontological notions of *continuity* becomes virtually incomprehensible.

It was emphasized, only a bit earlier, that we must now not view *spoken language, geometry,* and *music* as three respectively distinct phenomena, but as *elementarily inseparable* facets of a *common substance.* Only in academic or kindred fantasy, can we imagine *vocalization* of spoken language, without the musical harmonics shown to be the *natural* one by both *bel canto* and the successful line of development of modern mathematical physics by Kepler.

To know this language, one must know it in an appropriate sort of historical way, in terms of reliving in one's own mind some of the most crucial, at least, among the valid creative discoveries elaborated in terms of language in general to date.

Thus, do we say, a viable nation-state republic could not be maintained by a population which does not share primary dependency upon a literate command of a literate form of common spoken and written language. Except by means of shared communication and dependency upon such a common literate form of language, a people cannot truly reason together, and therefore could not become sovereign, as long as this defect were not remedied.

For the same reason, in principle, that an individual person's creative processes are sovereign, the nation's reaching of agreement to a development policy-conception, through means of deliberation in the medium of a literate form of common language, is also a sovereign (e.g., *indivisible*) act. A process of self-government so defined, is, therefore, a sovereign quality of self-government. Hence, for that latter reason, such a process of deliberation must define the scope of a sovereign political process, a sovereign nation-state republic.

The qualification for a sovereign form of nation-state republic, must include, absolutely, the efficient use of a common literate form of language in all matters of policy-deliberation; that is indispensably necessary, but not sufficient. The state must be founded upon a common principle expressed efficiently in all use of a literate form of common language. Otherwise, if there were divergence in respect of principle, the policy-deliberations could not have a sovereign character. That common principle of a true republic, is the (Christian humanist's) *natural law*.

B. A Community of Republics

It may be said fairly, in summary, that, under the highest fully intelligible authority which the Christian humanists know as natural law, modern mankind as a whole ought to be nothing differing from a community of such natural law, a community of respectively sovereign, anti-oligarchical, anti-usury national republics. The desired clarity of principled conception in this matter is aided by referring to the notion of cardinality of a transfinite ordering.

We review briefly, the notion of such a cardinality.

We have situated a notion of a transfinite ordering dialectically in respect to the axiomatically nonlinear sequence of states representing higher levels of potential population-density, achieved successively under the continuing impetus of a society's investment in the generation, communication, and efficient (productive) assimilation of scientific and technological progress. In this case, the same causal principle is generating the next term of a series, ostensibly from the immediately preceding term in each and every part of a series of terms.

Thus we have:

1) The generating (ordering) principle is always *equivalent* to itself.

2) The generating (ordering) principle in each locality is *equivalent* to the same principle as the characteristic of the series as a whole or in any part.

3) The ordering-principle, in each and every *equivalent* form, is always absolutely indivisible in every interval and in respect to the process as a whole.

So, modern mankind as a whole or any community of principle based upon natural law, in any anti-oligarchical sovereign nation-state republic, or the sovereign person, are each and all sovereign processes, which are definite (discrete) in respect to the self-bounding character of self-similar equivalence and indivisibility of determining transfinite cardinality.

This overview treats the collection of modern, mortal mankind as a whole as both a *Becoming,* in the Platonic sense, and also approximately, a *One.* The nuclear families of which the most viable portion of the mortal collection is composed, are each distinct as a definite kind of nuclear family, by means of a reproductive function of such a family which is *indivisible,* thus definite, implicitly a transfinite process in development of the new individuals. The sovereign individual is, by virtue of the functions of the divine spark of creative reason, also transfinitely definite. And thus, the relative ones and manys of that process which is society are arranged.

Take the relationship of *Many* sovereign national republics to *One community* of principle containing them in that light. What defines that community as relatively a Platonic *One* among *Many,* is, for example, the transfinite principle of natural law, by which the community is defined. *Natural law* thus displays, in respect to the functioning characteristic of community as a coherent community, transfinite qualities of self-similarity, equivalence, and indivisibility. This overlaps the similar role of a continuous creative process, in respect to such indispensable forms of manifestation as valid fundamental scientific progress. As the principle of creative reason is the means by which natural law is known efficiently, as scientific progress so ordered is the means by which scientific knowledge exists, so the two facets, commitment to creative progress and natural law, cohere as two facets inseparable, as they come to form a principle of community which is in form itself indivisible.

C. The Controversy

1. Empiricism
During modern centuries, the principal advocates of these cohering views have been the modern Augustinians, typified by Nicolaus of Cusa and Gottfried Wilhelm Leibniz, otherwise fairly described as the "Christian humanists." During a more or less equal period, the chief opponents of these principles have been the positivist gnostics (e.g., empiricists), including, most relevantly, Thomas Hobbes, John Locke, David Hume, Adam Smith, Jeremy Bentham, as well as John Stuart Mill and Mill's godson, Bertrand Russell.

It is relevant to stress, that during the most recent times some of these gnostics have followed the term which Thomas Huxley fabricated, "agnostics," or have termed themselves "secular humanists," indicating their devotion to hatred of Christian humanism. Respecting the issue of British neo-imperialist world-federalism, it is sufficient to put Hobbes and Locke together as at the center of our adversarial interest at this moment.

For both Hobbes and Locke, as for Adam Smith, Bentham (1748–1832), Malthus, Darwin, John Stuart Mill, et al., man is but, at best, an elegant variety of cultivated farm animal. Such a man, as he is closer to the wild predator species or dull-witted, domesticated vegetarians, is always governed by mere "instincts." So, for Hobbes and Locke, society is but a state of each individual implicitly at war against all others, and respecting impulses more sociable than the primeval heteronomic instincts, man begins as a *tabula rasa*. Hence, for them, the state, at best, is no better than a tyranny by the relatively few, or a tyranny, by social contract, by the majority. In consequence, for example, of such positivists, the nation-state, assumed by them as being composed of bestial beings, has also the instinctively inherent, alternate qualities of a carnivorous or vegetarian beast; the state is, in other words, a bestial "ego-state." "Hence," they agree, "away with the cause of war, the nation-state. On, with the absolute world-federalist tyranny of a one-world, imperial *Pax Romana*."

World federalism, in all those among its names which are legion, is a sophist's intellectual and moral fraud. War long antedates the first emergence of the republic. So, the world-federalist argument is a historical fraud. There are conditions far more murderous than war, such as International Monetary Fund "conditionalities"-induced spread of famine and epidemic disease; or a peaceful submission to a "new world order," implementing the racialist genocide of the Draper Fund, "Global 2000," and the Club of Rome. Most wars, such as the Thirty Years' War in ancient Greece (the Peloponnesian War), the Persian Wars, the wars of the Roman Empire, the usury wars of the fourteenth century, the 1618–48 Thirty Years' War, Marlborough's War, and the British-orchestrated 1912–45 "Thirty Years' War," were caused by oligarchism and, like the wars of Teddy Roosevelt's cronies on behalf of murderous, imperialist usury, in a form as crude as London's and Napoleon III's conquest and looting of Mexico.

"Is not anything better than war?" the sophists of the neo-Roman imperialism, the "new world order," argue. "Yes," the thoughtful Auschwitz slave replies, "there are worse conditions than war." The peace which the "new world order" provides, were an evil far worse than any war to free mankind from slavery to such a satanic world-rule.

Indeed, whence comes today's danger of war? As the unjustifiable U.S. butchery in Panama and Iraq illustrates the answer, war today is brought to crush, in the most mass-murderously, exemplary fashion, those who resist the spiritual heirs of Diocletian's use of famine and epidemic, as the means to reduce the world's population-level, especially the darker-skinned portion, over the next pair of generations or so, by approximately 80 percent.

It is not the nation-state which is the cause of modern war; the cause of war today is chiefly the satanic lust of oligarchs for one-world rule.

The picture of man painted by the evil Francis Bacon's evil protégé, Thomas Hobbes, appears to have been the self-image which the English-speaking oligarchy has adopted for itself. Such oligarchical bestiality is not the natural moral characteristic of mankind in general.

2. Goodness/Keplerian Negentropy

We have all experienced frequently the essential goodness to be found among the majority of men and women. Each time we reflect upon that fact, the thought may occur to the Christian: "God had His reason to love humanity, as the Gospel of St. John affirms this to be the case." Humanity is worth saving; we find evidence of this even among the proverbial cesspools of humanity.

For our uses here, it is sufficient to add now two distinct, although interdependent evidences of the quality which makes humanity lovable by God.

The one facet of this is natural law; the second is that quality manifest to us even among very young children, the which, upon deeper examination, locates for us the proximate cause of man's impulse toward living according to natural law.

Now, examine this indicated connection from the vantage-point implicit in Kepler's axiomatic approach to the first successful approximation of a comprehensive mathematical physics. Bring into consideration, in studying the apparent intuitive genius, especially, of Kepler's relatively most elementary discoveries, the warning supplied earlier here against the absurd "cyberneticist's" assumption, that the message "information," is contained statistically within the medium.[13] Remember, that the central feature of Kepler's discovery of the possibility of a comprehensive mathematical physics, is that same principle, earlier emphasized by Leonardo da Vinci et al., which Kepler addresses with relatively greater conciseness in his "Snowflake" paper, on, in fact, *analysis situs,* or "physical topology": that, on the ordinary scale of perception, all living processes are characterized, morphologically, as a class, by harmonic ordering congruent with the Golden Section; non-living processes are not.

Kepler's work as a whole, his astrophysics most luminously, is based on the courageous and fully accurate recognition of the fact, that if the universe contains living processes as proximate causes of physical effects upon the inorganic domain, the universe as a whole is axiomatically ordered in a manner not inconsistent with a Golden Section congruence of the harmonic congruence of the universe, a universe taken everywhere, always as a *One,* as a sovereignly indivisible, transfinite unity as a whole.

Compare this with Professor Bostick's "L'chaim" characteristic of the photon, and so forth.[14] Compare this with the work of Prof. Dan Wells, a long-time collaborator of Bostick et al., on the "Keplerian" characteristics of the atom. The *negentropic* characteristics of living organisms (or, the relevant remains of such living forms), are not some super-Turing-like configuration of dead inorganic building-blocks; the tiniest singularities of material processes already show such embedded *hylozoic* characteristics. These are the characteristics of the curvature of the physical space-time in which the existence of the photon, etc., is a determined singularity of a continuing process.

So, can we be properly surprised if the principle of living processes asserts itself, even in defiance of the philosophical dogma of that most efficiently tyrannical, anti-life state? Can we rightly protest ourselves to be incredulous, at the fact that this principle of life is not only in accord with *natural law*, but that biological substrate of our mental processes is in apparent accord with our mind's peculiar capabilities for constructing ever-less imperfect, intelligible representation of that natural law?

As an individual personality locates his or her social identity in that personal contribution which makes one's completed mortal life to have been historically necessary to mankind to have existed, the difference between a poor quality of nation and the personality of a truly honorable republic is, as France's President Charles de Gaulle warned his nation's citizens, that a true republic defines its distinctive national self-interest as in the continued success of some essential function it provides to the effect of defending, maintaining, and improving civilization as a whole.

"Of what good is the existence of your cruel nation to me?" the citizen of a looted African or South American nation, who dares to speak frankly, speaks bitterly, as he rebukes the, unfortunately, typically arrogantly chauvinistic, morally shallow, and callous representative from the citizenry or officialdom of the United States of America. Shame upon the United States and shame upon those citizens who defend the evils of monetarist usury, and genocidal Malthusianism, which the U.S. government over the past 25 years has imposed upon the developing-sector nations increasingly and generally.

What U.S. citizen can rightly claim any honest self-respect and not do better than merely wish, that the foreign policies of his nation's government and financial establishment might become, at the very minimum, civilized behavior?

There are today those general tasks of mankind as a whole, around which all the persons of good will of all nations, ought to be united, tasks in respect to which each nation might find its necessary place in the general division of labor for the common good.

1) To establish on this planet no oligarchical sort of world-federalist, utopian tyranny, but rather an expanding community of anti-oligarchical, sovereign nation-state republics, a community committed to increasing the potential population-density of all mankind, by the included indispensable means of the fostering of investment in scientific and technological progress, progress made effectively available to all republics of this community. To this purpose, to ban the practice of usury from relations among nations, and to establish a just international monetary order, fostering the expansion of trade and related credit.

2) To end and to eradicate the effects of that monstrous injustice typified by the recent, Malthusian, pro-usury "conditionalities" policies of the International Monetary Fund, the World Bank, and other relevant institutions.

3) To begin to move mankind beyond the limits of this planet Earth, into expanding programs of colonization and exploration of intra-solar and interstellar space.

The importance of the first two listed of these three missions is virtually self-evident, at least in light of relevant matters taken up at earlier points. The third requires some clarification; we treat the subject as such "Gaullist" kinds of "dirigistic" mission-orientation in respect to the crucial exemplary feature of a space-colonization orientation.

3. Smaller and Further

The indefinitely extended general increase of the per capita value of mankind's potential population-density, correlates with both an increase in the per capita and per hectare *power* (*action→work∝power*). This correlates with an extension of both the astrophysical and microphysical limits of man's currently effective range of reach of effective comprehension of physical processes. In smallness, we progress from the cubic millimeter, toward the micron, to the Angstrom unit, to the scale of characteristic molecular, then atomic, then nuclear, etc. action—scales corresponding to ranges of increasing frequency of simple electromagnetic radiation. So, at the same time, the realm of the stars is reached by the simple nighttime's eyesight, by simple and improved optical and radio telescopes, followed at last by man's ventures into space.

As we travel on Earth and into space, we meet the obstacles of ratio of range of effective power per units of weight and volume of fuels. This translates into the succession of chemical, fission, fusion, and subnuclear sources of power: absolute distances reached, during what lapsed time, in respect to the ratio of fuel weight to total weight, and rate of power generated per unit of fuel weight consumed, and so on and so forth.

This pushing back, more or less simultaneously, at more or less coordinated rates of scale of advancement, of the microphysical and astrophysical limits of our useful action, correlates with the emergence of those successively successful (e.g., decreasingly imperfect) advances in scientific conception, and with potential increases in per capita and per hectare generation and application of *power* to accomplish useful work. Thus, to sustain progress in this way, it is not sufficient to extend merely contemplation of the universe; we must also extend man's range of practice, down into the microphysical and outward, toward beyond the stars.

This view of the matter just portrayed suggests, that if we choose practical missions of scientific exploration which are in accord with the correlated directednesses just identified, we shall force scientific progress along those lines of fruitful inquiry which generate valid scientific revolutions more rapidly, with a greater rate of fruitful result to relevant effort applied. Thus, on condition society is committed to give priority to capital-intensive, power-intensive modes of investment in scientific and technological progress, the kind of coordinated microphysical and astrophysical state-promoted "crash programs" implied here, represent "science-driver" programs, as a sort of effort which supplies society in all its facets the highest rate of fostered increase of potential population-density per ration of society's available effort applied.

We should mean to include emphatically in an appropriate form of coordinated microphysical/astrophysical "crash aerospace program," a program in *extended optical biophysics,* extended to the limits of the notions of electromagnetic forms of "optical."

Such commitments by a republic and community of republics to a microphysical, "optical biophysical," and "crash aerospace" program, become, first, a manner for locating the identity of each republic as a necessary personality for mankind as a whole. This assists in elevating the individual sovereign person within each such republic, to access directly, practically, to an intelligible representation of oneself as both a patriot and a world-citizen, and locating one's practical reflection of higher self-interest along such pathways.

Those scientific and economic considerations have their correlate reflections in the realm of classical-humanist art-forms. All taken together, define implicitly a "level" of literacy required of the current form of literate popular language.

4. Democracy?

The case of Meletys's wicked, then-ruling Democratic Party of Athens, warns us of the evil and onrushing tyranny which mankind incurs whenever a people embraces longer than briefly a radical version of "faith" in the populist principle of "a Jeffersonian-Jacksonian democracy." By "radical," one signifies the model of British liberalism otherwise known as British philosophical radicalism, the model of David Hume, Adam Smith, Jeremy Bentham, and John Stuart Mill.

The crux of that matter of a liberal's "blind faith in democracy," is the agreement with the fascist-tending, amoral positivism in law of John Locke's tradition. (See Appendix XIV.) This kind of radical democracy spawns fascism in the manner typified by the Democratic Party's jurors of the trial of Socrates; the irrational tyranny of a perceived "democratic majority in opinion," in crushing its opposition. The issue of fascist *philosophy* is the positivist's irrationalist advocacy of a political equality of virtually "value-free" (e.g., amoral, immoral) opinion, as mere opinion.

The remedy for such a fascist-tending faith in democracy, is the notion of *a republic under natural law,* as the Christian humanists have supplied, succeeding Plato, the correct, exemplary definition of *natural law.* Without the higher authority of natural law, which often finds a few in the right, against the impassioned sincerity of wrong-headed majorities, a democratic majority is morally no better than a fascist lynch mob. The laws enacted by such a majority are no proper laws at all.

Hypothetically, it were better for all men, and more advantageous to the individual true freedom of all persons, to be ruled by an autocrat, whose conscience is awed by that natural law's higher authority, than by a perfect democracy of the "New Age." The fascist epidemic of "political correctness" invoked among many leading university campuses of 1990–91, illustrates the evil of radically populist democracy on this account.

Yet, as the history of monarchism attests, after the good king, we were likely to suffer several or more corrupted successors. The remedy is, as Schiller's Posa in *Don Carlos* says to the drama's Philip II, a state in which the king is one among a million kings. In short, a democratic republic, under natural law, based upon a classical-humanist, compulsory, universal

secondary education, in turn based upon a truly literate, oblig-
atory form of popular speech.

A sovereign democratic republic under natural law, were
the most secure and highest known form of government. The
question, as the young U.S.A. federal constitutional republic
was considered by its Founding Fathers, Benjamin Franklin et
al., was how "to keep it." Without a general, compulsory
classical-humanist form of secondary education, *in terms of
reference to* one's own adequately literate form of common
language, what occurs is the probable erosion of general qualifications for citizen, as witness most emphatically, the past 25
years' widespread degeneration of U.S. language, morals, and
intellect, of the under–50 strata of adults in the U.S.A. today.

II. Economics and Natural Law

A. The Example

For the purpose immediately before us, now let us select two
examples as the cornerstones of reference for our discussion.
Let us focus at relatively greater length, upon some leading,
crucial policy-shaping problems respecting a successfully
guided development of a new, durable, peaceful, and productive relationship among the peoples of Eastern and Western
Europe. First, let us focus briefly upon the second exemplary
case, the impossibility of a "purely political" solution for
the half-century conflict between invading Israelis and the
indigenous Palestinian Arabs.

During a period of approximately 15 years to date, for
example, there have been several periods of relatively more
promising—or, if one prefers, "less unpromising"—attempts
to begin a process of serious peace discussions between Israelis
and Palestinian Arabs. One of the principal contributing reasons for the pre-assured failure of these tantalizing moments
of hope, has been the delusion expressed in such form as, "We

326 The Science of Christian Economy

must concentrate on seeking a political solution; discussion of economic development must wait, until a political solution establishes the basis for negotiating economic cooperation."

Take the maps of the physical and physical-economic geography of that portion of the Near East. Put a canal and tunnel, cutting below Beersheba, leading down to the fabled Dead Sea, approximately 1,300 feet below sea level. The salt waters of the Mediterranean, rushing toward the evaporation-basin, which, among other things, that Dead Sea represents, augment the mining and related potentials along the Jordan, West Bank, and Israeli shores.

Along the portion of this new waterway devoted to a canal, a series of the latest model of high-temperature gas-cooled fission-power plants is constructed, producing, among other useful output, electrical power, a liquid-chemical trans-ported power, and, aggregately, a river's worth of fresh water processed from the Mediterranean influx.

This promotes new, dense agro-industrial development in the area through which the canal cuts. Piped fresh water from here supplies Jordan, Gaza, and the West Bank, as well as Israel's territory.

This canal-tunnel typifies a general commitment to pro-vide added fresh water supplies equal to a new river in that Israel-Palestine-Jordan region. Water and power are the indis-pensable, interdependent, added ingredients upon which such a sustainable, rational exercise of the per capita and per hect-are physical wealth of the region depends.

This approach toward mission-oriented economic-devel-opment cooperation for that region, creates, in that develop-ment itself, a vital interest in common among the participating nations. That vital interest becomes, in turn, the basis for a common "political" interest, and that, in turn, supplies the motive for a "political settlement."

The opposite approach, to postpone economic coopera-tion pending a "political" settlement, must almost certainly fail in the short term, and fail more assuredly over the medium to longer term. Simply, there is no true common interest.

Our comprehension of this difficulty is enriched, if we inquire: Which portion of each national grouping—say, of Israelis and Palestinians—is pro-usury? That pro-usury current in either camp is inherently—"objectively"—the adversary of the vital interests of virtually every other family household, whether Jewish, Muslim, or Christian, in the region as a whole. Consequently, for as long as Israeli unity against the Arab, or Arab unity against the Jew, prevails on either of the respective sides of the quarrel, a toleration of the pro-usury interest's veto-power is virtually the certain death of any proposal for a durable Middle East peace negotiated among the principal nationalities themselves.

Once an indivisible economic development mission, as illustrated by the cited Dead Sea canal, is adopted in the manner indicated, that mission becomes the shared interest which acquires the form of a common or mutual interest. It "acquires the form of," is a crucial nicety. The interest lies not within the acquired objective wealth, but the use of the production, maintenance, and operation of that useful object, to foster a significant rise in the development of the sovereign, creative potential of the members of nearly all among the region's affected family households.

Much of the inability shown among educated persons, the inability to grasp the concept just illustrated, is derived from the unfortunate success of the British liberals in spreading the empiricist/inductive philosophical poison of John Locke and so on. Usually, the proposed, "non-economic political solution," echoes the empiricist's definition of a "social contract." The brainwashing of Middle East political-science students, at London and elsewhere, in Adam Smith, Karl Marx, J.M. Keynes et al., has polluted the intellectual bloodstream of the Jewish and Arab intelligentsia alike. They are thus conditioned to the notion of a "peace" achieved through the Kantian mechanisms of negativity. As in Kant's *Critique of Practical Reason,* the "positive" (e.g., "peace") appears to your imagination only pathetically, negatively, as a "negation of the negation" (e.g., of the "horrors of war").

Apply the foregoing illustrative case's lessons to the vaster and vastly more complex issues of, first, Charles de Gaulle's continental Europe "from the Atlantic to the Urals"; and, extend that further, to the vastness of the issues uniting Eastern and Western Europe in the urgent economic development of Eurasia as a whole.

The Soviet Union, like czarist Moscow's imperium before it, is a quilt of nations and of smaller quasi-autonomies. It is at this moment a crumbling domain of numerous languages and many dialects. In size of area and population alone, it is most nearly comparable to the U.S.A. It lacks the kind of "melting-pot" tendency for integration around a common language, which was formerly a leading characteristic of the U.S.A.; the comparison, on this and other leading counts, shows us the inherent instabilities of Moscow's present domain, and so shows us implicitly, the more clearly, in this way, the kind of forces which have held this assemblage together under a central authority for seven preceding decades, and, also, the similar case for the old czarist Moscow earlier.

If one attempts to resolve the crises of the former Comecon domain, or, more narrowly, within the Soviet Union's borders, by means of "political solutions" alone, the entire latter region of this nuclear-armed superpower were likely to converge upon civil war, a development of incalculable global implications.

This poses implicitly a point central to any effective programmatic understanding of the situation. To put the point in a suitably startling form: The inherent, chief source of potential civil warfare within the territory of today's Soviet Union, is identified by the simple statement of fact: The very notion of "racial equality" is an affirmation of the blood-strewn evil of racism.

1. Racialism

Whoever chooses to describe himself or herself as of a different race than some other persons, is inherently, axiomatically a racist and a—possibly dangerous—fool. Thus, to speak of "racial equality," is to draw certain biological distinctions among classes of persons, analogous to the distinctions rightly made among breeds ("races") of dogs, cats, horses, pigs, cows, and cockroaches. Once such liberal nonsense is established as official opinion, along come the liberal racists, such as the notorious liberal perverts Jensen and Shockley,[15] to remind us why the assertion of "racial equality" is to concede defeat of the struggle for *individual personal equality* to the "genetical racialist."

Christians rightly emphasize the mission of the Apostle Paul. As was stressed earlier in this present location, the only quality which defines a person as human, is that which sets all persons axiomatically apart from and absolutely superior to all species of beasts: the *divine spark* of each and every person's innately sovereign capacity for creative reason; there is but one human race; there is but one feature, one demonstrable singularity, that *divine spark* of humanity, which defines, elementarily, absolutely, each person as a person; one such defining distinction; one race.

This, as will be elaborated, is programmatically crucial for solving today's Eurasia crisis. Before coming to that practical application, we explore the issues associated with the distinction itself.

Consider the relevant implications of the Jensen-Shockley case.

Shockley, associated with a singularly important accomplishment in the field of engineering,[16] brought into and out of that accomplishment an increasingly bloated, fanatical quality of overconfidence in the commonly taught, but axiomatically defective positivist version of excessively algebraic classroom mathematics. He shifted away from his field of relative usefulness and competence, to deploy his defective mathematical learning in service of a purely arbitrary, irrationalist, "social Darwinist" sort of racialist prejudice. Out of this came the atrocious, Nazi-like dogma, which won 1969 public endorsement by then-U.S. Rep. George Herbert Walker Bush (R-Tex).[17]

Recognize the efficient, central role of something *hereditary* in those 1969 racialist utterances of Congressman Bush. Here, "hereditary" is employed in the same general sense one speaks, narrowly, of a "hereditary principle" in deductive theorem-lattice systems, or, more profoundly, more generally, of a true, Cantorian transfinite ordering. (See Appendix VII.)

In the Shockley-Bush case, we are referencing Shockley's affinity for a positivist current of excessively deductive mathematics. As some might read the current U.S. government's economic reports, former Congressman Bush does not impress us as exactly a mathematician. Shockley's defective mathematical heritage, yes, but only as that is congruent with a flaw also central to Congressman Bush's mind-set.

This is to focus attention momentarily upon the common, hereditary roots of Shockley's and Congressman Bush's converging racialist policies. That common root is chiefly the modern British tradition of *gnostic* cults, as typified in modern history by sixteenth- and seventeenth-century "Oxbridge" *cabalism*,[18] and also by the permeating influence of the *Rosicrucian* cults upon the empiricism of Sir Francis Bacon and such followers of his as Isaac Newton.[19]

In the case of Shockley, we trace the hereditary influence of gnostic cultism from the introduction of the anti-scientific principle of *induction*,[20] into one influential, reductionist faction in mathematical physics. In the case of Congressman Bush, we are tracing the same gnostic tradition as Shockley's, in such forms it is transmitted, from Bacon, down to the 1960s, by aid of such notable Anglo-American names as John Locke, David Hume, Adam Smith, Jeremy Bentham, Thomas Malthus, John Stuart Mill, Thomas Huxley, William James,

John Dewey, Walter Lippmann, and such myth-makers as Thorsten Veblen and R.H. Tawney.

2. Descartes and Kant

Not only does Bush's Yale baseball-diamond empiricism have, predominantly, the same British origins as radical positivist Shockley's engineering-school classroom reductionism. Any positivist statement, if sufficiently rigorously so, if issued first in the medium of spoken English, can be restated in mathematical or formal-logical quasi-algebraic form. On both counts, first, common religious (gnostic) roots, and, second, linear equivalence of positivist statements in different choices of forms, there is a simple—i.e., linear—kind of functional congruence between the 1969, country club locker-room's "social Darwinism" of a Bush and the stiff formalism of race-theory crank Shockley.

The extra annoying feature of dealing with British empiricism, is that the British empiricists lard their utterances with irrelevant sophistries, usually relying more often upon an appeal to the irrelevant bit of rhetoric, than force of argument on the issue debated, to persuade the dupes in their audiences. For that reason, it is often desirable and also admissible, to attack a British empiricist proposition, by two successive steps. The first such step, is to address the content of the British empiricist's argument, or as the same conclusion is argued in a relatively less turgid, more rigorous form, by French or German notables. The second, following step is to prove that underneath the Oxbridgean lard,[21] the British empiricist has actually offered nothing more of substance than the relatively more translucent French or German case considered for comparison.

Although neither Descartes nor Kant should be termed an empiricist, most of the crucial propositions of British empiricism are included with more compelling logic among the work of these two continental neo-Aristotelian gnostics; for related reasons, where the indicated sort of comparison is appropriate, these two are usually the modern continental sources to be preferred.

Refer to a point underscored in the preceding chapter. Newton's "clock-winder" paradox is a constructed paradox which rests upon nothing different than Descartes's case for his *deus ex machina*. Without further ado, it should be sufficient at this point, to call to the reader's attention, that the notion of *deus ex machina* relegates to the domain of, if not the nonexistent, the unintelligible, both all in the universe which reflects *negentropy*, and all in the powers of the human mind by means of which negentropy might be comprehended.[22]

Kant is more important to us than Descartes on this specific point, for two principal, historical reasons. Not overlooking the development of those differences with the more radical turn Hume took later in life, as Kant's *Prolegomena* indicates: Prior to the appearance of his *Critiques*, Kant had chosen to become the chief disciple of Hume's empiricism and opponent of Leibniz, in the German language. Despite the issue with the aging Hume, referenced in the *Prolegomena*, Kant remained a gnostic defender of empiricism's quarrel with Christian humanism to the end of his life. During the nineteenth century, Kant's work and so-called "neo-Kantianism," contribute an indispensable part to the survival of fledgling radical positivism in France and the German language.

Examining briefly once again Kant's restatement of Descartes's *deus ex machina* argument, leads us now to the needed fresh view of that paradox of Eurasian development being treated here. To show the roots of the Anglo-American-dominated policy-conflict, we must begin our summary of the Kant case with a glance toward the English roots of former Congressman George Bush's policy today.

The summary begins with the accession of the wicked first Duke of Marlborough's political ally, George I, to the newly established throne of the United Kingdom. This was a triumph for Marlborough's British liberals, otherwise known as the "Venetian Party," the pro-usury party, over that pro-development party which included Leibniz's British admirers.[23] Under the long prime ministership of Walpole, a prolonged orgy of moral, intellectual, and economic decadence produced the curious phenomenon of Scottish apologetics for the moral degeneracy among their wealthy English neighbors to the south. This curiosity was advanced under the perverse title of "moral philosophy," as concocted by an alleged lunatic, David Hume, and his emulator, Adam Smith.[24]

The crux of this "moral philosophy" is summed up in two principal books of Adam Smith, his 1759 *Theory of the Moral Sentiments,* and its sequel, the 1776 work known best by the abbreviated title of *The Wealth of Nations.*[25]

Smith argues, that since man is, in his view, incapable of anticipating the longer-term consequences of policy of practice, the individual must forget such concerns and limit himself to pursuit of the simplest, instinctual sense of narrow individual self-interest. That, at least, is a fair summation.[26] In *The Wealth of Nations,* this Nazi-like argument ("all is permitted") of Smith, serves as the defense of Smith's employers, the British East India Company, Baring's Bank, conducting the opium-trade against China at that time. It serves also as the sole apology for the infallibly ruinous, irrationalist Smith cult-doctrine, "the invisible hand"—"free trade." It is the same argument used later by Jeremy Bentham in his own "In Defense of Usury," and "In Defense of Pederasty," in addition to his book, *An Introduction to the Principles of Morals and Legislation.*[27]

Kant later applies a more challenging sophistry in defense of Hume's and Smith's immorality. This sophistry is a central feature of Kant's *Critiques,* as summed up in relatively more popular language in his *Critique of Judgment.* This sophistry is essentially a fresh defense of Descartes's *deus ex machina* and implicitly, therefore, also of the Newton "clock-winder" copy. Although Kant, in the Preface to the first edition of his

Critique of Pure Reason, features a devastating attack upon (British) philosophical (moral) "indifferentism"—a kind word to employ as euphemism for the satanic abomination of Adam Smith's apologetics—Kant himself supplies the theorem upon which the nineteenth-century positivism depends for a mere show of philosophical credibility.

Kant denies categorically the possibility that human beings might develop an intelligible representation of those processes of mind by means of which a valid creative discovery is generated as hypothesis.[28] He derives from this theorem the corollary assertion, that there exist no possible, rational criteria for defining artistic beauty. These featured, failed aspects of his *Critique of Judgment,* represent the relatively most rigorous among known extant efforts to justify theorems equivalent to Descartes's *deus ex machina.* For related reasons, Kant's failed theorems are congruent with any rigorous form of formalists' attempted proof of Smith's "invisible hand" dogma.

To the point immediately at hand, the entire systems of empiricist or positivist theorems depend upon an assumption equivalent to Kant's failed attempt. This is underlined by a fact, cited earlier, that the fledgling, nineteenth-century positivist movements of France and Germany, invoked the neo-Kantian authority of Kant, in the attempt to fill up gaping epistemological holes in their systems.

Thus, we have such a qualified congruence among the Cartesian *deus ex machina,* the central Kantian theorem (of the *Critiques*), and the elementary assumptions of empiricism. The mind-set underlying these relatively more rigorous, mathematical and other formal representations, is the same empiricist mind-set transmitted across the centuries since the appearance of Oxbridge cabalism and Rosicrucian gnostics' empiricism, as reflected in the referenced, 1969 racialist utterances of Congressman Bush.

Before a final bit of tidying up significantly relevant loose ends on the history of empiricist gnosticism, consider a significant aspect of both the Israeli-Palestinian and Eurasian paradoxes to which this line of inquiry is addressed. In short, how do issues of philosophy, as philosophy, exert an efficiently direct, overriding influence on strategic processes?

Earlier, the fact was stressed,[29] that despite the significant number of what have been, in some among these instances, rather radical successive changes in U.S.A. economic and monetary policies, domestic and foreign, the succession of change is, with rare deviation, in a constant direction. That direction is summed up as three doctrinally regulated policy-trends: the objective of an Anglo-American-ruled world-federalist order; the objective of an "Aquarian" "cultural-paradigm shift"; and a global, Malthusian "post-industrial" order, the latter modeled as a matter of historical fact, upon those notorious "socialist" decrees of the Roman Emperor Diocletian (the, de facto, "Malthusian" doctrine upon which the subsequent Byzantine order was based).[30]

The case of Congressman Bush is exemplary of the philosophical determinism of the 1963–1991 "cultural paradigm-shift" in the United States of America.

Bush is derived from a Yale "Skull and Bones" chapter cult-circle, of such moderns as Averell Harriman (Bush's father's employer), Henry Stimson, McGeorge Bundy et al.[31] This circle produced the Eugenics Society of America, an overt supporter of the "racial purification" dogmas of Adolf Hitler's Nazi Party during the early 1930s. Congressman and President Bush's affinities for Malthusian racialism have been openly associated with the Draper Fund,[32] since the period of his 1960s terms in the U.S. Congress.

This is not to single out Mr. Bush. Quite the contrary. One may quip that there are three functional categories of Anglo-Saxon racism appearing significantly in the U.S. population. Category "A" is the country club or barroom loudmouth stratum. Category "B" includes the punctured pillowcase set. Category "C" includes those patrician establishment figures, like Britain's Bertrand Russell, who may be classed fairly as representing the "gas oven," or "famine-and-epi-demic" set. The Draper Fund, like the Club of Rome, the Carter administration's *Global 2000*, or International Monetary Fund and World Bank "conditionalities," belongs to those who, like Bertrand Russell, prefer "the more efficient" means of famine and epidemic to "gas ovens." The important thing is *not* to single out Congressman Bush, but rather to show that Bush's referenced, shameful political utterance echoes the prevailing philosophical mind-set in the relevant Harvard-Yale patrician elements of the U.S. part of the Anglo-American Liberal Establishment as a whole.

Thus, did persisting such establishment-centered philosophical influence exert an erosive influence upon what was taught by positivists in universities, what seeped from such university and think-tank circles into government, news media, Establishment media, and political parties, into the shaping of most policy-reshaping actions.

B. History

So, in general, history is made. It is but rarely that decisions on crucial events shape history. Usually, the accumulation of decisions which appear to shape history, are reflections of the influential philosophical, religious, and other "mind-sets" which determine what the prevailing trends in decisions will become. This connection is roughly analogous to the effect of the "hereditary power" of an integral set of axioms and postulates in determining the theorems of a corresponding deductive theorem-lattice.

To effect a real change of direction in current history, we must focus efforts upon the "integral sets of axioms and postulates" which define a "mind-set," or "cultural paradigm." In the two illustrated cases referenced here, there are two or more, respectively distinct, cultural "mind-sets" to be addressed.

In these cases, as the case of the Dead Sea canal-tunnel

project illustrates the point, the proposed approach to solutions gives us a practically much-needed physical-economic program to catalyze the needed shifts in "mind-sets."

Any much-needed economic-development program which fosters emphasis upon conscious employment of the sovereign individual's creative powers of reason, tends to shift the "cultural paradigm" toward inclination for agreement with natural law. On the contrary side, any policy of practice which suppresses emphasis upon scientific, technological, and related progress, is an affront to the individual's potential for creative reason; the result is a tendency to "bestialize" the members of that society.

Thus, the empiricist—e.g., British-style liberal—mind-set is inherently a racist one, a perverted view of mankind, which, like Britain's Thomas Huxley, cannot distinguish effectively between the breeding of cattle and dogs and the reproduction of the human species.[33] The necessary reasons underlying the causal relationship of positivism and racism (of the Shockley-Bush type) are already identified implicitly. Identify those connections and then apply the lesson of the connection to the Eurasia case.

The Cartesian *deus ex machina* has two common, noted, relevant, interdependent effects. It relegates creative reason, as Kant does, to an unknowable spiritual domain, outside the physical domain and human flesh. To consistent effect, all that is suggestive, empirically or otherwise, of a "Keplerian" negentropic physical space-time curvature of the universe as a whole, is banned from neo-Aristotelian mathematical physics.

On the first account, Descartes is to be compared with the Manichean gnostics, and also with the Cathar-Bogomil roots of Rosicrucian gnosticism, the gnostic Percival/Parsifal myth, and so on. Take, for example, the celebrated "clock-winder" admissions of Newton,[34] already noted earlier, and Maxwell's (1831–79) similar emphasis, in a letter supplementing the introduction to his famous published work, that his falsifications of certain known crucial evidence[35] was done out of a governing determination of Maxwell's own work, "to exclude any geometries but our own." The early Bertrand Russell publication of his assignment to attack and defame the work of Gauss, Riemann, and Georg Cantor, among others, attests to the same feature of English empiricism.[36]

The neo-Aristotelian form of gnostic mind-set being addressed here, is thus typified for our presently immediate uses, by the three cited landmark examples: Descartes's *deus ex machina,* the echoing, "clock-winder" theses of Newton, and the two corollary theses of the Kantian system as featured in Kant's *Critique of Judgment.* These are, each and all, equivalent to all those varieties of explicitly gnostic mind-sets, which, like Manicheanism, postulate a more or less hermetic separation of and mutual hostility between, a spiritual and physical universe, which are supposed to oppose, more or less fanatically, the concept of *consubstantiality.*[37] These include the Bogomil-Cathar cult-tradition. Cartesianism's hostility to Kepler et al., is thus fairly described as the *Cathar* cult[38] disguised as mathematical physics.

The forms of gnosticism, most conspicuously when expressed as an ideological imprint upon a mathematical physics, deny the existence of an intelligible mental-creative power capable of being necessarily an efficient cause within physical processes. In the same way, gnostic pseudo-Christian cults deny the existence of a *necessarily efficient* "divine spark" of creative reason in the individual person.

This has two included hereditary effects to be underscored here. The notion of the sovereign individual person does not exist as a theorem for such a cultist ideologue; nor does there exist a theorem which specifies a necessary, fundamental distinction between man and beast. This either leads to racism, or, for an obsessed racist, this gnostic denial of a "divine spark" is sought out and embraced as an axiom necessary to provide the racist a suitable mind-set.

The same cult-ideology allows the practice of usury. Either the society's increase in per capita wealth is the result of the sovereign, mental-creative powers of persons, or it is not. If not, then we have the theses of the physiocrat, the theses of a gnostic worship of the "Mother Earth" whore-goddess, Ishtar-Gaia-Cybele-Isis. Similarly, there is no sacredness of individual human life.

Conversely, whoever denies systematically the theorem of the sacredness of an individual human life, is neither a Christian nor a respecter of natural law.

We can now leap directly from the foregoing to the point in view.

C. Dealing with Moscow

In dealing with Moscow, currently (1991), from "the West," one approach will assuredly produce nothing but disaster for all concerned: Continue to insist that Moscow et al. submit to the disastrous "Polish model" of International Monetary Fund, Group of Seven, Schacht-like "conditionalities," as a "precondition" for this or that. The second approach to be considered, is the more complex correlative of the cited Arab-Israeli case: the political solution, the demand for sovereign independence by nationalities which have been under decades of Moscow's rule.

The case of pre-1989 Moscow trade-relations with such crucial Comecon trading partners as Czechoslovakia and East Germany (G.D.R.), illustrates a principal included feature of the matter to be considered. Focus upon the transition from 1988–89 to 1990–91 in trade relations between Moscow and the part of a now-united Germany which was formerly the G.D.R.'s "Land of Mielke and Honi."[39]

First, prior to the political change, East Germany and Czechoslovakia were suppliers of crucial products to the Soviet economy; without a continuing flow of such trade, on the Soviet side, the resulting bottlenecks are crippling for Soviet industry as a whole. Without such trade, a very significant segment of the former G.D.R. economy has no suitable source of orders to keep its production going.

A similar situation confronts not only all of the newly reformed, former Comecon states of Eastern Europe; the avowedly or prospectively independent states from within 1989 Soviet borders, such as the Baltic states, Georgia, Ukraine et al., each and all have acute interdependencies with what has been the Soviet economy as a whole. The nearly disastrous effects of a 1990 cutoff of former lines of such trade between eastern Germany and Moscow illustrates the general problem.

This aspect of the matter overlies the military-strategic problems.

Moscow's Red Army (in a larger sense) continues to be a thermonuclear superpower. Worse, the recent behavior of the Anglo-American forces, in the enunciation of "the Thornburgh Doctrine," actions against Panama, actions in the Persian Gulf, as otherwise, put lower limits on Moscow's willingness, or, indeed, political capacity to retreat as far, strategically, as the legal, morally legitimate, national aspirations of the Balts and others obviously desire and demand. "Two steps backward," thinks the Voroshilov Academy's General Staff group, "but not three and never four."

1. The SDI

In 1979, as part of his own U.S. 1980 Democratic presidential nomination campaign, the author published a personal "Campaign Platform Plank,"[40] which later became known as President Ronald Reagan's Strategic Defense Initiative (SDI) announcement of March 23, 1983. The point on which emphasis is to be placed, for the purposes of the matter immediately under discussion, is the special offer to Moscow which President Reagan included in that March 23 address and repeated at least several times after that.[41]

Consider the following relatively very compact summary of the "SDI" proposal as this writer came to see it, over the

period 1977–1979 and later. The autobiographical accounting given in published locations elsewhere, is largely omitted
here for sake of brevity.[42]

The summary given in text above is a repetition of the
author's conception of the problem-area during 1977–78.
However, some of the facts used here to represent aspects of
that conception, were not documented in the writer's proposal
until some point during the 1979–1982 period.

As Bertrand Russell reflects this in his famous, Churchillian contribution appearing in the October 1946 edition
of the *Bulletin of the Atomic Scientists,* the original British
strategic goal for the post-World War II period, was to use
the United Nations Organization as a vehicle for establishing
a global, new Roman Empire of the principal victors of World
War II. Essentially, this signified a global Anglo-American/
Soviet condominium, the Soviets a junior partner, and the
virtual Anglo-American arrangement, according to the trans-
atlantic watchword of that time, "British brains, American
brawn."

As Russell emphasized in that October 1946 piece, and
in later published writings and published interviews on the
same theme,[43] the temporary postwar Anglo-American monopoly on nuclear arsenals was a key feature of the proposed
world-federalist forms of "new world order" at that time.
That 1946 piece was the first of a series of occasions, during
the post-1945 Stalin period, that Russell delivered to Moscow
his Churchillian "Iron Curtain" threat of "preemptive nuclear
war," should Moscow continue Stalin's postwar rejection of
the proposed Soviet junior partnership in the world-federalist
scheme.[44]

To his Western readers, beginning with that 1946 piece,
Russell warned, that he believed that the Anglo-American
powers lacked the courage to go to the brink of preemptive
nuclear war with Moscow, in time to force Moscow to submit
to the world-federalist arrangement on terms relatively most
favorable to London and Washington, i.e., at some point

342   The Science of Christian Economy

prior to the inevitable Soviet acquisition of nuclear arsenals.[45] Russell predicted, essentially, that because of the West's lack of nerve, the new world-federalist arrangement would emerge only after Moscow had such weaponry.

So, as if Russell had predicted it, the first step toward such an Anglo-American/Soviet global condominium occurred under Nikita Khrushchov, after Stalin's death, beginning with the appearance of four Soviet representatives at the 1955, London meeting of Russell's own World Association of Parliamentarians for World Government.[46] Out of this came the Fabian-sponsored (Cyrus Eaton's) Pugwash Conferences, which, at the second, Quebec Pugwash Conference of 1958, set forth the first arms-control arrangements, detailed by Dr. Leo Szilard, preparatory to world-federalist government.[47]

Put aside the ups and downs of 1958–82 relationships between U.S. Presidents, on the one side, and Khrushchov and Brezhnev on the other. Essentially, supported by the Council on Foreign Relations's New York City branch of London's foreign intelligence organization, Henry A. Kissinger's Chatham House,[48] the U.S.A. and Soviets reached agreement on Pugwash Conference terms under Henry A. Kissinger's terms as national security adviser (1969–75) and secretary of state (1973–77) for Presidents Nixon and Ford. The most prominent features of Kissinger's role as a Pugwash Conference agent, for which many suspected him of being a Soviet agent,[49] was in dealings with Moscow and Beijing. The arms-control negotiations, including the crucial 1972 ABM (Anti-Ballistic Missile) Treaty, are the most directly relevant for examining SDI policy.

Already in 1958, 14 years before Kissinger rammed through the 1972 ABM Treaty, Bertrand Russell's accomplice, Dr. Leo Szilard,[50] had proposed to outlaw anti-ballistic missile weapons, as a way of ensuring that both thermonuclear superpowers remained in a state of pristine vulnerability to intercontinental thermonuclear warheads of the other. Why? To

force a world-federalist sort of Anglo-American/Soviet imperial condominium upon the world as a whole.

Kissinger, trained by British foreign intelligence's Chatham House, under Prof. William Yandell Elliott at Harvard,
and at Tavistock in London, was a hardened follower of the
Castlereagh of "Masque of Anarchy"[51] notoriety, before being
assigned to work on Russellite Pugwash dogmas, under
George Franklin, John D. Rockefeller III, McGeorge Bundy
et al., during the mid-1950s, at the New York Council on
Foreign Relations.[52] During the interim years, from the time he
was booted out of his consultant's position with the Kennedy
administration, until he became virtually "acting President"
during the years 1969–77, Henry A. Kissinger's principal
association was with the ostensibly left-wing co-thinkers of
Bertrand Russell, at Pugwash.

By the middle of the 1970s, the Russellite Pugwash
dogma had put the world on a short nuclear fuse. So this
author found the situation, in launching his 1976 campaign
for the U.S. presidency.

By the mid-1970s, the introduction of increasingly accurate, medium-range, MIRVed thermonuclear land-based and
submarine-based missiles, such as the conspicuous Soviet SS–
20, had put the world *potentially* on a hair-trigger. The reduction of preemptive missile-attack warning-time, from more
than 20 minutes, to the order of five or even less, meant that
the detection of close-in submarine launch of a relatively few
Soviet missiles against U.S. territory, or analogous targeting
of Soviet territory, could even probably mean a full-scale
launch, in reply, by the threatened party. So much for Szilard's
"balance of terror," and the McNamara-Kissinger "Mutually
Assured Destruction" (MAD).

If, however, *both* the U.S.A. and U.S.S.R. possessed an
anti-ballistic missile defense (BMD) capable, in the 1963
words of Soviet Marshal V.D. Sokolovsky,[53] of eliminating "a
strategically significant" ratio of missiles launched against it,
the hair-trigger effect could be brought under control. During

the early 1960s, Sokolovsky's *Soviet Strategy*[54] had rightly deprecated what 1980s convention came to term "kinetic-energy weapons" of strategic ballistic missile defense; Sokolovsky had emphasized the emerging alternative, which, later, the addenda to the U.S.A.-U.S.S.R. 1972 ABM Treaty defined as anti-ballistic missile defense based upon "new physical principles."

During the mid-1970s, the chief of U.S. Air Force intelligence, Maj. Gen. George Keegan, noted the Soviets were working on a "new physical principles" BMD, and proposed that the U.S.A. match this. Defense Intelligence Agency head Lt. Gen. Daniel Graham was only one prominent figure among those influentials who shot down Gen. Keegan's findings and proposals at the time. On the basis of an independent scientific audit of Gen. Keegan's report, in the fall of 1977, this writer publicly supported that report at the time and also went further to develop what became the "SDI" plank in his own 1980 Democratic presidential nomination campaign, and, in a larger form, the author's 1981–82 "SDI" proposals to the Reagan administration. This was also the subject of the author's 1982–83 White House back-channel discussions with official Soviet representatives.

What this author proposed during 1981–83 to the Reagan National Security Council and other relevant U.S. institutions, represented in U.S. back-channel discussions with the Soviet government, to institutions of U.S. allies et al., was a precursor to what he projects now as a basis for working discussion on the Eurasian crisis of 1991. Now, review the mere highlights of the LaRouche 1982 "SDI" proposal in that light.

The 1982 LaRouche "SDI" proposal was first brought prominently to international attention before several hundred participants, at a two-day seminar held in Washington, D.C. for this purpose, on Feb. 17–18, 1982.[55] This public announcement was followed by the issuance of a published version of the same announcement.[56] This proposal had three leading components: military, technological, and political, representing, taken altogether, a *war-avoidance policy.*

1) *Military:*

The military element of this war-avoidance package, was the reliance upon introduction of a high rate of technological attrition in strategic and tactical methods of warfare, centered around a "crash program" employing so-called new physical principles, to construct a global ballistic missile defense capable of destroying assuredly a strategically significant ratio of an adversary "first strike" missile-launch.

This design was premised upon the feasibility of early deployment of a new generation of electromagnetic weapons systems, with an estimable, inherent design-principle advantage of approximately ten-to-one cost of destruction advantage over (relatively) lumbering intercontinental missiles and their warheads and busses. The same family of "new physical principles" technologies was extended to the tactical battlefield (e.g., Europe) and the seas.

2) *Technological:*

The apparatus which is developed to effect a relatively perfected form of a crucial experiment is, as a matter of geometrical-physics principle, the model of reference for designing a corresponding family of weapons and *machine-tools*. The machine-tool developed in conjunction with a weapons program, is the means by which the physical advantage of the weapon-design becomes the device introducing a greater or lesser degree of technological revolution and quality of products and productivity into production in general.

Thus, insofar as military production is an applied reflection of high rates of scientific progress, etc., and on condition that military technologies are encouraged adequately to spill, via the machine-tool interface, into high rates of *capital-intensive, energy-intensive* investment in technological progress in the economy in general, a "breakeven point" is implicitly projected, above which level of rate of such latter investment, a large military program may be maintained at a *net negative cost* to the economy as a whole. This became known as the "spill-over" principle.

This reflection of the principles of Leibnizian physical economy, was the point of the proof of both military and economic feasibility of what later came to be known as the "Edward Teller" version of the SDI.[57] That is: a) the U.S. could afford whatever a proposed BMD program required, and b) the "spill-over" principle allowed the U.S. to go as far as necessary in the direction of advanced technology, to achieve the performance required.

2. The Economy

This military-technological package was also conceived as a "science-driver" form of "jump start" for the world economy. In this respect, during 1982, the author conceived and presented his BMD package as complementary to a package of global economic-recovery packages including his famous *Operation Juárez* of August 1982.

The general perspective was to combine a science-driver "jump start" industrialization boom in the industrialized nations, with a general international monetary reform. The intended result, as *Operation Juárez,* and the 1983 LaRouche "Indian/Pacific Basin" reports typify the point, was to unleash a self-sustaining, growing capital-goods export boom from the industrialized to developing sector.

The other distinctive feature of the 1981–82 LaRouche proposals for the Reagan administration, was that the U.S.A. must propose the new BMD program-package to Moscow as a basis for cooperation between the two strategic blocs.

Why not? *The two adversary-blocs were already cooperating militarily,* along Pugwash lines. Medium-range rocketry had proven what should have been apparent all along: e.g., Bertrand Russell is perhaps the most evil man of the century and Dr. Leo Szilard had been arguably insane; his "Rube Goldberg" scheme was leading rapidly toward the very thermonuclear war it was alleged to prevent.

Some concrete features of the LaRouche BMD "crash program" addressed aspects of the 1982–83 U.S.A.-U.S.S.R. SDI negotiations, which bear upon the solution for the Eurasian crisis today.

Approximately eight weeks prior to President Reagan's first public announcement of the SDI, the following three-point response was relayed from Moscow to the U.S. National Security Council by way of this writer: 1) We agree that your BMD (based upon "new physical principles") is feasible; 2) We agree with the feasibility of technological economic "spillover"; 3) However, we will reject any such proposals from your government, because, under "crash program" conditions, you will race ahead of our economy.

When President Reagan did announce the SDI, the Yuri Andropov government in Moscow reacted as the three-point message had indicated about two months earlier. Instead, Andropov ordered the package-proposal publicized through his interview with *Der Spiegel*'s publisher, Rudolf Augstein.[58] The U.S.-Soviet negotiations, since some time during 1984, until the beginning of 1990, generally followed the outline of that *Der Spiegel* interview with Andropov.

Today, in retrospect, Moscow's reaction to the offer of cooperation in deploying BMD based upon "new physical principles," appears to have been more or less a tragic error.

At that time, 1982–83, both the Soviet and Anglo-American economic systems were sliding near to the brink of that collapse which erupted to the surface, on the Anglo-American side, in the October 1987 financial crisis. By 1982, both the Anglo-Americans' radically Malthusian monetarism and accumulated effects of Soviet "socialist primitive accumulation," were converging asymptotically upon the collapses we are witnessing today.

At that time, 1982–83, the joint U.S.A.-U.S.S.R. adoption of a "crash program" to escape a worsening of the MAD-caused "hair-trigger" threat of the late 1970s, relying chiefly upon "new physical principles," would have initiated a desperately needed, global economic renaissance, with proportionate benefits on both sides of the "thermonuclear divide."

This writer's design for a "BMD based upon 'new physical principles,' " developed and deployed, in separate, successive phases,[59] in open coordination among the powers, represented the combination of, first, a uniquely effective, real-life solution to the indicated military crises,[60] and second, an urgently needed "cultural-paradigm shift" in political and economic thinking on both sides. It was understood by this writer, at the time, as an initiative in imitation of Gottfried Wilhelm Leibniz's eminently successful reforms proposed to Czar Peter "the Great." It was also, in fact, an echo of the Eurasian development projects of France's great statesman Gabriel Hanotaux.[61]

It was not a "peace proposal." It was, rather, something far less ambitious, far more realistic, something effective. It was proposed as nothing more ambitious than *a necessary means, by means of which the temporary avoidance of war might be significantly prolonged and that avoidance otherwise enhanced.*

3. The Question of Peace

"Peace," as the term is used customarily, has merely a *negative* meaning, as the term "negative" is employed in the setting of Kant's "dialectic of practical reason," which is the same general quality of meaning "peace" has when the idea of "peace agreement" is referenced to the romantic/empiricist notion of "social contract."

The virtual worthlessness of such popularized, negative usage of the term "peace," is as a description of a *symptom,* the mere absence of "non-peaceful" conditions.[62] Whenever this *negative* meaning is misused, to treat negative peacefulness as a positive condition to be constructed, politics acquires the hues of a possibly dangerous delusion.

The delusional character implicit in popular attribution of rapture to the mere sound of the word "peace," ought to remind us how deservedly contemptuous is this century's experience with other such mere words as "a war to end all wars," "League of Nations," "Kellogg-Briand," or "non-aggression pact." Kant's "perpetual peace"—a social contract for peace—by negation, is a bloodstained folly which we must not repeat.

Peace in the positive sense exists only in that sense of truth, beauty, and charity which is characteristic of a natural law's community of principle among nations. It is a positive state of affairs, which must be built, as an Indian parent plants mango trees whose fruit will nourish his children and grandchildren.

If one were instructed to describe this positive, true, *agapic* peace in strictly formal terms of deductive approximation, one would say that such peace is a constantly regenerated, necessary theorem of practice, affecting all dimensions of social life within and among the nations comprising a community of principle. This "hereditary" determination is rooted, one would say, "axiomatically," in the shared confidence of each such nation, that all the others are committed *truthfully* to be self-governed according to the natural law.

In the language of the "Tavistockians,"[63] it is by building up among all of a certain prospective community of nations, an appropriate "cultural paradigm," that we bring about the state of affairs represented approximately by such a formalist attempt at description.

Apply now, in somewhat greater detail and depth, what was said of the Dead Sea project, to the image of a project of physical-economic cooperation, to develop a community of principle "from the Atlantic to the Urals," within Europe—and beyond.

D. Eurasia's Great Projects

If one accepted the low standard of personal political "success" popular among most of the North American and European mass news and entertainment media, it would be said that Soviet General Secretary Mikhail Gorbachov's bad luck was to have his patron, Yuri Andropov, die prematurely and thus leave poor Gorbachov to receive the blame for the inevitable failure of Andropov's *perestroika* economic and monetary reforms. So, today, Soviet power is disposed to attach itself to whatever leading political faction is credited with having put "meat and potatoes" more or less regularly on the table for the Soviet people.

Unfortunately for a public afflicted with today's popular opinion, there are no simple, distributionist, or so-called "free market" solutions for this problem of hunger and other current, or imminently threatened, grievous material want. The presently functioning levels of employment and productivity in basic economic infrastructure, agriculture, and manufacturing, are variously underdeveloped and also collapsing rapidly, so much so, that a general catastrophe of spreading material want is the preponderant reality globally, until an essentially global, "dirigist" form of economic-recovery program reaches the level of net effect, at which the presently downward trend in physical economy is reversed.

History

Let us now consider, once again, summarily, the degree to which twentieth-century world history was determined chiefly by certain global events unleashed during the 1860s. The latter was centered around the relationship which emerged between U.S. President Abraham Lincoln and Russia's Czar Alexander II.

The so-called U.S. Civil War and the Union victory, became key to the British motive for causing World War I, and also, thus, implicitly, World War II. This is contrary to what is popularly believed, of course, but the documented truth is overwhelmingly contrary to the vastly popularized mythology.

The British plot to create the Civil War began, in approximation, with the successive U.S. victories in the 1776–83 U.S.

War of Independence and the War of 1812–15. London, to this day, has never given up its determination to re-take, and keep, all of North America. Following the 1812–15 "War of 1812," the British and their Scottish Rite freemasonic agents (such as the 1814 Hartford Convention crowd) inside the United States, adopted a new strategy. To establish a branch of the New England Scottish Rite, which became the pro-slavery "Southern Jurisdiction," while the New England free-masons, although profiting, like Friedrich Engels's family British firm, from cheap, slave-produced cotton, became the "abolitionist" backers of John Brown et al. As the letters of British agent and treasonous head of the U.S. Democratic Party, August Belmont, revealed, the British intent, behind such figures as August Belmont and British spy Judah Benjamin, was to tear the United States apart, into a "balkanized" set of quarrelsome, tyrannical baronies, easily controlled from London.[64]

Thus, the leadership of the Confederacy, around London agent Judah Benjamin, was not a collection of bravely independent Southerners; they were slaveholding oligarchs in the worst sense of human rights violations *en masse*. These proud families were purely and simply British-controlled traitors of the lowest sort. In fairness, their freemasonic, "abolitionist" brethren of New England, were not much better.

The plot was coordinated from London, by the opium-trading circles around the Mazzinian libertarian, Lord Palmerston and Palmerston's confederate, the same Lord Russell who is the grandfather of super-racist Bertrand Russell. So, Palmerston and Russell planned to rescue their Confederate agents, as they directed Britain's agent of influence, Napoleon III, into a Suez-like operation against Mexico.[65]

At Lincoln's front, were his enemies and London and the Confederacy's freemasonic Southern Jurisdiction. At his back, were the Democratic Party "Copperheads," whose darling of the day was General McClellan, and also the "abolitionist" New England freemasonry.

Into this situation, during 1862–63, intruded the shadow and then the military substance of Russia's Czar Alexander II. The Russian Navy deployed *en masse* on friendship visits to New York City and San Francisco; the czar warned London and Paris that Russia would unleash war in Europe, should Britain and Napoleon III attempt to do against the U.S.A.[66] what they did do in full at that time against Mexico.[67]

Then, the British intelligence services assassinated anti-carpetbagger President Lincoln, bringing into power President Andrew Johnson, who set back the United States a whole half-century, by establishing usurious "carpetbagging" against the region of the former Confederate states.[68] Meanwhile, Czar Alexander II re-freed Russia's serfs, at least to the degree of lifting Russia out of the barbarism into which it had been returned over the course of the preceding 100 years.

It was in the context of these Russian developments, that France's Hanotaux launched his efforts of aid of Eurasian economic development. It was to defeat the natural tendency for the cooperation of economic-leader Germany in this Eurasian perspective, with Hanotaux's France and Sergei Count Witte's Russia, that the British corrupted France (by circa 1900) with the Entente Cordiale, and organized World War I.[69]

The symptomatic evidence is plain enough and crucial; the relevant British lies on these matters prevail in global policy-shaping today. Does France's leading opinion have the courage, even 90 years later, to accept the truth, that the Entente Cordiale, was not only France's shameful, virtually catamite, strategic submission to Milner's Fabian London, but was the crucial folly by France's corrupted government, which made World War I almost inevitable? More than 70 years after World War I, how many credulous people still tolerate the popular lie, that Germany, not Britain, sought and caused that war?

The persistence of the falsehoods inherent in the popularized, and also official Anglophile myths, betrays, in a crucial way, the existence of corresponding elements of "axiomatic" assumptions of belief in most relevant public and private, national, and international institutions. These myths reflect also an aggravation, as well as persistence of those "axiomatic" assumptions of institutionalized belief, which permitted the British to corrupt 1890s France against Hanotaux, successfully, and to bring about the monstrous combined direct and radiating effects of World War I. In short, most of us appear thus to be greater fools today, than our grandparents or great-grandparents at the beginning of this century. They made their horrible mistake; we appear to insist upon repeating it.

The 1989 developments which brought the subsequent reunification of Germany, evoked the vilest anti-Germany propaganda outbursts from such circles of Britain's Prime Minister Margaret Thatcher as Nicholas Ridley and Conor Cruise O'Brien. There were supporting echoes of this irrationalist hate-propaganda from leading circles in France, and France's and Moscow's support for a Thatcher-ordered, 1956 Suez-modeled U.S.A. Middle East adventure, the latter of which was plainly unleashed to target the economies of Germany and Japan, and to erode as much as possible the possibility of a Germany-led, vigorous economic recovery in Eastern Europe—and also the Soviet Union.

Echoes of 1900–14! The British Empire was up to the old "geopolitical" war-mongering tricks of those scoundrels Mackinder, Milner, and H.G. Wells.[70] Mitterrand's France of 1990 had rejoined the Entente Cordiale, was joined once more with London in a new "Suez" adventure, and a rewarming of the old Anglo-French Sykes-Picot atrocity. Meanwhile, the neo-Bukharinist "cosmopolites" of Russia were also up to their old tricks. The events which the British-led cabal unleashed in the Middle East, blended with the simmering Balkan crisis to echo the 1900–19 breakup of the old Ottoman Empire; the pattern of Entente Cordiale-like policy action in Europe echoed the British efforts to organize World War I.

Yet, history is not "repeating itself." On the contrary, it

is but displaying, that the cultural paradigm set into place over the 1900–90 period still prevails. Men are not making history; history is dangling entire nations and continents by its puppet-strings.

As long as nations refuse to recognize how a lunatic "cultural paradigm," such as that whose outlines we have just reviewed, controls their consistently foolish behavior, and does so again, and again, and again, over spans of a century or longer, the tragedy will continue its bloody course up to the disastrous end, which brings down the closing curtain on such an effort of mass folly.

"I refuse to accept such conspiracy theories," an objecter retorts from onstage.

From off-stage, the mocking, Delphic voice of the puppet-master is heard: "Then die, you poor fool of a nation, which refuses to show sufficient intelligence to be qualified to survive."

Look at this history, this British-led cultural paradigm, from the standpoint of economies. Start with British hatred against Lincoln's U.S.A.

Under President Lincoln's leadership, principles adduced from the American System of political-economy were applied to generate the investment credit, the investment, and the production needed to win the war, and to prepare to defend the U.S.A., if needed, against a British and French military aggression like that conducted against Mexico during that same period. Thus, the U.S. emerged from the most ruinous war in the history of the federal republic, vastly more powerful in economy and military capabilities than at the outset of the British-directed Confederate insurrection.

The kernel of Lincoln's postwar reconstruction policy is summed up in his last public address, shortly before his assassination at British hands.[71] Had this Lincoln policy, instead of Andrew Johnson's, prevailed, the ruined Southern states would have become immediately a center of a nation-wide "infrastructure-building boom," led by railroad development, establishing the mandatory basis for a great agricultural and industrial growth throughout the United States as a whole. President Johnson prevented that. With British success in corrupting the U.S. Congress of the 1870s, the London-designed U.S. Specie Resumption Act was passed, an act which made the U.S.A. economically a semi-colony of London, and kept the growing U.S. economy in a state of depression or near it, from 1877 through 1907.

With the assassination of U.S. President William McKinley by a transient from New York City's and Emma Goldman's Henry Street Settlement House, the leftist and Anglophile Teddy Roosevelt became President, thus putting the U.S.A. fully in the British Fabian camp of Mackinder, Milner, and H.G. Wells, for a war against Germany. Roosevelt established the U.S. military as the British collection-agent in the Americas,[72] and made war against the American System of political-economy in general.

Despite a threat of a London-directed British-Japanese war against the United States during the 1920s, with Teddy Roosevelt's accession to the U.S. presidency was born the later watchword of the century's Anglo-American partnership, "American brawn, British brains."

Teddy Roosevelt was the creator, through his attorney general, the nephew of France's Napoleon III, Charles Bonaparte, of a national political-police agency to control political opposition, the National (later Federal) Bureau of Investigation. He was crucial in the process of putting the United States under a plainly anti-constitutional, British form of oligarchical (usury-based) central banking, the Federal Reserve System. He ensured that Taft would be defeated,[73] bringing Harriman-House dupe, Woodrow Wilson, into the presidency for 1) ramming through the Federal Reserve Act, 2) ramming through the Federal Income Tax law, and 3) for the case of an expected war against Germany.

Why should 1890s Britain regard Germany as a strategic threat? Were not the royal families cousins? Had the Hohenzollerns not been Anglophiles since the Napoleonic Wars or even earlier?

The British of the 1890s were even more clear than Mrs. Thatcher's cabal on this matter: The prosperous growth of Germany's economy was the *casus belli*. We have an analogous situation today, as Washington, D.C. voices threaten Japan and Germany for "unfairness." How are the latter nations unfair? Simply, they have refused, thus far, to be as self-destructively stupid in their economic policies of the past 25 years, as the U.S.A. and Britain have been. The 1897–1900 Britain might have resolved to gain the benefits of initiating policies already proven then successful in Germany; instead, they elected to create an Anglo-French-Russian alliance to destroy Germany, rather than correct the insanity of their own economic policies at home. That is the issue in a nutshell.

The Policy for the Great Projects

The British of 1897–1900 were still the liberal oligarchs they had been during their 1763–1814 efforts to crush economic development in the English-speaking American colonies. The issue is defined by Schiller's view of the conflict between the oligarchical model of Sparta's Lycurgus and Athens's Solon. The leading expression of these fundamental philosophical differences was and is physical-economic policy. This is so, just because Physical Economy is essentially the mode of social reproduction and development of the society and of the individual personality within it.

The area of Europe east of the former, pre-1990 eastern border of the Federal Republic of Germany, is a desert of a previously, already insufficient development of basic economic infrastructure, which has been ruinously depleted subsequently, by 50-odd years of "socialist primitive accumulation," by 40 years of war and of deep economic depression, and of more war before that. Talk of the "miracles of free trade" is worse than infantile babbling in such circumstances.

There must be a mobilization of all otherwise idled or wasted productive resources of labor, to create rapidly the trunk lines of a network of modern forms of basic economic infrastructure from the Atlantic to the Urals, and beyond. The market defined by this massive infrastructure-building provides the base-line for the development of agriculture, high-technology small entrepreneurships, and modern manufacturing operations.

The mobilization of this region's population for such a great undertaking, in common interest *of Europe as a whole,* is the practical foundation for conditions of durable, just peace among all of the rightfully sovereign nationalities of that continent. Conversely, to allow the described geopolitical syndrome of World War I to rule, by default, would ensure the worst possible outcome as the probable one.

The crux of the matter is the specific way in which the *Becoming* of a physical economy, based upon investment in scientific and technological progress, reflects *natural law.* That *Becoming* does not contain the *Good,* but, like the instructions in the message which is a crucial historic source-document in the history of revolutionary scientific progress, it bestirs the divine spark of creative reason in the individual mind, to find the echo of the *Good* within itself.

Since we have emphasized science and physical economy so much, this is a most appropriate point to give credit to the creative role in classical humanist art, in this case classical tragedy. We reference the manner in which certain kinds of messages—such as a historically crucial scientific source-document or masterful tragedy—unlocks the mind of the recipient to knowledge generated from within the recipient's own sovereign, creative-mental processes. In such ways do creative minds employ mediation by inferior means, to address one another's innermost voices directly.

Contrary to Wiener, Shannon, Von Neumann et al., in such exemplary cases of scientific and classical-artistic communication, what is transmitted to the recipient is far greater than might be estimated as the statistically significant content of the transmission itself.

To illustrate the principle most simply: "Remember that day in __, 19__?" All significant scientific communication of ideas is broadly analogous to such a query. However, instead of invoking the recollections of a finite experience, as the illustrative message suggests, in statements describing a process of scientific discovery, we invoke the transfinite generative capacities of the recipient's mental-creative powers. Within the relatively brief statement of an important problem, are months of justified labor by the recipient of that statement, to explain adequately the proper solution to that problem. Such also is all great artistic composition.

Consider a Shakespeare tragedy, *Hamlet,* for example. Or, Schiller's *Don Carlos,* for example. Is the power of the drama in any of the utterances—even in Posa's "king of a million kings"? The passion is located in the juxtaposition of essentially simple, more or less stylized words and movements, to force upon the audience a conception of something which might be said to "lie between the cracks" of anything said or done onstage. Hence, the form of a dramatic composition is as essential as the form of a non-Euclidean constructive geometry is to creative thinking in mathematical physics.

So it is with a configuration of individually simple tasks of labor, when those tasks are an essential part of a useful process of increase in the productive powers of labor (increase of potential population-density). It is not the acts per se which define what is special in this case. What is crucial is that the basing of the meeting of elementary household needs of consumption upon a process of production governed by generating, communicating, and efficiently receiving valid scientific and technological progress, defines the relationship of person to person, in terms of those activated qualities of sovereign, creative reason which are the resonators of natural law.

A family, a nation cannot live safely in a Christian household, while we permit the devil to reign in those economic processes to which the material existence of the household is kept hostage.

Let it be clear, the attempt led by the Anglo-American, liberal, imperialist Establishment, to establish now, irrevocably, their neo-Roman, world-federalist "one world order," impels an increasingly brutalized, increasingly immiserated world into a kind of global "Thirty Years' War."

In this set of circumstances, as long as it appears to be the hegemonic trend, the tendency of Moscow, and elsewhere, is, in Kant's language, predominantly *heteronomic,* and that with increasing propensity for violence. Moscow, for obvious reasons, will prepare for the likelihood of global war, if, indeed, its military is not already doing so, as slyly as is manageable under presently difficult circumstances.

In this circumstance, respecting nearly all of the territories recently within Soviet or Comecon borders, Soviet doctrine will be, in effect, *two steps backward, one step forward.* This would be, under that circumstance, the underlying, Muscovite strategic view of the Baltic states, Georgia, Ukraine, and so forth.

This strategic horror is the result of longstanding Anglo-American oligarchical (liberal) imperialist policy, as the foolish U.S. President Woodrow Wilson, London's Lord Lothian, Chatham House, Bertrand Russell, and so forth expressed this. This liberal, neo-Roman, neo-Malthusian imperialism, is the correlative of a pro-usury, oligarchical economic policy, synonymous with the "free trade" dogma. Thus, "free trade" means global tyranny and global warfare; the conditions in Eastern Europe would be determined accordingly.

If, instead, we unleash a general economic-development approach of the characteristics indicated here, a different state of affairs dominates Eastern Europe, and Europe's central position in today's depression-wracked world as a whole becomes a positive one for all humanity. Relations among nations, political as well as economic, would be susceptible to a corresponding sort of creative initiative.

Appendices

| Conical versus
Cylindrical Action

The qualitative difference between cylindrical and conical action is seen in the projections of elliptical cuts through the cylinder and cone (**Figure 1a**). The cut through the cylinder projects as a circle; that is, cylindrical action does not transform the universe. The conic section, however, projects as an ellipse, whose perihelion is the radius of the cone's circular cross section at the base of the cut and whose aphelion is the radius of the circular cross section at the top of the cut. The ellipse demonstrates the transformations produced by conical action.

The shift from one to the other is characterized by a transformation from one to two singular characteristics (singularities) (**Figure 1b**). Instead of a center, the ellipse has two foci; instead of every radius being of equal length (as in the circle), the ellipse's radii vary in length with a minimum (perihelion) and maximum (aphelion); instead of one diameter, the ellipse has major and minor axes.

A self-similar series of expanding circles (**Figure 1c**) represents the Riemannian transformation from N to N+1.

Source: *21st Century Science & Technology*

FIGURE 1. Conical versus Cylindrical Action
(a) Projection of elliptical cuts through the cylinder and cone

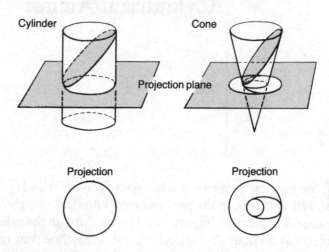

(b) Transformation produced by cylindrical and conical action

(c) Series of self-similar expanding circles on a cone

Circle
Center: f
Radius: r
Diameter: d
Constant curvature

Ellipse
Foci: f_1, f_2
Perihelion: r_1
Aphelion: r_2
Major axis: d_1
Minor axis: d_2
Inflection points in maximum and minimum curvature occur at the end points of the major and minor axes

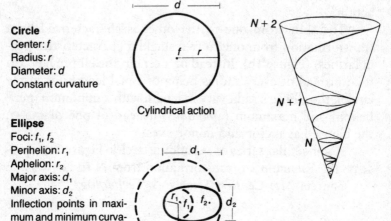

APPENDIX II # The Maximum-Minimum Principle

"Now if the curvature of the curved line decreases as the circle whose circumference it is increases, then the circumference of the greatest possible circle is the least curved, thus completely straight. The smallest thus coincides also with the largest. . . ." (**Figure 2**) Nicolaus of Cusa, *De Docta Ignorantia* (On Learned Ignorance), Vol. 1, F. Meiner (Ham-

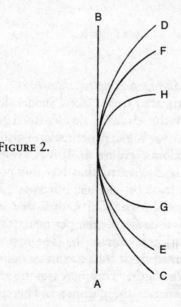

FIGURE 2.

burg, 1979), p. 49. *In this way Nicolaus of Cusa discusses the coincidence between the individual human being (Minimum) and God (Maximum).*

Nicolaus of Cusa's Circle

Nicolaus of Cusa, in his 1440 book *On Learned Ignorance*, showed geometrically that human reason is not attainable through mere logical thought. If we attempt to approach a circle (reason) through construction of polygons with more and more sides (logical thought), it might be thought that we would actually get closer and closer to a circle (**Figure 3**). Nonsense! A circle has no angles; the more angles we add to the polygon, the further we are from a circle.

FIGURE 3. Nicolaus of Cusa's Circle

Least-Action: The Isoperimetric Principle

About 400 years after Cusa, Jacob Steiner devised the following proof that the circle is the figure that encompasses a maximum area for a given perimeter—also without the use of algebraic axioms (**Figure 4**). If it is assumed that another figure has been discovered that has this property, then this figure must at least be convex; otherwise, a connecting line could always be drawn from A to B that increases the area of the figure and decreases the perimeter (a).

Take an arbitrary figure (b). The first step—if it is concave—is to transform it into a convex figure by wrapping a string around the figure. This increases the area by the amount shown but decreases the perimeter. Therefore, the last step

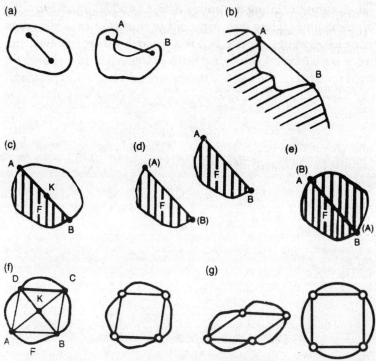

FIGURE 4. Least-Action: The Isoperimetric Principle

here is to expand the figure by a continuous amount along its entire edge to bring the perimeter back to its original length.

The second step is to make the figure symmetrical. To do this, divide the perimeter into two parts of equal length, AB and BA (for example, by measuring the perimeter with a string and then folding the string in half) (c). Then the figure can be divided along the straight line that joins A and B. Choose the larger of two halves (d). Cut the other half out and rotate the chosen half 180 degrees from A to B (e). Then a symmetrical figure is produced with the perimeter of the original figure and possibly with a greater area. If the new figure is no longer convex, it can be made so by application of the first step.

Next, fold the resulting figure in half twice (f) (as in the illustration), creating the points A, B, C, and D. Join them with straight lines. They will form either a square or a rhombus parallelogram as shown. If it is a square, we are finished and have transformed the figure into a circle. If it is a rhombus, then the area of the figure can be increased by "straightening" the rhombus into a square, while the perimeter does not change (g).

If this procedure is repeated, then the figure will get closer and closer to a circle. The circle is the only figure whose area cannot be increased in this way.

APPENDIX III | # The Golden Section

An Algebraic Construction of the Golden Section

The Golden Section, or Golden Mean, divides a line into two segments, such that the ratio of these segments is proportional to the ratio of the whole length to the larger of the segments.

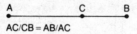

A———————C————B

AC/CB = AB/AC

This being the case, when the length AB is extended by the segment AC, the ratio of the original to the new length, A′B/AB, will also be proportional to the Golden Section ratio.

AC/CB = AB/AC = φ
(φ is the traditional symbol for the Golden Mean)

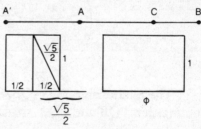

The Golden Section ratio is $(1+\sqrt{5})/2$, which is approximated by the number 1.61802. A simple construction of the ratio $(1+\sqrt{5})/2$ can be determined from the Pythagorean The-

369

orem. Construct a square on an extended line. Draw a diagonal through one-half of the square, and mark this length on the line. The extended line will be in the Golden Section ratio to the length of the side of the original square.

A Geometrical Construction of the Golden Section

The Golden Section can also be constructed directly from a circle, as follows: Take any circle, and determine the length of its diameter by folding it in half. Now produce a tangent from any point on the circumference of the circle, which is extended so that it has the same length as the diameter. Connect the endpoint of the tangent to the center of the circle, and continue this new line until it reaches the opposite half of the circumference. This line will be cut in the Golden Section proportion (ϕ) by the diameter.

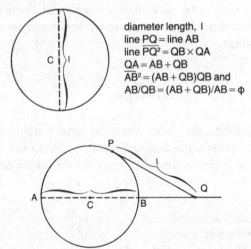

diameter length, l
line PQ = line AB
line $\overline{PQ}^2 = QB \times QA$
$QA = AB + QB$
$\overline{AB}^2 = (AB + QB)QB$ and
$AB/QB = (AB + QB)/AB = \phi$

The relationship $\overline{PQ}^2 = QB \times QA$ can easily be shown by noting that PQB and PQA are similar triangles.

Formal Deductive Systems

Strict formal-logical deductive systems were developed for the first time at the end of the nineteenth century by various mathematicians and philosophers and attained their definitive form in the first three decades of this century. Essential aspects of such systems, however, as they apply to this work, can be explained most suitably with reference to deductive Euclidean geometry.

Euclid's *Elements* (ca. 300 B.C.) presents in 13 books the known mathematical and geometric science of the period, and contains the first attempt in history to present a whole area of knowledge, that of three-dimensional metric geometry, in the form of an axiomatic-deductive system.

In that system, the initial assumptions were five postulates (e.g., "Prove that it is possible to draw a line from any point to any other"), and five axioms or "universally valid concepts" (e.g., "if A equals B and B equals C then C is also equal to A"); and it was asserted that, based solely on these, it was possible to deduce *all* the valid and *only* the valid laws of geometry.

Today we distinguish these "universally valid concepts" more sharply from specific axioms of geometry, since they underlie every formal-deductive system of mathematics, not just geometry. Purely logical axioms, which codify certain permitted logical conclusions, do not occur in Euclid in explicit form; however, already before Euclid, specific cases were presented by Aristotle in the *Organon*.

Euclid's assertion, that *all* and *only* the valid laws of geometry as theorems are derivable from his axioms, corresponds to the modern formal-logical concepts of the deductive *completeness* and *freedom from contradiction* of axiomatic systems.

Assertions of completion actually allow only a proof by contradiction: They are proven false by the discovery (construction) of a new, apparently true law, which was, however, not derivable as a theorem from the axiomatic system. Such new laws are usually not in contradiction to the given axioms; they might be absorbed into the axiomatic system or can by made derivable by an appropriate transformation of the axioms.

Freedom from contradiction naturally is a characteristic that is the *sine qua non* of any set of axioms. Otherwise every assertion and its contradiction might be deduced as theorems from the system. The famous Göttingen mathematician, David Hilbert, in his book *Foundations of Geometry* (1899) presented a new, detailed system of axioms for Euclidean geometry, which satisfies the most stringent formal-logical demands. Hilbert's program, in the elaboration of which he was helped by John Von Neumann, led to the conclusion, that arithmetic's freedom from contradiction was valid only with respect to the methods of the so-called "finite" (not to the perfect infinities), roughly corresponding to operations that can be carried out by electronic calculators. This program was definitively wrecked in the year 1931, thanks to Kurt Gödel's famous *Indeterminacy Theorem* and certain consequences arising from that.

Gödel shows that, in a complete formal-logically constructed mathematical system, the statement which demands that the system be free from contradiction, is itself in principle impossible to derive from the system.

That should have meant the end of formal-logical deductive systems or at least their application to mathematics, mathematical physics, etc. But the precise opposite occurred in "pure" mathematics, and especially after the Second World War, under the influence of the French "Bourbaki group." The attempt to subject all domains of mathematics to axioms and formalisms is being forcefully extended, not least at the expense of constructive-geometric methods, the only ones apt for creative work.

Especially destructive in the course of this development was the effect of the introduction of "New Math" within the framework of the so-called educational reforms of the 1960s. Geometric tasks and constructions, which constituted the vestiges of synthetic geometry in the curriculum, were eliminated and replaced by brainwashing set-theory, which no longer had anything to do with Cantor's concept of the manifold, and limited itself only to the memorization of axioms and definitions, all of which could only remain completely incomprehensible to the student.

How Newton Parodied Kepler's Discovery

Johannes Kepler (1571–1630) published the laws named after him as "Kepler's Laws" in his *New Astronomy* in the year 1609. Isaac Newton (1643–1727) published the *Principia* in the year 1687. Newton's greatest achievement is supposed to be the law according to which the gravitational force of a body decreases with the square of the distance. The idea that all the paths of the planets might be explained by the attractive force of the Sun, which is but a special case of the universal mutual attraction between all bodies, did not spring fully formed out of Newton's head when the famous apple fell on him. G.P. Roberval had already publicly asserted this in 1644. Newton's collaborator, Robert Hooke, in 1666 in a letter to the Royal Society had explained the curvature of the planetary orbits as the consequence of the attraction of the Sun, and demonstrated it with respect to research he was doing on pendulums. The idea of gravity also was not new. What was supposedly new in what Newton formulated, was the law, that gravitational force decreases with the distance r as a ratio of $1/r^2$. In reality, however, this relationship is also already contained in Kepler's laws of planetary motion and in Nicolaus of Cusa's work in 1450.[1]

Kepler's First Law establishes, that the planets move in ellipses, of which the Sun is at one focus. (To make the explanation simpler, we will treat the paths as circular. The same argumentation is nonetheless precisely valid also for ellipses.)

Kepler's Second Law establishes that the radius vector between the Sun and a planet sweeps out equal areas in equal intervals of time. As **Figure 5** shows, the planet moves more quickly when it is closest to the Sun (perihelion) than at the greatest distance (aphelion) from the Sun.

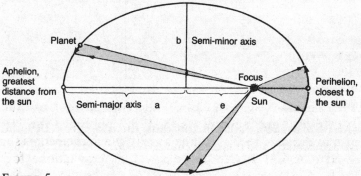

FIGURE 5.

Kepler's Third Law establishes that for all the planets there exists a coherence between the radius of the orbit and the time it takes to go around it. For all the planets a value K holds for the third power of the radius r divided by the square of the period T:

$$\frac{r^3}{T^2} = K$$

In **Figure 6,** the points A, B, and C represent the place where a regularly rotating planet around a midpoint S, will find itself after several seconds. According to Kepler's Second Law, the areas MAB and MBC are equal. The arrow from A to B gives the velocity v of the planet in the first second, the arrow from B to C the velocity in the second second. The change of velocity Δv from the first to the second second is the arrow from B to L. Since in our simple example the radius

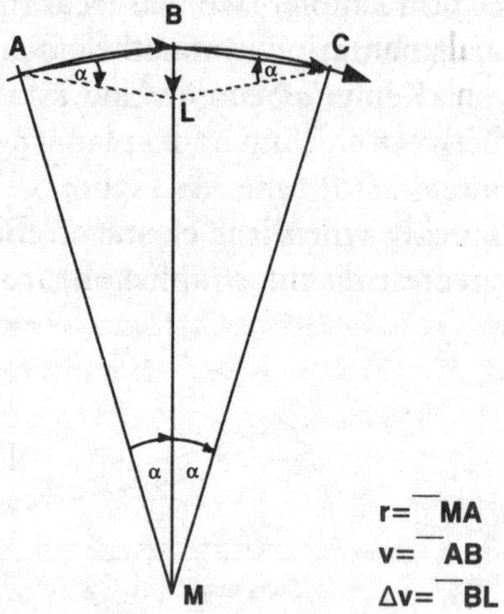

FIGURE 6.
Radius r=MA.
Velocity v=AB
Δv=BL
The triangles ABM and ABL are
similar

r= ‾MA
v= ‾AB
Δv= ‾BL

r=MA=MB=MC equals a constant, the triangles MAB and ABL are similar, and it is clear that $\Delta v/v = v/r$. Hence it follows:

$$\Delta v = \frac{v^2}{r}$$

Yet the velocity is nothing other than the relationship of the circumference $2\pi r$ to the period T. Substituted into the above equation that gives:

$$\Delta v = \frac{4\pi^2 r^2}{T^2} \times \frac{1}{r} = \frac{4\pi^2 r}{T^2}$$

If one substitutes the relationship $r^3/T^2 = K$ of Kepler's Third Law to eliminate T, you get

$$\Delta v = \frac{4\pi^2 K}{r^2} = k \times \frac{1}{r^2}$$

in which the product $4\pi^2 K$ is a constant, which, for the sake of simplicity, is represented by k.

Up to now, no mass of any kind has come into consideration. Since Newton defined Force F as the product of the mass m with the acceleration a, thus $F=ma$, and the change in velocity Δv is precisely the acceleration, then $a = \Delta v$, by multiplying by m on both sides of the equation, you get "Newton's" Law of Gravitation:

$$F = ma = km(1/r^2)$$

The celebrated achievement of Isaac Newton, namely the discovery that the gravitational force of a body decreases with the square of the distance, is nothing more and nothing less than an immediate consequence of Kepler's laws.

The Art of Forming Hypotheses: Kepler's Harmony of the World

There is hardly a greater antithesis than between Newton, who asserted "I do not make hypotheses," and Kepler, who became a master in this typically human art. For the making of hypotheses rich in consequence there is no better example than the path Kepler took, which led to the discovery of the three planetary laws. Kepler draws his genius from his belief in the creative God, who lent mankind the capacity of at least divining his thought in ever-increasing approximation. The unity of geometry and music, theology and observation of nature, scientific rigor and poetic expression, makes Kepler's *Harmony of the World* into both a work of art and a ground-breaking work in astronomy.

In the *Harmony of the World*, Kepler describes the development of his hypotheses about the planetary system. Kepler's first geometric hypothesis about the distance of the planets of our solar system, bases itself upon the five Platonic solids, the fundamental geometric forms of visible space. Since between each of the elliptical paths of the planets there is nested a Platonic solid, a geometric ordering is attained, which is a very close approximation to the actual average distance of the planetary orbits from one another (**Figure** 7). Kepler had already developed this hypothesis in his youth and published it for the first time in his *Mysterium Cosmographicum*.

Kepler's work inspired the scientists J.D. Titius and J.E.

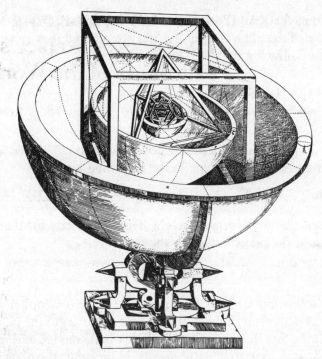

FIGURE 7. Kepler's Model of the Planets
The cube lies between Saturn and Jupiter; the tetrahedron between
Jupiter and Mars; the dodecahedron (12 faces) between Mars and
Earth; the icosahedron (20 faces) between Earth and Venus; and the
octahedron between Venus and Mercury.

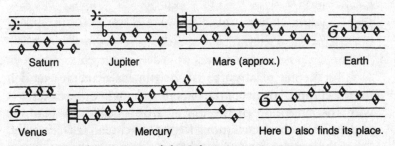

FIGURE 8. Kepler's 'Music of the Spheres'
The original illustration from the German edition of the *Harmony of
the World.*

Bode in the year 1766 to make the hypothesis, that the distances of the planets from the Sun are ordered to correspond to the following number series:

0+4,	1(3)+4,	2(3)+4,	4(3)+4,	8(3)+4,	16(3)+4,
Mercury	Venus	Earth	Mars	Asteroid belt	Jupiter

This contains a series of powers of two (2^n), if we solve the numbers in bold type in the following way:

$$0=2^{-n}, \quad 1=2^0, \quad 2=2^1, \quad 4=2^2, \quad 8=2^3, \quad 16=2^4, \quad \text{etc.}$$

Out of this they formulated the Titius-Bode law for the distance of the planets a^n from the Sun:

$$a^n=0.1[2^n(3)+4]$$

The following table shows the degree of agreement between Kepler's planetary model, the average distances of the planets from the Sun demanded by the Titius-Bode model, and more recent and accurate measurements. In the table, the planetary distances from the Sun are taken relative to the distance of the Earth from the Sun taken as 1.

	Mercury	Venus	Earth	Mars	Asteroid belt	Jupiter	Saturn
Kepler	0.429	0.762	1	1.440	—	5.261	9.163
Titius	0.400	0.700	1	1.600	2.800	5.200	10.000
Modern	0.390	0.720	1	1.520	—	5.200	9.500

But Kepler was not satisfied with his planetary model, for the elliptical orbits of the planets did not fit quite exactly— with the exception of the tetrahedron—inside the Platonic bodies. From these deviations Kepler concluded self-critically, "that the regular fundamental forms do not suffice to derive the distances, . . . for the Creator never deviates from his original thinking." Yet he was in no way irritated by this, rather and much more, he wrote in the *Harmony of the World*, "It is a joy to contemplate my first steps to discovery, even when they erred."

From the Danish astronomer Tycho Brahe, Kepler obtained a complete set of astronomical tables, which are until today the most exact data made without the help of a telescope. According to Kepler, Tycho Brahe's measurements meant, that all previous hypotheses had given a false value for the path of Mars. (The error becomes the clearest in the case of Mars, since its elliptical orbit has the greatest eccentricity of the then-known six planets.)

Kepler, therefore, set aside his hypothesis based on the Platonic solids: "Hence we have again torn down the structure. We had to go through this since we had followed some plausible, in reality false, assumptions in copying earlier teachers. What a great effort did I waste in copying the early teachers!"

With the help of Tycho Brahe's observational data, as well as quite new hypotheses of a higher order, Kepler discovered quite incidentally the three laws of planetary motion valid until today. (See also Appendix V.)

The planets thus moved not in circles, but in ellipses. In the process they change their velocity, which is at a maximum nearest to the Sun (perihelion) and a minimum farthest from it (aphelion). Kepler then compared the angular velocities (W, expressed in angular minutes) of the planets to one another, and with respect to one planet between its aphelion and its perihelion, and with that, discovered relationships which correspond to musical intervals:

Saturn	W at aphelion 1'48"	relationship 4/5
	W at perihelion 2'15"	major third
Jupiter	W at aphelion 4'35"	relationship 5/6
	W at perihelion 5'30"	relationship minor third
Mars	W at aphelion 25'21"	relationship 2/3
	W at perihelion 38'1"	relationship fifth

Earth	W at aphelion 57'28"	relationship 15/16
	W at perihelion 61'18"	half-tone
Venus	W at aphelion 94'50"	relationship 25/25
	W at perihelion 98'47"	sharp
Mercury	W at aphelion 164'	relationship 5/12
	W at perihelion 394'	octave=minor third

This is how Kepler lists the harmony of the planets in his *Harmony of the World*. Of course, all the planets pass through all the other notes that pertain to their interval as they travel their daily path around the Sun. Kepler adds an illustration of the music of the spheres (in **Figure 8**), and says this about it: "Thus all these heavenly movements are none other than an eternally wonderful, many-voiced song, which transcends— only in thought, not recognizable from actual notes—discordance of tension." With a sense of humor Kepler adds a footnote: "The Earth sings Mi Fa Mi, so that one can already from these syllables discern, that our home is ruled by *Mi*seria and *Fa*mes (misery and hunger)." This he wrote in the middle of the Thirty Years' War.

Thus did the Creator picture "the infinity of the duration of the world within the short fraction of an hour through an artfully structured musical work." "Try to follow for me, you musicians of today," Kepler writes, "and form for yourselves a judgment according to your rules of art, which were not yet known in antiquity. You have finally brought forth in the last centuries as the first [law], in which the universe is truly mirrored, Nature that is ever abundant, after 2,000 years of brooding. Through their many-voiced melodies, through the mediation of their ears, she [Nature] has whispered to the human spirit, the favorite child of divine Creator, her innermost being." And in a footnote he calls upon the composers of his time to compose an "artistically just motet" in six voices. "Whoever best expresses the heavenly music presented in my work, to him Clio may give a wreath of flowers and Urania pledges him Venus for a bride."

With the hypothesis of the elliptical orbits of the planets, Kepler rose to the higher geometry of polyphonic music. It is also in this domain of geometry, that spiral-formed action upon the surface of a cone belongs. This is the simplest way of presenting processes in which negentropic changes take place, where net work is being attained. Stimulated by Kepler, Titius-Bode, and LaRouche, Dr. Jonathan Tennenbaum made a model which presented the elliptical planetary orbits as conic sections on one cone, with the Sun at its apex. A second cone with the same apex angle, whose apex is formed by the innermost planet Mercury, clarifies the consequence of the powers of two, that is, the doubling or "steps by octave" in the Titius-Bode law (**Figure** 9) relative to the average distance of the planets from the Sun. From planet to planet the self-similar spiral completes one turn, during which the length traveled along the axis is doubled.

Between Mars and Jupiter there lies the asteroid belt, for which the Titius-Bode law gives the orbital distance. For reasons of geometric harmony, Kepler already suspected be-

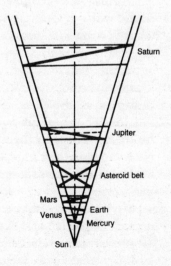

FIGURE 9. Conical Model of the Planets according to Titius-Bode
The average distances of the planets from the sun show self-similar proportions, as they do in organic growth processes and in music (the octave jumps: $2^0, 2^1, 2^2, \ldots, 2^n$.)

The eccentricity of the orbits corresponds to actual values.

tween Mars and Jupiter another, unknown "wandering star" (planet). (See also Appendix V.)

Taken together, one might say: When forming hypotheses, it is not a matter of discovering from the very beginning an eternally valid "formula"—and whoever pretends to have done that, is likely more spy than researcher—rather that a hypothesis points in the right direction and brings man a step closer to the recognition of the laws of the ordering of creation, even if this order might not be fully known. In this process of approximation lies true knowledge, lies transfinite truth, the characteristic of the creative human spirit.

Transfinite Ordering

Georg Cantor's (1845-1898) discoveries between the years 1870-1883, which show that the domain of the infinite allows itself to be ordered with the same rigor as the finite, belong to the most beautiful and important questions of mathematics.

Today in school and at university, Cantor is called the creator of *set theory;* he himself preferred the expression *Theory of Manifolds,* and a great deal of what is today taught as axiomatic set theory in the framework of "New Math," would have filled him with disgust. For his goal was not to reduce various domains of mathematics to primitive ideas of sets; rather to conceptually penetrate the hitherto far-reaching, unelaborated, and controversial idea, philosophical as well as mathematical, of the *actual-infinite.* In this process he dealt with the problem with the same rigor as had Gauss or his student Weierstrass.

Cantor began from the problem dealt with by Bernhard Riemann in his paper "On the Representability of a Function with a Trigonometric Series," submitted toward his Göttingen habilitation (1854), of the representability of arbitrary functions as trigonometric arrays (Fourier), and at the same time let himself be guided philosophically by the thinking of Plato, Augustine, Leibniz (and in part also Thomas Aquinas) about the actual-infinite.

The problem consisted of discovering whether there exist infinite manifolds which are different in magnitude. For example, Galileo Galilei in his *Discorsi* in 1638, presented the riddle as to whether or not the square numbers 1, 4, 9, 16, etc., be just as many in number as the natural numbers 1, 2, 3, 4, etc.—for were they not more thinly distributed with increasing magnitude? He wrote the rows next to each other:

$$1 \quad 2 \quad 3 \quad 4 \quad \ldots$$
$$1^2 \quad 2^2 \quad 3^2 \quad 4^2 \quad \ldots$$

and thought, that between infinite aggregates it were impossible to compare magnitudes. Although this is correct in the case of numbers that are squares, this cannot be asserted in general, as Cantor showed. Yet the method used by Galileo was useful, in that the one-to-one ordering of the two infinite aggregates, which in this way were able to be mapped onto each other, showed them to be of "equal magnitude."

The same question, more difficult than with numbers that are squares, posed itself with regard to fractions and rational numbers. For between 0 and 1 there already lie infinitely many rational numbers, such as for example the array ½, ⅓, ¼, etc.; and between each natural number and the following one, there lie again infinitely many fractions. One of the first great achievements of the young Cantor in the domain of the infinite was the proof, that there are precisely as many rational numbers as natural numbers. Cantor invented a special method, the "diagonal method," for proving that the rational numbers also are in a one-to-one relationship with the natural numbers 1, 2, 3, 4, . . . despite the fact that there is already a whole infinity between 0 and 1.

The proof in simplified form goes as follows: Cantor reduced all rational numbers to fractions, found a system for ordering them, and put that into the following form:

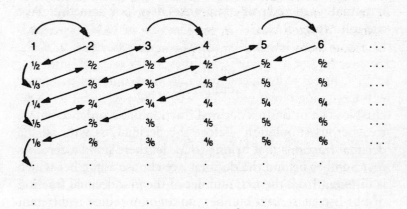

Then departing from 1, he takes the path shown in the above array, so that he reaches the linear sequence 1, 2, 1/2, 1/3, 2/2, 3, 4, 3/2, . . ., etc., which contains all possible rational numbers.

As also happened with the numbers that are squares, in this process the fifth axiom of Euclid ("the whole is greater than its parts") was decisively confuted. Cantor writes in 1887: "There is no contradiction when, as often happens with regard to infinite aggregates, two of them—of which one is a part or a constituent of the other—have [the same number of elements]. In mistaking this matter, I see the main obstacle which from antiquity has hindered the introduction of infinite numbers."

Cantor, however, did not believe that different infinite aggregates of points could always be simply mapped one-to-one to each other. This was, for example, as he showed, not the case with infinite decimal fractions. Cantor's proof (simplified) goes like this:

The aggregate of all infinite decimal fractions between 0 and 1, thus numbers such as 0.1213. . . or 0.4999. . . we denote with M. Now I assert, that with *each* attempt to simply order the natural numbers with the decimal fractions in M, at

388 *The Science of Christian Economy*

least one decimal fraction must remain unmatched. A list of
a mutually ordered pair of numbers may look something like
the following:

> 1 and 0.397 . . .
> 2 and 0.216 . . .
> 3 and 0.752 . . .

etc. Now let us build a decimal fraction in accordance with a
treacherous stipulation: Before the decimal, as with all the
decimal fractions that belong to *M,* let there stand a zero. As
first number behind the decimal, we choose a number which
is different from the *first* number of the *first* decimal fraction
of our list; as *second* number, an amount which is different
from the *second* number of the *second* decimal fraction; as
third number, an amount which is different from the *third*
number of the *third* decimal fraction, etc. This new decimal
fraction thus is not equal to the first decimal fraction of the
list (for it distinguishes itself from it in the first decimal); it is
also not equal to the second decimal fraction of the list (for
it distinguishes itself from it in the second decimal), and so
we see that it is not equal to *any* of the decimal fractions of
the list. . . . With that it is proved, that it is completely impossible to marry all the decimal fractions between 0 and 1 with
the natural numbers. The aggregate of this decimal fraction is
so large, is in such a strong degree infinite, that it by far
surpasses the infinity of the natural numbers.

Thus Cantor proved that there are gradations of the
infinite. An infinite aggregate, which can be matched one-by-
one with the aggregate of natural numbers, he gave the ordinal
number *w.* That was now the first *infinite number.* To this
aggregate there belong among others, the even and the odd,
the square and the rational numbers.

After Cantor had already proved in 1873, that a) the
rational numbers (fractions) can be mapped one-to-one onto
the natural numbers, but that b) such a one-by-one ordering
between the real numbers (points of the linear continuum, for

example, decimal fractions) and the natural numbers is not possible; that thus in the domain of the infinite there definitely exist aggregates of two, fundamentally different *powers*, he asked himself the question whether perhaps the concept of space might lead to a further differentiation of the infinite. To his own and Dedekind's (with whom he corresponded about this) great surprise, and to the great chagrin of other mathematicians (especially Kronecker), in 1877 Cantor was able to deliver the proof that the answer to this is "no": All the points of a surface (for example the unit square) can be mapped in one-to-one correspondence to the points of a line (for example, the closed interval [1,0]). Spatial dimension thus does not allow itself to be defined by a determinate gradation of the actual-infinite.

Cantor wrote about this to Dedekind: "We ought to look for the distinction between the mapping of various dimensions, much more to other features than those numbers held [by Riemann to be] characteristic of independent coordinates." He was referring with this to Riemann's habilitation thesis, *On the Hypotheses Which Underlie Geometry*.

Now this provided the proof of the fruitfulness of Cantor's concept-formation also for the domain of geometry and topology; and at the same time in the idea of *power* of (infinite) aggregates he obtained the principle that allowed him to formulate his theory of the *transfinite* in his *Foundations of a General Theory of Manifolds* (1882-83), independently of the concept of derivation from point-aggregates, and to bring this to a preliminary crowning conclusion.

The *actual-infinite* has a determinate ordering and possesses lawful gradations, constructed according to self-reflexively acting hereditary principles, and hence intelligibly presented: The *first hereditary principle* Cantor defined as "adding unity to a given, already-formed number." The *second hereditary principle* is "that when any determinate succession of defined, real whole numbers is presented, of which no largest one exists . . . a new number is created, thought of as the *boundary* of those numbers, that is, defined as the next greater number after all of them."

The first principle describes the genesis of the sequence of positive whole numbers, the second assures the transition to the transfinite. A third, so-called "limiting principle"—"to embark upon the creation of a new whole number with the help of one of the two [above] principles, only [then] when the totality of all preceding numbers possesses, with respect to their domain, the power of an already available, defined class of numbers"—divides the domain of transfinite ordinal numbers into lawfully sequential, well-ordered classes of numbers, which turn out to "offer themselves, in a unified form, as the natural representatives of the lawful sequence of increasing powers of well-defined aggregates," as well-defined ordering of the transfinite (*Hypotheses*, Section 1).

This lawfully increasing ordering defines the transfinite ordering of the *actual-infinite*, as distinct from the *bad infinity* of dully adding still another 1 to the end. There is nothing to be done with bad infinity; on the contrary, the concept of the transfinite is the key to the laws of the universe and of the creative human mind.

Non-Algebraic Curves and Negative Curvature

There are basically two different kinds of curved surfaces, those with *positive* and those with *negative curvature*.

With the first type (often called convex-convex or concave-concave), all curves (normal sections) which arise from the cutting of the surface with a plane perpendicular to it, are curved in the same direction. The centers of curvature all lie upon the same side of the surface. Typical examples are the sphere and the outer half of a torus. Such surfaces have *positive curvature*.

The second type (often called convex-concave or concave-convex) have at every location two, contrary directions of curvature; the centers of curvature are located on different sides of the surface. These are surfaces of *negative curvature*. The most familiar example is the saddle form.

Among curved surfaces, those of *constant* positive or negative curvature are especially interesting. The typical surface with constant positive curvature is the sphere. In the nineteenth century Eugenio Beltrami, especially, concerned himself with thoroughly investigating surfaces of constant negative curvature.

There are three basic kinds of these (see **Figure 10**, *In Defense of Common Sense*), which all arise from the rotation of different segments of the *tractrix* (literally "drag line"): a) the elliptical-pseudospherical surface of rotation, which also corresponds to the surface of rotation of a *caustic* (literally "burn line"); b) the parabolic-pseudospherical surface of rotation or "pseudosphere"; and the hyperbolic-pseudospherical surface of rotation.

Both curves, the caustic and the tractrix, are non-algebraic curves. These curves were thoroughly investigated by Fermat, Pascal, Leibniz, Huygens, Bernoulli and their collaborators, as well as later at the Ecole Polytechnique under Gaspard Monge. Descartes, on the other hand, wanted to exclude them from geometry, since they could not be constructed either with ruler and compass or by simple algebraic equations. Leibniz thought that was insane, since these curves belong after all to the real universe. How often they occur in nature and in art, we will soon see.

The most simple of these non-algebraic curves is the *cycloid* (literally "wheel line") (see **Figure 9**, *In Defense of Common Sense*). It originates by the movement of a point upon a circle (wheel), which rolls on a plane. Other cycloids can be created when you roll a circle along the inside or the outside of another circle.

The cycloids have characteristically optical properties. The cycloid, which originates by rolling a circle inside a semicircle, whose radius is double the diameter of the rolling circle, is called in optics the caustic (**Figures 10a and 10b**).

The catenary originates, as the name tells us, when a chain hangs between two fixed points (**Figure 11a**). We get the same curve when we dip two rings parallel to each other into soapy water and pull them out. The soapy water forms as minimum surface the surface of rotation of the catenary (**Figure 11b**). We can, however, also construct it by rolling a parabola upon a flat plane while marking off the focus of the parabola.

Any child can make a tractrix. We use a string to attach an object lying next to the track to a toy train traveling along the level. When the train pulls on this object, the path it travels will be a *tractrix* or "drag line."

FIGURE 10a. A Hypotrochoid

A hypotrochoid is a cycloid in which a circle rolls along the outside of a bigger circle. The same hypotrochoid originates when the smaller circle rolls inside the greater semi-circle. In this case the radius of the semi-circle corresponds to the double of the diameter of the rolling circle and the *caustic* is formed. Figure 10a corresponds to Leonardo's presentation of a caustic in Figure 10b.

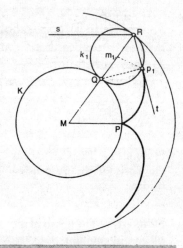

FIGURE 10b.

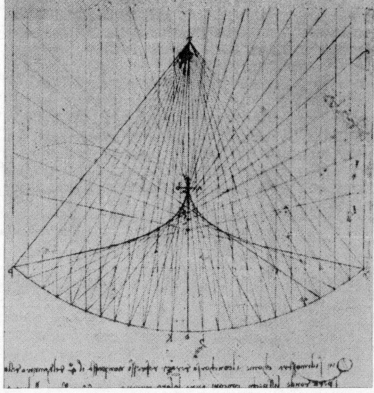

11a.

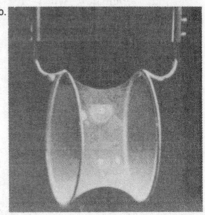

11b.

FIGURE 11a. The Catenary Is the Basis of the Suspension Bridge

FIGURE 11b. The photo of the surface of rotation of a catenary: a catenoid from soapy water as a boundary curve between two parallel rings.

The tractrix also originates as *involute* of the catenary. The *evolute* is the locus of its centers of curvature. In a construction it occurs as the *envelope* of normals of a curve (**Figure 12**). Each curve is itself an evolute of its involute. The evolutes of a cycloid are again cycloids.

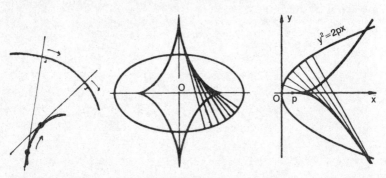

FIGURE 12. Involute, Evolute and Envelope
a) Relation between involute and evolute; b) evolute of an ellipse and parabola, constructed as envelope of the family of normals to the curve. Each curve is the evolute of its involute.

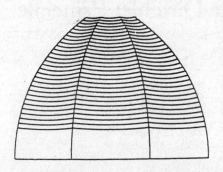

FIGURE 13. Brunelleschi's Dome

The dome of Florence designed by Filippo Brunelleschi between 1404 and 1420 and completed in 1436. The negatively curved surfaces between the ribs were formed by families of catenaries.

(Drawing by Leandro Bartoli)

Cycloids, caustics, catenaries, and tractrices are all real curves, geometric-physical curves, which occur in nature and play a decisive role in the building of bridges and other structures. They cannot be constructed with ordinary algebraic methods.

The catenary occurs, for example, in the building of the dome of Florence (pictured on the cover of this volume). At first sight, Brunelleschi's dome seems to have positive curvature. Yet, as the drawing by Prof. Leandro Bartoli (**Figure 13**) makes clear, the surfaces between the ribs of the cupola are bent in towards the inside. They are surfaces of negative curvature. They originated from catenaries, which were formed by hanging actual chains between the ribs.

The Dirichlet Principle

In the year 1857, three years after his Göttingen habilitation thesis, Bernhard Riemann wrote and published his most important and influential mathematical work, his *Theory of Abelian Functions,* which further develops in particular the concept of "Riemann surface" and its applications.

These hyperelliptical functions, named after the Norwegian mathematician Niels Henrik Abel (1802-1829), are generalizations of elliptical functions, which were achieved by reversing elliptical integrals. Kepler already ran into such an integral with the problem of calculating the elliptical curve, and in his *New Astronomy* had made a call to all European mathematicians to come to his aid in solving this problem. Only through the work of Gauss, Abel, Jacobi, and finally Riemann, was Kepler's wish fulfilled and the problem of elliptical functions definitively solved, exactly in the way he wanted it.

In order to show, that even such general functions as the hyperelliptical (Abelian)—which Jacobi initially held "contrary to reason" because of their possible infinite meanings— are susceptible with the help of Riemann surfaces to being completely intelligibly presented, Riemann made use of what he called the *Dirichlet Principle* (defining it at the same time).

This principle is borrowed from potential theory and permits us to conclude the existence of a desired solution-function from the fact that it is demonstrably single-valued. Specifically: If certain boundary values are given (e.g., temperature gradients at the edge of a disk), then there exist within the domain under concern precisely one (constant and differentiable) function, which a) corresponds at the edge with the given boundary values, and b) makes a specific integral into a minimum (brings the stationary temperature gradient to expression). Fundamentally, this principle is only a version of the *principle of least-action* adapted to specific conditions.

Since Riemann conceived of all complex (analytic) functions as conformal mappings, and since he abrogated the "irrationality" (polyvalence) of Abelian functions by creation of a many-leafed Riemann surface, which is the form of both a multiply- (but with respect to the leaf, simply-) connected, mapping/structure/surface, he could now use the Dirichlet Principle (in principle only applicable to a simply-connected surface) and apply it to the Riemann surface, and guarantee in this way a single-valued function, integrate it, etc. This made a complex, polyvalent, algebraic structure/mapping/surface, possessing singularities that could not be ignored, accessible by means of an ingenious topological construction, and the application of the *Maximum-Minimum principle* to intelligible representation and simple calculation.

| # Scientific Tuning: Middle C=256 Hz

A t a Schiller Institute conference on April 9, 1988 on the above theme, Dr. Jonathan Tennenbaum explained why middle C=256 is the only acceptable tuning in music, which might not be arbitrarily raised or lowered without negative consequences. In this, Tennenbaum drew upon a working group of physicists, biologists, musicians, and instrument builders, who had pursued the urgings and hypotheses of Lyndon LaRouche. What follows is a selected summary of this presentation.

The human voice, the fundamental musical instrument, is a living process. Leonardo da Vinci and Luca Pacioli proved, that all living processes are characterized by a special geometry, whose most visible manifestation is the morphological proportion of the Golden Section. Since music is a product of the human voice and the human spirit, it must be lawfully coherent with the Golden Section. And it is. The classical well-tempered system is based upon the Golden Section. The American mathematician Carol White showed this with respect to two sequences of notes, whose musical significance ought to be clear to every musician:

C, E-flat, G, C and C, E, F-sharp, G.

In the first sequence, the different frequencies of successive notes form a self-similar array in relation to the Golden Section. The different frequencies of the second sequence decrease according to the Golden Section (**Figure 14**).

Where now lies the particular meaning of the frequency middle C=256 Hz as the true value of musical tuning?

Kepler derived the musical intervals from the division of a circle (so to speak, a circular string) by an inscribed regular polygon:

Interval	Figure	Relation
Octave	Folding circle	1:2
Fifth	Triangle	2:3
Fourth	Square	3:4
Major third	Pentagon	4:5

Thus, going one octave down corresponds to doubling the length of the string, a series of such steps by octave, an array of powers of two: $2^0, 2^1, 2^2, \ldots 2^n$.

The frequency middle C=256 corresponds to the circular action (rotation) of 256 rotations per second, or one rotation in 1/256th of a second. Now let us take the time for the Earth to rotate on its axis. We divide this by 24 ($2 \times 3 \times 4$) and get one hour. Divide by 60 ($2 \times 3 \times 4 \times 5$) and get one minute. This again divided by 60 gives us one second. These seconds we divide now by 256 ($2 \times 2 \times 2 \times 2 \times 2 \times 2 \times 2 \times 2$). Now we can easily calculate that the Earth's rotation corresponds to a G that lies precisely 24 octaves below middle C=256 Hz! Scientific tuning is thus anchored in the solar system. The

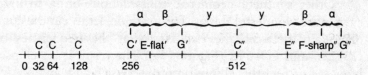

FIGURE 14. Proportion of the Golden Section Relative to the Musical Scale: a:b=b:y

A=440 Hz, on the contrary, has a purely arbitrary value, which absolutely cannot be justified in physical geometry.

Today we might still add some essential points to this, for our universe can only be imperfectly described using circular action. At the beginning of the nineteenth century, Carl Friedrich Gauss introduced, instead of circular action, spiral-formed or conic action into synthetic geometry. Spiral-formed action combines the isoperimetric principle of the circle with the principle of self-similar growth, such as expressed, among other things, by the Golden Section.

In our **Figure 15a**, the axis of the cone presents the frequency. During the frequency shift of one octave, the spiral completes one rotation upon the surface of the cone, e.g. from C'=256 to C"=512. One octave thus corresponds to one complete turn of 360° of the spiral upon the cone.

Figure 15b shows the projection of the cone onto the flat plane below. The surface of the cone is divided into 12 parts. Each slice of the circle represents a *bel canto* half-tone. A 30° turn around the spiral corresponds to a half-tone interval. The radial lines (measured out from the mid-point) correspond precisely to the frequencies of the scale of the even-tempered system. The fifth corresponds to a turn of 7/12ths of the circle, the minor third corresponds to a right angle, etc.

It is important that the note F-sharp lies after each half-turn of the spiral starting from C. This interval C–F-sharp, the minor fifth of C, is known as the "devil's interval." It lies precisely at the geometric mean of the spiral-formed action between C'=256 and C"=512 Hz (**Figure 15a**).

Other synthetic-geometric constructions bring to light even more wonderful things. For example, let us cut the cone diagonally between the two circles at the frequencies of C'=256 and C"=512 (**Figure 16a**). The result is an ellipse. Now we project this ellipse onto the plane below and obtain precisely the frequency relations between the most important dividing points of the octave (**Figure 16b**). Let us imagine that the ellipse were a planetary orbit with the Sun at the left focus.

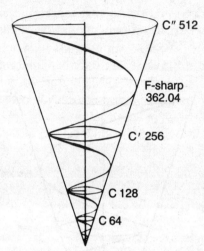

FIGURE 15a:
The spiral on the cone fulfills complete turn of 360° for the octave. The length upon the cone's axis indicates the frequencies.

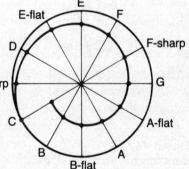

FIGURE 15b.
A projection of the same conic spiral upon the plane. If the complete 360° turn is divided into 12 equal angles, each one will correspond to a half-tone interval.

Thence in **Figure 16b,** we call the shorter segment of the major axis, perihelion (closest distance to the Sun), and the longer segment, aphelion (greatest distance from the Sun).

C'=256 corresponds to the perihelion of the ellipse
C"=512 Hz corresponds to the aphelion of the ellipse
F corresponds to the perpendicular at the focus
F-sharp corresponds to the semi-minor axis
G corresponds to the semi-major axis

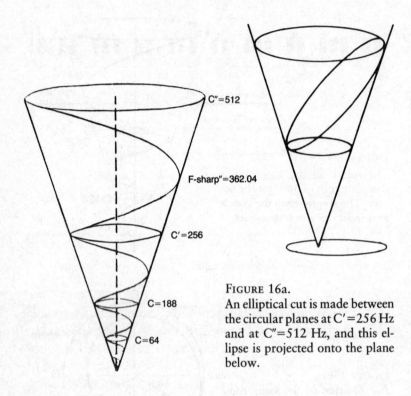

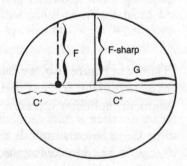

FIGURE 16a.
An elliptical cut is made between the circular planes at C'=256 Hz and at C"=512 Hz, and this ellipse is projected onto the plane below.

FIGURE 16b.
If we imagine that this ellipse is a planetary orbit and that the Sun is located at the focus on the left, then the shorter distance corresponds to the perihelion (that point of the orbit which is nearest to the sun) C'=256 Hz, the distance of the aphelion (that point of the orbit most distant from the sun) C"=512 Hz. The frequency F corresponds to the perpendicular dropped to the focus, F-sharp the semi-minor axis, and G the semi-major axis.

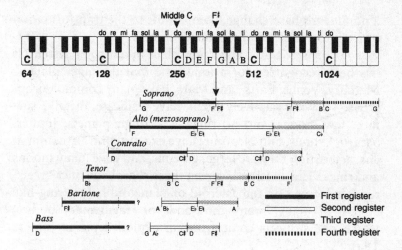

FIGURE 17. Registers of Human Singing Voice

At the same time, F, F-sharp, and G correspond to, respectively, the harmonic, the geometric, and the arithmetic means of a turn around the spiral. These three means form the foundation of classical Greek architecture, of perspective, and of music.

The same notes, F, F-sharp, and G, mark the fundamental lawful division of the C-major scale. It consists of two congruent "tetrachords" CDEF and GABC. The dividing note is F-sharp.

Precisely at the F-sharp is where the register shift of the soprano takes place (**Figure 17**). The first tetrachord CDEF is sung in the first register, while GABC is sung in the second. The register shift hence divides the scale precisely at the geometric mean or after a half-turn of the spiral upon the cone. The same thing repeats itself in the next-higher octave, where the change from the third register again occurs at F-sharp.

The register shift of *bel canto* is a physical property of fundamental lawful significance, not only something regarding voice technique. The register shift is a physical singularity, a nonlinear phase change, comparable to the transformation of ice into water, or water into vapor.

A "register shift" also takes place in our solar system. It has been recognized for a long time that the inner planets, Mercury, Venus, Earth, and Mars, have many common properties. They are relatively small, have a silicate, metallic surface, few moons, and no rings. The outer planets, Jupiter, Saturn, Uranus, and Neptune, have exactly contrary common characteristics: they are large, gaseous, and have many moons and rings. The division between these sharply distinct "registers" is formed by the asteroid or planetoid belt, a ring-like system of many, many thousands of fragmentary bodies, which one suspects come from an exploded planet (**Figure 18a**).

Now it can be shown that the "register shift" of the solar system also takes place at the F-sharp, at the geometric mean of a spiral turn, just as with the soprano voice.

If, starting with the outer surface of the Sun, we make a self-similar spiral to the orbit of the innermost planet Mercury, then the continuation of this spiral from Mercury to the intersecting orbits of Neptune and Pluto will travel through one full turn, an "octave." The geometric mean of the spiral is reached precisely at the outer limits of the asteroid belt!

And it gets even more precise. If we compare the planetary spiral with the spiral of the even-tempered system (**Figure 18b**), where the interval from Mercury to Neptune-Pluto corresponds to the C'-C'' interval, then the orbits of all the planets are intersected exactly at the place where we find the most important notes of the scale. And the asteroid belt occupies precisely the angular position, corresponding to the interval F to F-sharp (**Figure 18c**).

Hence there exists a perfect harmony between the human voice, the solar system, the musical system, and the synthetic geometry of spiral-formed action upon a cone.

If the musical tuning is set arbitrarily, say to A=440 Hz, usual today in many places, then this either ruins the soprano

FIGURE 18a.

In the large distance between Mars and Jupiter, there is to be found, which Kepler already had suspected, a relevant orbital region—the asteroid belt. It has a "singular " character, for it divides our solar system into two fundamentally different arrays of planets, so that we might speak of a "register shift."

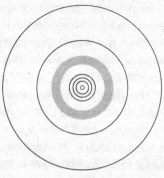

FIGURES 18b AND c.
From Mercury to the intersecting orbits of Neptune and Pluto, the self-similar spiral carries out one full turn, which here is presented in its projection in two halves. Because of the elliptic orbits, we have presented for each planet two concentric circles. The intersections of the spiral with the orbits, projected as presented upon the outer circle of the 12 halftones, hence provide us with a range of frequencies. Now, the asteroid belt occupies exactly the angular position of the register shift for a soprano from F to F-sharp.

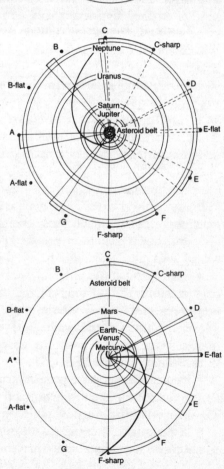

voices or it shifts the register shift to the interval from E to F, instead of from F to F-sharp. With that, however, the octave is divided in the wrong place, which destroys the geometry of the musical system, and the harmony between music and the laws of the universe.

If we should in this way wish to arbitrarily alter the "tuning" of the solar system, it would explode. God does not make mistakes. Our solar system functions very well with its just tuning, which is in exact harmony with the musical tuning of middle C=256 Hz. This is hence the only scientific tuning.

Source: *Ibykus (special edition), August 1988, Böttiger-Verlag, Wiesbaden.*

APPENDIX XI | # Euler's Fallacies on the Subjects of Infinite Divisibility and Leibniz's Monads

*L*eonhard Euler (1707-83), renowned Swiss mathemati-*cian, astronomer, and natural scientist, studied mathematics for 11 years under Jean Bernoulli. Bernoulli had collaborated with Gottfried Wilhelm Leibniz, the German philosopher, statesman, and universal genius who invented the calculus, on various problems of mathematics and physics. But, in his 1761 Letters to a German Princess, Euler attacks the followers of Leibniz, who had died 45 years earlier, in a manner revealing his own lack of understanding of Leibniz's notions of space, time, and substance.*

He was an opponent of the Newtonian reductionist method in mathematical physics. In an attempt to refute Newton's bowdlerization of Kepler's great discoveries, Euler tried to show that Newton's theory did not correctly account for perturbations of the Moon. While Euler was absolutely correct philosophically in his criticism of Newton's axiomatic barbarism, this could not be demonstrated for the case of the Moon's orbit.

LaRouche, in a three-part essay dictated by telephone from prison in the third week of January 1990, demonstrates the fallacies in Euler's argument and revives the standpoint of Leibniz's Monadology. Following LaRouche's critique, we publish two of Euler's letters, which present the essentials of his argument.

407

A Critique by LaRouche

Author's note: Euler's material was sent to my attention by Larry Hecht.

Let me deal first with the core argument by which means Euler introduces the subject. (I'll deal later with the second part of his argument, which is more specific, on the subject of monads.)

Euler obviously starts with a very simple proposition, winds up to it, then gets into monads, and premises the entire discussion which ensues on a certain fallacy. I shall now just summarily address that fallacy, specifically because it is very interesting to do so, as well as profitable.

He argues simply for the case of infinite divisibility and I need not replicate his argument; it is clear enough. Simply by asserting infinite divisibility, he comes up against a problem which he ignores, a problem which was recognized implicitly as early as Leonardo da Vinci, in respect to physics qua physics.

All through the discussion of this subject, there's been the question: If we divide all observation into three categories, can we attribute the same sensory properties of phenomena to all three categories in the same fashion, without some qualification as we move from one to another?

The three categories are the following:

First is the level of simple visual observation, simple sensory observation, a physical space-time as it appears to our senses by virtue of the limitations of our senses. The second is astrophysics, the macro-scale, that which is accessible in a sense to our senses, but which involves things which are far beyond our senses' immediacy. The third, of course, is microphysics, that which is so small, that it is beyond the capacity of direct observation by means of the senses.

Now, from early times up through Riemann, those of my persuasion have insisted, that when we come to the extremes of astrophysics and microphysics, we can no longer make the simple projections which might be suggested by observation, or successful observation, within the realm of visible and kindred phenomena, on that scale.

This begs a third question: What is the nature of the boundary separating each of the extremes, i.e., the large, astrophysics, and the very small, microphysics, from the ordinary scale of observation?

Generally I think we accept the notion, or those of us do who ponder this matter, that we speak of microphysics as that which lies in the vicinity of such a boundary, as in microphysics, the very small. You might say an Angstrom unit, or two or three Angstrom units, might not be that boundary or might be that boundary, but that when you get down into micron and similar kinds of areas of measure, you are in a troublesome area, relative to projections simply of the ordinary rules of visible observation and visible phenomena. Similarly, when we deal with matters on an astro-scale or astrophysical scale, for various reasons, having to do largely with time and so forth, we can no longer trust the simple rules of observation, of visible related phenomena. So, we are not concerned, generally, when we speak of astrophysics or microphysics, with knowing, at least for preliminary purposes, the exact boundary which separates the classes of phenomena. But we say, "When we get in the vicinity of those, a certain area, a certain scale, we have to be alert for sudden changes, abrupt changes hitting us."

We would say the boundary, of course from the standpoint of physics, is not a wall, but is rather a singularity. An example of that would be satisfying, since this was already addressed by Leonardo da Vinci in respect to sound, for example, and light. When we project a body under power to a supersonic speed, velocity, it is not in this case impossible to have supersonic velocities; but certain changes occur within the realm in which this occurs, the trans-sonic, supersonic phenomena occur, changes associated with phenomena which are not otherwise evident on the scale of observation of events at the lower speeds. So the speed of sound is a singularity. A trans-sonic area is a singularity, such that we cannot generalize what appears to be adequate interpretation of phenomena at lesser speeds, as we move through the trans-sonic to the higher speeds.

So that's what we mean, generally, when we say there is a change in the rules for observation of physical space-time as we encounter a boundary condition in the form of a singularity, as we continue to venture into the ever-smaller and the ever-larger scale.

The way we generally would approach this, particularly in the present century, is in respect to the limiting factor of the speed of light. As we approach the speed of light, we speak of a boundary area, which we call relativistic conditions. Generally, this is applied to the scale of astrophysics. But, ingenious minds will promptly attempt to reflect what is true of astrophysics, even as a consideration, back onto microphysics. That is, it is the common tendency in mathematical physics to treat the infinitesimal as an inverse of the infinite. Thus, if the speed of light is a boundary condition in the one scale, we must expect that there is a complementary boundary condition, i.e., a singularity, in the microphysical scale. That is essentially the way this should be approached.

What this would mean, of course, is that there is no infinite divisibility, in the sense I just implied. That is, we are not talking about an impossibility of some kind of divisibility on the microphysical scale below the scale of this boundary, this singularity, but we are implying the singularity as such.

This whole business, in both instances, is associated with the issue of the proper definition of physical space-time itself. Is physical space-time, in respect to physical cause and effect, a matter of simple linear extension, or is it not?

Kepler's astrophysics says it is not a matter of simple linear extension: that the available planetary orbits are not only limited in number, in the sense of being enumerable, but that this enumerability is defined by a very definite, intelligible principle, a principle susceptible of intelligible representation, which is the harmonic ordering; and that in the values of a special kind of Diophantine equations, if you like, in the values which lie between these harmonically ordered, enumerable values, there are no states of a similar nature, or precisely similar nature, at least, to be found.

Now, this introduces a kind of discreteness into physical space-time per se. That physical discreteness is the first aspect of a monad in the micro-scale.

Let me skip a bit and go ahead to another consideration respecting both astrophysics and microphysics. What about the large monads? The very large monads belong, not necessarily, immediately, to the microphysical scale, but rather to the astrophysical scale. Ahaaa! Right? Now there is a second consideration.

This goes to what I treated under the title of the Parmenides Paradox: the immediate relationship between the infinitesimal and infinite, say in the case of a human being. In this case you will see that it leads to the second point, on the monad.

We, in a sense, are, on the scale of astrophysics, an infinitesimal. Our mortality makes us all the more so. Nonetheless, we can affect the universe as a whole, at least implicitly so. We do so by an agency; that agency is creative reasoning.

We are capable of discovering, less imperfectly, the laws of the universe and doing this by creative reason. By activating and acting upon those discoveries by means of the agency of creative reasoning, that is, by acting on them by means of creative reasoning as well as discovering them by that means, we are able to influence the course of behavior of society as a whole and society as a whole is able to act on the universe, on an ever-larger, implicit scale of chains of cause and effect. By that agency, in terms of discovering universal principles, less imperfectly, and by discovering more powerful and more efficient means of acting upon the universe in the large by these means, we show that the human individual, this mortal ephemeral creature, we, the individual, actually have an implicitly direct relationship to the universe at large.

Similarly, we come to the second principle. Not only is the monad, so-called, something which is defined in respect to scale, but it is defined in respect to an active principle. Now here we come to the crucial matter, as treated by Leonardo da Vinci and treated explicitly by Kepler, as in the small paper *On the Six-Cornered Snowflake*.

On the ordinary macro-scale of observation, it appears to us that we have two harmonic orderings: one characteristic of living processes and the other characteristic of non-living processes, as Kepler treats this matter in *The Snowflake*. Thus, is the universe bifurcated in this way, or do we find some reflection of this question in the microphysical and macrophysical, or astrophysical, scale which removes the apparent paradox, or which makes comprehensible the apparent anomaly of the division of visible space-time and physical phenomena of observation into these two, living and non-living parts?

We find it just so. We find it implicitly required, for example, that the monads, in the scale of the small, in the microphysical scale, be implicitly negentropic, rather than entropic. That is, since negentropy, as a phenomenon, is characteristic of living processes, and entropy of non-living processes, then we must find, what might be considered by some, the simplest aspect of the non-living, the simple physical monad, to be implicitly negentropic—that is capable of showing negentropy or entropy, but being primarily negentropic. This again bears upon our relationship to the universe as a whole through creative reason, that is, our individual relationship to the universe as a whole as creative reason.

This goes to the simple *Parmenides* paper, to that little, beautiful irony, which is the center of that artistic composition, rightly called artistic. Amid all of these antinomies, this elaborate, quasi-deductive array of antinomies, Plato inserts a touch of irony: that after all, the problem here is, that the transition between these qualities which seem paradoxical, is defined by change and if we introduce, implicitly—Plato says, not explicitly, but implicitly—if we introduce change as having the primary ontological actuality, in this case, then the mystery of the antinomies dissolves and vanishes.

The problem here is, that when we say, that this divisibility of physical space-time in its linear aspect is elementary, we get into precisely the problem which Euler creates here. So, by assuming that simple extension, in that sense, is the property of matter, we create all the chimeras which haunt Euler's dream in this instance.

We recognize the implications of the speed of light as a singularity of the astrophysical scale, and recognize that the speed of light has a reflection in terms of a singularity in the microphysical scale; then we see where the fallacy of Euler's argument lies respecting physical geometry. If we recognize that the connection between the micro- and the macro-, the maxima and the minima, is expressed by change, where change is the quality of negentropy generalized, as typified by creative reason—as I have, I think adequately, defined at least in the preliminary degree, in *In Defense of Common Sense* and locations to the same effect, earlier—then the problem vanishes.

Só, the problem for Euler lies in his definition of extension and in the use of a linear definition of extension. In principle, Euler excludes, thereby, the realm of astrophysics and of microphysics from physical reality. This is where Leibniz did *not* fail and where Euler, at least in this case, did. That is my preliminary observation.

One thing added, as a footnote: Microphysics and astrophysics do not simply stand independently of the universe of the scale of simple observation; but, there is a point of scale at which, in the vicinity of whatever boundary condition is defined, we must *change*. We must recognize that we can no longer rely simply on simpler elementary methods of observation, but must change our view to accommodate the fact that we are approaching a singularity. Thus, in practice and in fact, as we get into the very small, divisibility in the ordinary sense *vanishes,* as it does as we get into the astrophysical scale, where the relativistic considerations remind us, or should remind us, that we are approaching a boundary condition in that respect.

Thus, as we get to certain areas of scale in practice, we no longer trust infinite divisibility. What that exact boundary condition might be, as, say, from the standpoint of the eighteenth century, we might not know. But we must know that one does exist, as Leibniz recognized. We must also recognize, as Leibniz recognized and Euler *does not,* that there is a qualitative change in the immediate implications of phenomena, of existence, as we get into the microphysical scale, i.e., that that which seems to be entropic non-living processes, on the scale of simple observations, can no longer be treated as simply entropic, but as a negentropic existence susceptible of generating ostensibly entropic phase spaces.

Not only is Euler wrong—and it is important to find Euler wrong, because of how otherwise useful he is—but, I think he has made what we might call a *strong* error, which has tremendous pedagogical value.

Letter 12, on the Subject of Monads

I address the content, in part, of Letter 12 of Euler's letters on the same subject of monads.

Euler introduces a fallacious argument of some significance, an argument whose foundation is a simplistic reading of the *Monadology* by some critics of Leibniz's work. This pertains to the magnitude of monads. Are they greater or lesser? Since they cannot be greater or lesser by the method which Euler imputes, then the whole thing is absurd. He also, therefore, says that relative to magnitude, they are absolute nothings.

It is interesting to look at this from the standpoint of the method we associate with the early work on integration by

Roberval, L'Hôpital's accounts, and so forth: the primitive view of infinitesimals, as Roberval et al., define them, which is the result of the conventional reductionist view, or quasi-reductionist view, prevailing in mathematics and mathematical physics today.[1] Nonetheless, it is not the point of view of the *Monadology*.

For example, the simple demonstration of the fallacy of Euler's argument here, from the standpoint of geometry, to which we can hold Euler accountable, is that it is easily demonstrated, beginning with nothing but the circular action of constructive geometry and hence multiply-connected circular action, that we generate discontinuities or singularities *out of continuity*. These singularities pertain to the nature of monads, at least in respect to the question of magnitude.

Now, the singularities are not generated by division. They are not generated according to the principle of extension which Euler in these letters demands be the standpoint of examining the *Monadology*. Rather, they are generated with precisely the geometrical qualities which may be attributed to monads by a continuous geometry, which takes no regard of infinitesimals generated by division.

Let us take the case of the simple fallacies which arise from the calculus by the simple method associated with L'Hôpital. If we use L'Hôpital's approach, we cannot equate an infinitesimal to virtually anything; but, in the case where we are trying to get the slope of a discontinuity, this infinitesimal becomes wildly indeterminate in a ponderable degree. That is, the indeterminacy is not infinitesimal, is not marginally infinitesimal, but the indeterminacy is of a very large order of magnitude relative to the function itself. Thus, there is no problem of the type which Euler attributes.

Thus, this is another way of looking at the boundaries of geometrical division, that is, in respect to scale, micro-scale and astrophysical-scale versus the ordinary scale of observation. What we call the micro-scale, the microphysical-scale, or the astrophysical-scale, is associated with the boundary conditions, which are associated in turn with the generation of singularities. What all of this involves, more specifically, is something which is made clearer successively by the work of Leonardo da Vinci, Kepler, Huygens, Leibniz et al. in the seventeenth century into the eighteenth century.

Huygens, for example, in his treatment of the pendulum clock, shows the role of the cycloid and of course this extends throughout the entire period, the tautochrone, the isochronic, the brachistochronic, functions, this shows that universal lawfulness and determination of time with respect to universal lawfulness is determined in respect to these non-algebraic functions. The implication of that is that the Cartesian notion of extension, of space, time, and matter, does not exist. Rather, that physical space-time, which has a definite curvature, is what *does exist,* and thus the significance of astrophysics and of microphysics and of the boundary conditions which ostensibly or putatively, or what not, separate the three domains from one another (or each of the two extreme domains from the domain of simple observation) and involve the generation of singularities.

The other aspect of this I stated before and must emphasize again: The characteristic of a monad, in Leibniz's setting and as I have situated it in the previous little oral memorandum on this subject, is that it is a universality; it is the minimum in which is embedded implicitly the maximum, or the minimum in which the maximum is implicitly embedded. This relationship of minimum to maximum is demonstrated immediately from the standpoint of the *Parmenides* dialogue, by the demonstration of the negentropic character of the monad. This we know, from the standpoint of human reason, from examining the nature of human reason itself, or its efficient and therefore existent nature. The fact that we are able to change the potential population-density of mankind through scientific and technological progress, i.e., through negentropic processes, nonlinear processes of creative discovery, demonstrates that this process of efficiently expressed discovery is existent and is thus *reason*.

Thus, when we look at man as a monad, as embodying *reason* in this efficient-existence sense, we thus define a relationship between the mortal individual, a monad, and the universe as a whole and with the Creator—the reflection of the Creator, the *imago viva Dei*. This negentropic monad, us, the creative reason, individual creative reason, becomes the standpoint from which we understand the monads in general. That is Leibniz's point of view.

Letters 13-15

Here we are dealing with Euler's attack on the principle of sufficient reason.

Now, the first thing to look at in Euler's criticism as a whole, particularly when, most to be emphasized, when we come to this issue of sufficient reason, is the question of ontology: It is not accidental that Euler starts this entire discussion on extension with the issue of ontology and affirms infinite divisibility, as a corollary of extension, to be a quality of substance, a necessary condition, a universal requirement, a universal property, of ontological actuality.

The best vantage-point from which to view this, critically, is to recognize the point made by Plato in the *Parmenides* dialogue. Plato anticipates, in effect, this entire argument of Euler's and of others, by showing through antinomies the inexhaustible absurdity of the idea of simple extension—and does so by showing that simple deductive methods, which are linear methods and hence the method of simple extension, cannot define substance. He does this in the beautiful, ironical method indicated, by referencing *change* as the key to the whole business. Thus, not extension, but rather *change in the process of extension,* is the location of efficient ontological actuality.

What Euler does, is to deny the efficiency of monads, except as *deus ex machina*—the Cartesian argument. He says, for example, in the Brewster English translation: "In this philosophy everything is spirit, phantom, and illusion; when we cannot comprehend these mysteries, it is our stupidity that keeps up an attachment to the gross notions of the vulgar." And then again (this is in 14) and in 15, he extends this to include the powers of the soul: that ideational properties are the mechanism which the monadologists profess to be efficient ideas, efficient principles. But, we know precisely that, in respect to change, ideas, insofar as they are limited to images of linear space, are not efficient.

So, therefore, by agreeing with Euler on this point, which he asserts, we thus demolish his argument, because that is not the issue. It is the creative processes through which valid scientific principles are discovered; and it is in changes in human behavior resulting from these ideas, that the monad expresses its efficiency. Therefore, it is not simply an abstract idea of movement, that is the idea in this case, that is the creative idea, as distinct from the simple mental image of an object, which is at issue. This, therefore, he assents to, by saying it would be to descend into obscurity to see efficiency in a mere image idea; he avoids the fact, that it is not the image idea that is the question here, but as Plato says in the *Parmenides,* it is *change.* The change, in this case, is the change effected by overthrowing an entire set of assumptions controlling human behavior, through discovery of a valid, crucial principle of natural law and thus changing human behavior to the effect of increasing the per capita power of the human species over the universe.

The sufficient reason in this case applies to the discovery and the elaboration of the discovery of this negentropic characteristic of individual human mortal existence. The fact that human beings have this capability, is sufficient evidence of the existence of this capability within an individual existence within the universe. The fact that this capability within an individual existence expresses a coherence of the maximum and the minimum—that is, the maximum in the minimum and the minimum in the maximum—is sufficient to demonstrate, against Euler, that this nature of existence is a general, i.e., Maximum, within the universe. General, not in the sense that all existence is immediately manifested, but that it is general in the universe and defines existence.

The *Parmenides* dialogue comes back into play here, by showing the absurdity of any notion of efficient existence from a linear standpoint, the absurdity of the notion of efficient existence from any other standpoint but change.

Selections from Euler's Letters

From *Letters of Euler on Different Subjects in Natural Philosophy, Addressed to a German Princess* (David Brewster, ed., New York: Harper & Brothers, 1840).

Letter 8: Divisibility of Extension in Infinitum
The controversy between modern philosophers and geometricians, to which I have alluded, turns on the divisibility of body. This property is undoubtedly founded on extension; and it is only insofar as bodies are extended that they are divisible and capable of being reduced to parts.

You will recollect, that in geometry it is always possible to divide a line, however small, into two equal parts. We are likewise by that science instructed in the method of dividing a small line, as *ai*, (**Figure 19**), into any number of equal parts at pleasure: and the construction of this division is there demonstrated beyond the possibility of doubting its accuracy.

You have only to draw a line AI parallel to *ai* of any length and at any distance you please, and to divide it into as many equal parts AB, BC, CD, DE, etc., as the small line given is to have divisions, say eight. Draw afterward, through the extremities A *a*, and I *i*, the straight lines A*a*O, I*i*O, till they meet in the point O; and from O draw towards the points of divisions B, C, D, E, etc., the straight lines OB, OC, OD, OE,

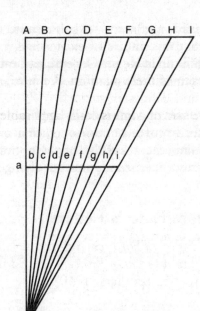

FIGURE 19.

etc., which shall likewise divide the small line *ai* into eight equal parts.

This operation may be performed, however small the given line *ai*, and however great the number of parts into which you propose to divide it. It is true that in execution we are not permitted to go too far; the lines which we draw have always some breadth, whereby they are at length confounded, as may be seen in the figure near the point O; but the question is, not what may be possible for us to execute, but what is possible in itself. Now, in geometry lines have no breadth, and consequently can never be confounded. Hence it follows that such division is illimitable.

If it is once admitted that a line may be divided into a thousand parts, by dividing each part into two it will be divisible into two thousand parts, and for the same reason into four thousand, and into eight thousand, without ever arriving at parts indivisible. However small a line may be supposed, it is still divisible into halves, and each half again into two, and each of these again in like manner, and so on to infinity.

What I have said of a line is easily applicable to a surface, and, with greater strength of reasoning, to a solid endowed with three dimensions,—length, breadth, and thickness. Hence it is affirmed that all extension is divisible to infinity; and this property is denominated *divisibility in infinitum*.

Whoever is disposed to deny this property of extension is under the necessity of maintaining that it is possible to arrive at last at parts so minute as to be unsusceptible of any further division, because they cease to have any extension. Nevertheless, all these particles taken together must reproduce the whole, by the division of which you acquired them; and as the quantity of each would be a *nothing* or *cipher* 0, a combination of ciphers would produce quantity, which is manifestly absurd. For you know perfectly well that in arithmetic two or more ciphers joined never produce any thing.

This opinion, that in the division of extension or of any quantity whatever, we may come at last to particles so minute as to be no longer divisible, because they are so small, or because quantity no longer exists, is therefore a position absolutely untenable.

In order to render the absurdity of it more sensible, let us suppose a line of an inch long divided into a thousand parts and that these parts are so small as to admit of no further division; each part, then, would no longer have any length, for if it had any it would be still divisible. Each particle, then, would of consequence be a nothing. But if these thousand particles together constituted the length of an inch, the thousandth part of an inch would of consequence be a nothing; which is equally absurd with maintaining that the half of any quantity whatever is nothing. And if it be absurd to affirm that the half of any quantity is nothing, it is equally so to affirm that the half of a half, or that the fourth part of the same quantity is nothing; and what must be granted as to the fourth must likewise be granted with respect to the thousandth and the millionth part. Finally, however far you may have already carried in imagination the division of an inch, it is always possible to carry it still further; and never will you be able to carry on your subdivision so far as that the last parts shall be absolutely indivisible. These parts will undoubtedly always become smaller and their magnitude will approach nearer and nearer to 0, but can never reach it.

The geometrician, therefore, is warranted in affirming that every magnitude is divisible to infinity; and that you cannot proceed so far in your division as that all further division shall be impossible. But it is always necessary to distinguish between what is possible in itself and what we are in a condition to perform. Our execution is indeed extremely limited. After having, for example, divided an inch into a thousand parts, these parts are so small as to escape our sense; and a further division would to us no doubt be impossible.

But you have only to look at this thousandth part of an inch through a good microscope, which magnifies, for example, a thousand times and each particle will appear as large as an inch to the naked eye; and you will be convinced of the possibility of dividing each of these particles again into a thousand parts: the same reasoning may always be carried forward without limit and without end.

It is therefore an indubitable truth that all magnitude is divisible *in infinitum;* and that this takes place not only with respect to extension, which is the object of geometry, but likewise with respect to every other species of quantity, such as time and number.

28th April, 1761

Letter 10: Of Monads

When we talk in company on philosophical subjects, the conversation usually turns on such articles as have excited violent disputes among philosophers.

The divisibility of body is one of them, respecting which the sentiments of the learned are greatly divided. Some maintain that this divisibility goes on to infinity, without the possibility of ever arriving at particles so small as to be susceptible of no further division. But others insist that this division extends only to a certain point, and that you may come at length to particles so minute that, having no magnitude, they are no longer divisible. These ultimate particles, which enter into the composition of bodies, they denominate *simple beings* and *monads*.

There was a time when the dispute respecting monads employed such general attention and was conducted with so much warmth, that it forced its way into company of every description, that of the guard-room not excepted. There was scarcely a lady at court who did not take a decided part in favor of monads or against them. In a word, all conversation was engrossed by monads—no other subject could find admission.

The Royal Academy of Berlin took up the controversy, and being accustomed annually to propose a question for discussion and to bestow a gold medal, of the value of fifty ducats, on the person who, in the judgment of the Academy, has given the most ingenious solution, the question respecting monads was selected for the year 1748. A great variety of essays on the subject was accordingly produced. The president, Mr. de Maupertuis, named a committee to examine them, under the direction of the late Count Dohna, great chamberlain to the Queen; who, being an impartial judge, examined with all imaginable attention the arguments adduced both for and against the existence of monads. Upon the whole, it was found that those which went to the establishment of their existence were so feeble and so chimerical, that they tended to the subversion of all the principles of human knowledge. The question was therefore determined in favor of the opposite opinion and the prize adjudged to Mr. Justi, whose piece was deemed the most complete refutation of the monadists.

You may easily imagine how violently this decision of the Academy must have irritated the partisans of monads, at the head of whom stood the celebrated Mr. Wolff. His followers, who were then much more numerous and more formidable than at present, exclaimed in high terms against the partiality and injustice of the Academy; and their chief had well-nigh proceeded to launch the thunder of a philosophical anathema against it. I do not now recollect to whom we are indebted for the care of averting this disaster.

As this controversy has made a great deal of noise, you will not be displeased, undoubtedly, if I dwell a little upon it. The whole is reduced to this simple question, Is a body divisible to infinity? or, in other words, Has the divisibility of bodies any bound, or has it not? I have already remarked as to this, that extension, geometrically considered, is on all hands allowed to be divisible in infinitum; because however small a magnitude may be, it is possible to conceive the half of it, and again the half of that half, and so on to infinity.

This notion of extension is very abstract, as are those of all genera, such as that of man, of horse, of tree, etc., as far as they are not applied to an individual and determinate being. Again, it is the most certain principle of all our knowledge, that whatever can be truly affirmed of the genus must be true of all the individuals comprehended under it. If, therefore, all bodies are extended, all the properties belonging to extension must belong to each body in particular. Now all bodies are extended, and extension is divisible to infinity; therefore, every body must be so likewise. This is a syllogism of the best form; and as the first proposition is indubitable, all that remains is to be assured that the second is true, that is, whether it be true or not that bodies are extended.

The partisans of monads, in maintaining their opinion, are obliged to affirm that bodies are not extended, but have only an appearance of extension. They imagine that by this they have subverted the argument adduced in support of the divisibility *in infinitum*. But if body is not extended, I should be glad to know from whence we derived the idea of extension; for if body is not extended, nothing in the world is, as spirits are still less so. Our idea of extension, therefore, would be altogether imaginary and chimerical.

Geometry would accordingly be a speculation entirely useless and illusory, and never could admit of any application to things really existing. In effect, if no one thing is extended, to what purpose investigate the properties of extension? But as geometry is beyond contradiction one of the most useful of the sciences, its object cannot possibly be a mere chimera.

There is a necessity then of admitting, that the object of geometry is at least the same apparent extension which those philosophers allow to body; but this very object is divisible to infinity: therefore existing beings endowed with this apparent extension must necessarily be extended.

Finally, let those philosophers turn themselves which way soever they will in support of their monads, or those ultimate and minute particles divested of all magnitude, of which, according to them, all bodies are composed, they still plunge into difficulties, out of which they cannot extricate themselves. They are right in saying that it is a proof of dullness to be incapable of relishing their sublime doctrine; it may however be remarked, that here the greatest stupidity is the most successful.

5th May, 1761.

APPENDIX XII | 'Anthropomorphic Science'

Since we did not desire to narrow our audience to exclude non-specialists in the following matter, the author has chosen to relegate to this appended section the treatment of certain topics relevant to *The Science of Christian Economy* Chapter VI, "The Reproduction of Man." To indicate the fuller scope of relevance of the technical difficulties addressed in this Appendix, we excerpt two passages from a writing by Max Planck.

In his 1948 paper, "The Concept of Causality in Physics" (in *Scientific Autobiography and Other Papers,* New York: Philosophical Library, 1949), Max Planck writes (pp. 144–5):

It could be maintained that a relationship possessing such profound significance as the causal connection between two successive events ought to be independent by its very nature from the human intellect which is considering it. Instead, we have not only linked, at the very outset, the concept of causality to the human intellect, specifically to the ability of man to predict an occurrence; but we have been able to carry through the deterministic viewpoint, only with the expedient of replacing the directly given sense world by the picture of physics, that is, by a provisional and alterable creation of the human power

426

of imagination. These are anthropomorphic traits which ill-befit fundamental concepts of physics; and the question therefore arises, whether it is not possible to give the concept of causality a deeper meaning by divesting it as far as it can be of its anthropomorphic character, and to make it independent of human artifacts, such as the world picture of physics.

Now we come to a second quote (pp. 149–50):

The law of causality, which immediately impresses the awakening soul of the child and plants the untiring question, *"Why?"* into his mouth, remains a lifelong companion of the scientist and confronts him constantly with new problems. For science is not contemplative repose amidst knowledge already gained, but is indefatigable work and an ever-progressive development.

The fact, that a "non-anthropomorphic science" is a contradiction in terms, did not prevent that catch-phrase from gaining today a widespread and stubbornly persisting popularity within academic and other strata. In the chapter from which we have just quoted, Planck is much too generous with his positivist adversaries on this point. A more precise treatment of the issue bears directly on the material within Chapter VI, above.

First, a matter of terminology.

To define the word *science* in the first approximation, we restrict initial inquiry to the domain of so-called *physical science,* or, earlier, *natural philosophy.* It is useful, because of a relevant dispute between the followers of Leibniz and the Kantians, to equate physical science, in first approximation, to the nineteenth-century usage of the German term *Naturwissenschaft.* Later, we shall complement our initial case by integrating the remaining aspect of science in general: what is named in German, *Geisteswissenschaft.*

The term, *modern physical science,* covers the period of, initially, European history beginning with the early fifteenth century's Italy-centered *Golden Renaissance.* By modern physical science so defined historically, we signify what is better described as physical geometry, a study of physical principles from the standpoint of demonstrable geometrical constructions.

The essence of *Physical Economy,* and therefore also of *political-economy,* is subsumed in conception by the single fact of the human species' absolute separation from, and superiority and proper dominion over, all other species of organic and inorganic processes. Unlike the animal species, mankind exists by means of a process expressed as scientific and technological progress.

This fact, this process of scientific and technological progress, is tested in practice by the yardstick of human-reproductive requirements. As we have already indicated in the text above, these requirements are associated with the need for a rise in the average, per capita, physical-productive powers of labor, and also a corresponding increase in the physical standard of human consumption, longevity, and health combined. This requires coordinate improvements in nature, to the effect that those improvements, combined with a rise in per capita productivity, represent a durable, continuing rise in the potential population-density of the human species.

Those facts summarized, lead us to the following proofs respecting the essential characteristics of human scientific knowledge. These proofs bear directly upon the relationship between Christian principles and sound principles of economy.

As we have identified that policy in the text above, everything we say rightly, respecting the potential scientific-creative powers of the individual human mind, is also implicitly a statement respecting the role and activity of those same processes in the generation of classical artistic beauty. With that point so emphasized once again, we proceed as follows.

As is shown in other published locations, the ordering of scientific progress consistent with increase of mankind's potential population-density is an ordering susceptible of intelligible representation. This intelligible representation of the principle of that successive ordering, is itself of the character of a cardinal notion, a *transfinite cardinality*. Strictly speaking, the name of physical science ought to be restricted in definition by direct and exclusive reference to this notion of transfinite cardinality. (See Appendix VII.)

At this point, we ought to take our conscious processes, in progress here, Socratically, as objects of our consciousness. We have just shown, implicitly, that the idea of "objective science" is a contradiction in terms, an absurdity. We have just said, implicitly, that absolute scientific truth exists only *subjectively!* We have said, implicitly, that there exists no science, or possibility of knowledge by any person, apart from the subjective instrument, the individual creative reason, by means of which Socratic method, scientific knowledge of the transfinite cardinality, is acquired.

Let us describe this process as follows.

First, through either crucial-experimental or equally significant observation, we discern some *axiomatic* flaw in principles of established physical science. The identity of such a flaw is sought by means of the same method permeating Plato's Socratic dialogue. The *Parmenides* dialogue is a beautiful and relatively simple illustration of this method.

Second, this Socratic treatment of established physics implies hereditarily efficient axioms and postulates, points us toward a potential form of creative solution through the detected error. That solution is in the form of a hypothesis, as hypothesis is explicitly and implicitly defined by Plato's dialogues as a whole.

Third, this hypothesis is subjected to either crucial-experimental or comparably significant tests. This test is initially addressed to the particular case or cases which had led us to discover the axiomatic error in established physics. If the result of that is satisfactory, we must also test the appropriateness of the hypothesis for physics in general.

Fourth, if the latter shows the hypothesis not only to correct the prompting error, but to increase practically the power of physics in general, the new principle is established and the activity leading to the success is viewed as a successful revolution in physics.

This increase in the power of physics means a demonstrable sort of potential increase of the power of the human species over the universe as a whole. This measurement is implicit in terms of *rate of increase of potential population-density.*

Such a success is a reflection of the *divine spark of reason* sovereignly situated within the individual personality. In other words, this is that *Minimum,* the creative individual, the Leibnizian *monad,* which is in relationship to the *Maximum,* the Creator.

As is shown among my published locations treating this matter, the successive successful revolutions in physical science, insofar as they are cases rigorously in conformity with what we have illustrated by the step-wise form, just above, define within science historically a series of transformations which do satisfy this requirement. The revolutionary work of Cusa, Leonardo da Vinci, Kepler, and Leibniz is exemplary. This typifies the notion of succession of successful scientific revolution. That notion of succession implies the relevant notion of a governing, *transfinite* ordering. The notion of that self-developing ordering as a cardinality, is the proper notion of *science in general.*

That *science in general* is associated with man's potential power over the universe. Thus, as long as we adhere to this rigor, the idea of separating the subjective from the objective is absurd. There exists nothing "objective" outside the realm of this rigorous kind of "subjectivity."

There is no possibility of a true science which is not of this rigorously *subjective,* or "anthropomorphic" form. We see, in science, efficient forms of *subjective* certainty of the

Creator's universal natural law. By that means, we increase the potential population-density of our species *in this universe as a whole*. The implicit increase of potential population-density is the proof of the *anthropocentric* experiment on which even the mere possibility of science depends. Since this science is produced by the sovereign faculty on which account the individual person resembles the Creator, the potential creative reason, the only possible form of science, is in that image, that *anthropomorphic* image.

On the Subject of Christian Civilization

Lyndon LaRouche delivered the following keynote address via audiotape to the Sept. 1–2, 1990 conference of the Schiller Institute in Arlington, Virginia.

I. The Strategic Focus of This Global Crisis

Let us consider first the strategic focus of this present global crisis.

We assemble this day, under the darkening shadow of a global strategic crisis, a crisis which is reaching toward the remotest corners of our planet, and into the most jealously guarded, most private places, where deluded persons might seek physical and mental refuge from awareness of unpleasant truths.

We are sitting presently in a process leading toward the possibility of a new world war. On the surface, it is the heirs of Britain's evil Castlereagh who are orchestrating such a war, in the same geopolitical fashion they caused World War I.

Events in the Middle East cockpit are being orchestrated by British Intelligence and diplomacy, to the purpose of pitting France and Moscow against Germany, and against Japan: all leading toward a later, nuclear conflict between Moscow and the Anglo-Americans.

If such a war comes, it will degenerate, as the 1618–48 Thirty Years' War in Central Europe degenerated.

To be specific, it will degenerate into a form of total war, which history usually associates with so-called religious wars. The character of any future world wars of this present time-frame (the period ahead), would indeed be derived from the fact, that the root of the present, global strategic crisis is a presently most visible effort, by some, to eradicate Christianity from this planet. It is to that deepest, axiomatic feature of the crisis, that I address my present remarks.

Before proceeding to that specific undertaking, it is of more than a little practical importance, that I identify a few ground rules for the discussion which I am provoking.

We assembled represent an international philosophical association, ecumenical in its composition. Thus, whenever we address matters of religion, as we are obliged to do that here, we allow no proposition to be presented, either as premise or topic of discussion, unless the truthfulness or error of that proposition, is to be subjected to those tests of truthfulness, which I associate with the term "intelligible representation."

For convenience, I reference the definitions of such intelligible representation supplied in the texts of *In Defense of Common Sense,* and *Project A.*

That said, we reference the fact, that the essence of the present global crisis is typified by the fact, that the British royal household's Prince Philip, the Duke of Edinburgh, has taken a leading, public position, in his words and in his corresponding practice, in promoting causes amounting to the attempt to exterminate Christianity from this planet.

Strong words, but true words. *There is no exaggeration in that. The Prince's own words are clear.* That factual observation situates the following proposition which we shall consider here.

Why must an informed Vedantist, Jew, Buddhist, or Muslim, view Prince Philip's expression of pro-bestial hatred of Christianity as representing a threat to the continued existence of the human species?

It is implicitly obvious, that that form of proposition pertains directly to the concept of successful or durable survival, treated in my text, *In Defense of Common Sense,* and also treated more extensively in *Project A.* Thus, we are putting this question, although it is a religious question, in a rigorous, scientific setting. We are treating it as a scientific-setting question. Thus, what some religion says, whether Christian, Jewish, or others, or seems to say, according to some putative authority, *is irrelevant here,* except that that proposition is sustained on the same basis of method, which I employ as typified by the outline of method in the two texts referenced.

So when we say, as I shall, in conclusion of this report, that Christian civilization is the highest form of social order yet attained by man, and thus must be defended by all humanity, as in the vital interests of all humanity, I am stating a scientific proposition with conclusive scientific proof, which does not rely upon the arbitrary assertions of any interpretation of a religious text in the fundamentalist or kindred sense.

The essential proofs of Christianity, in any case, have always been contended to be, by the leading Christian theologians, truths which were evident, *even if no text existed to assert them.* As Christ says in the Gospel (Luke 19:40): "The very stones might speak." Indeed, the stones and stars, as we know, do sometimes speak, in their own way, as they bespeak perceivable natural law, susceptible of intelligible representation by aid of the creative powers of reason of mankind.

The conflict we face can be more broadly described in the following terms.

For the past 2600 years, European civilization has meant essentially, at foundation, the opposition of Athens, as well as the Ionian city-state republics, to the usury-practicing culture of Babylon, of Mesopotamia, and has meant the overthrow of the usurers at Athens, by the so-called constitutional reforms effected by Solon of Athens.

We trace European civilization thus from Solon, in those

terms of approximation. We trace that civilization through the exemplary work of Socrates and Plato. We thereafter treat Socrates and Plato as they would treat themselves, had they been converted posthumously to Christianity: as Augustine and as Cardinal Nicolaus of Cusa, for example, exemplify Christianity.

We thus treat Christian civilization as an anti-oligarchical, anti-usury culture, extended as Christian civilization, implicitly, from Solon of Athens's overturning of usury in Athens, through to the present time.

The chief adversary, over most of the 2600 years to date of Christianity, has been pagan Rome: the pagan Rome which we identify, sufficiently, with such names as the Anti-Christ, the Emperor Tiberius, the Emperor Nero, the Emperor Diocletian. These are the enemy.

In more recent times, the enemy of Christianity within Europe, or the chief enemy, has been an oligarchy, which is characterized by its promotion of the licensing or practice of usury, and which has recurrently turned to pagan Rome for models as to law, as to social custom, and as to relations among states.

The most relevant case, for our present purposes, is the rise of what is called Romanticism, together with British liberalism, which is the same thing as Romanticism, in Britain and on the continent, during the eighteenth century. Examples of Romanticism on the continent, of course, are Voltaire and all of his friends: Montesquieu, Rousseau. In Britain, David Hume and Adam Smith are examples of Romanticism, as well as Gibbon or Jacques Necker, the man who ruined France in the eighteenth century, or his daughter, who spread the virus of Romanticism into Germany so prolifically: the Madame de Staël.

Romanticism is the modern form of the enemy, which leads to a second form, to which I'll come in a moment, a second expression of Romanticism: the Dionysiac form.

Romanticism proposes, essentially, to uphold pagan Imperial Rome and the idea of a global one-world empire, a Pax Romana, so to speak, modeled upon pagan Imperial Rome, as the hero; and Moses and the Christianity associated with Moses, as the arch-enemy.

It is not Judaism as such which is the target of paganism, but rather, Mosaic theology: the ancient Judaism of Moses, rather than something which is mixed with Babylonian myths such as cabalism, a pseudo-Mosaic concoction, cooked up since.

Thus, in modern times, especially since the eighteenth century, since the time of the enemy of Christianity, the First Duke of Marlborough, and his success in enthroning liberalism under the new United Kingdom, it is Romanticism and its successor, modernism, which have been the enemies of Christianity, the enemies of Christian civilization.

Thus, we have the picture.

Within Europe itself and the Americas, of course, by extension, we have two European civilizations: One is the basic civilization, whose achievements all rest upon what we might call the republican current, or Christian current, as identified with Solon, Socrates, and Plato, in the manner I indicated.

The second is oligarchical Europe: the oligarchy, the aristocracy, the nobility, who practice usury or who promote it; and who turn to the model of pagan Imperial Rome, to the model of Tiberius, of Augustus, of Nero, of Diocletian, and their policies, as the antidote to anti-usury republicanism.

Now, in this time, we've come to a conflict which flows from that conflict within European civilization. The oligarchy, the pro-usury oligarchy, the pagans, represent the standpoint of the British Empire, for example. The British Empire was explicitly devised as a concept, during the eighteenth century, developed by the Romantics, as an empire based upon the pagan imperial Roman model. Napoleon Bonaparte, for example, later, was an instant of the pagan Imperial Roman model introduced as a cult idea into the politics of France.

The pagan Imperial Roman model was adopted by the Russians, as early as Philotheus of Pskov, in 1510 A.D., and so forth and so on.

But, out of this imperial design, typified again in 1815 and thereafter by the Holy Alliance, calling itself Christian, but actually based on a *pagan* model, pagan Imperial Roman model, we have emerging the idea of the management of the balance of power, as a way of crushing out of existence the form of statecraft which reflects the Christian republican tradition.

This led to World War I: the British, working against Gabriel Hanotaux of France, Sergei Witte of Russia, and others, connived with others, *to prevent* economic cooperation from developing among France, Germany, and Russia, among others, with the view that if these three powers collaborated and, in turn, collaborated with Japan and against British interests in China, that the Eurasian continent, so dominated by economic development, would become an unbeatable force, from the standpoint of Britain. Thus, Britain connived, in its so-called Great Game, to pit Russia and France against Germany, and to utilize the decay of the Ottoman Empire, with the attendant Balkan crisis, to create what became World War I.

Britain then acted, following World War I in the 1930s, to recreate that circumstance, with British interests, as well as the Harriman interests in the United States, working to bring Adolf Hitler to power, for that purpose: to launch and create World War II, which, however they may have regretted later, they caused.

So today, forces in Britain, seeing the rise of a reunified Germany, and a shattered Russian Empire's *dependency* upon economic cooperation with Germany for its own mere survival, fear again, that the continent of Europe, dealing with the crisis of the development of Russia and reaching out to nations of the rest of Asia and other parts of the world, would present a powerful economic force, which British imperialism, in its new form, or Anglo-American imperialism, could no longer dominate.

And thus, today again, Britain, through a certain faction in the tradition of the evil Castlereagh, has moved, with the Middle East crisis, to attempt to manipulate Russia through its oil-lever, against continental Europe, with Britain, and thus, to set France and Russia again against Germany, with the ultimate view that this must lead to, not a Germany-Russia war, but a nuclear war between Russia and the Anglo-Americans.

If this war were to occur, the result would be, as I've otherwise indicated, a degeneration of war as occurred in the Thirty Years' War, 1618–48. The proud armies of Wallenstein, coming into the field of battle, might prevail in the initial battle, as U.S. and other forces might prevail if they attack Iraq. *But, in the aftermath of that apparent success, there would be unleashed a form of total war, which we associate, as historians, with the worst and most ferocious and most embittered of religious wars.*

In that and related forms, the warfare ignited at one fuse, such as, say, Iraq, would spread across the planet: not all at once, but over days, over weeks, over months. And the days and the months and the years would pass ever more precipitously, as was the case in the Balkans, as in the period 1910 to 1914. But this time on a global scale, and more bitter and more profound, until a little spark—and the spreading conflict from that spark, uniting with other sites of conflict and wars—spread around the world, and aligned the whole world in a form of warfare, best described as total war, in which all kinds of weapons, ranging from fists and hands clenched at the throat, and rocks bashing skulls, to the most modern weapons, are deployed with man on man, nose to nose, and knife to back, throughout this planet.

That is the nature of the conflict we face.

So, in organization of the conflict, as I've already indicated, we have the geopolitical form, with the British and Anglo-American elements attached to the British, attempting to replay the continental Europe balance-of-power game, as it was played earlier during this century, and, indeed, since the founding, and pre-founding, of the Holy Alliance, back in 1815.

At the same time, these British forces are focused upon a North-South conflict: the attempt to shift (at least temporarily) the conflict from the Cold War conflict of East-West, to North-South: in effect, to conduct population and raw materials wars against those regions of the world, whose populations have skin colors somewhat darker than those the British most admire, to put the point bluntly enough. And thus, the Middle East becomes the cockpit for a world war: not merely because of oil, or because of any other reason, but precisely because, strategically, it is the crossroads between the East-West and North-South points of conflict. And, that must be prevented.

However, the struggle is not simply a struggle between the Roman pagan imperial idea and Christianity.

Toward the latter part of the nineteenth century, and then into the twentieth in a second phase, there was the rise of modernism, beyond Romanticism. The reasons generally were very obvious ones.

Romanticism, while it eroded and damaged the republican movement greatly, during the period of the eighteenth century and early nineteenth, nonetheless was unable to suppress entirely scientific and technological progress, and unable, thus, to abort the improvement of mind of the general population, an improvement of the mind which caters to political freedom, as it caters to the power of intellectual freedom. And thus, those behind the Romantic idea, had to resort to more desperate means to attempt to uproot Christianity.

We had, thus, the existentialists of the nineteenth century, in which one would include, properly, Ruskin of Oxford University and so forth. But more notably, people of the stripe of Friedrich Nietzsche and Aleister Crowley and that crowd.

These fellows said explicitly, we must develop a cult, a destructive cult, modeled upon the Phrygian cult of Dionysios or the Greek cult of Apollo, and counterpose that to Christianity, to use this form of Anti-Christ, to destroy Christianity: to use Dionysios, to use Wasserman (waterman, the cult of Aquarius), to destroy the era of Pisces, Pisces being the symbol in astrological doubletalk for Christianity and for Socrates.

We had a similar event occur right after World War I.

Bolshevism failed to conquer Western Europe, as it had taken over Russia. This disturbed the Bolsheviks and their sponsors very much. And, to that effect, a fellow called Georg Lukacs appeared in Germany, around circles associated with Max Weber's tradition, on an occasion in which Lukacs laid out what became the program of the Frankfurt School, the Frankfurt Institute for Social Research.

Lukacs said, essentially, that Bolshevism had failed to conquer Western Europe, because Western European civilization had an inoculation, an immunological potential, against the virus of Bolshevism: that, essentially was Christianity, the Christianity in the tradition of a Socrates and Plato, converted posthumously to Christianity. Therefore, Lukacs proposed, we must *destroy* this Christian immunological trait, this Platonic trait of Christianity, as a precondition for effectively infecting Western Europe with the Bolshevik virus.

Out of that came the projects of the Frankfurt School: out of that came the rock-drug-sex Malthusian counterculture, which has erupted with such increasing force, since the inauspicious year, or auspicious year, 1963.

Since that time, there has been an outright, increasing effort to destroy Christianity per se. In the United States, this erupted to the surface most conspicuously with the work of Supreme Court Justice Hugo Black, in using the mythical argument of Jefferson's supposed moral separation between church and state, to create a vacuum to the effect that, while Christianity is outlawed from our public schools, satanism is invited in, under law. And, by these attacks upon Christian morality, and the attempt to substitute Roman-style, pagan ethics for Christian morality, we have at least two generations of young Americans, for example (and in other countries, similar conditions), who are essentially morally destroyed or disoriented; who have *lost the immunological potential* to resist such viruses as Bolshevism, fascism, and so forth and so on.

And thus, when Prince Philip says, that man must give way to the rights of the beast, that the human population must be curtailed for this account, one finds that today, what would have been impossible two generations ago, erupts: that we have animal rights movements, wild terrorists, completely irrational, insane, as insane as maenads, preparing to tear the society down, for the sake of a spotted owl, or a red squirrel, or even some lower variety of species.

And thus, that is the nature of the danger to civilization, on the negative side.

But, on the positive side, there's something else.

Christianity contains something superior to any other form of culture, objectively speaking, which is not a property of Europe, in the strict sense, or of the Americas. *What has been contributed by Christian civilization is the rightful property of every person on the surface of this planet.* And to that, let us turn next.

II. The Map of the Human Mind: Rendering Policy-Making Intelligible

Before we look at the specific qualities of Christian civilization, which make it so superior, as well as unique, we have to glance briefly at matters which are covered in some length in the two texts already referenced, but which should be restated at least summarily here, for the benefit of those who may not have read or may not have studied adequately, the two texts involved.

Very few people, unfortunately, know what the term *mind* ought to mean; at least, they don't know what it ought to mean in any scientific sense.

Most people, for example, would tend to accept, at least as a proposition, the idea that *rational* means *logical;* and by logical, they would mean formal deductive logic.

But, this is not true.

We know today, of course, the embarrassing fact that *machines* can perform deductive logic: computers, for example. At least, a very crude form of deductive logic, and we're able to do more and more in that direction. Not exhausting all possibilities in deductive logic, but going further and further, to the point that the initiates are rather awed by what can be done.

So, it does not seem that deductive logic is very much the quality of the mind, if it's the quality of a machine. Or, perhaps machines will replace men. Such are the things you get into if you don't take into account the fact that there *is* a difference.

But, there is a difference, very easily demonstrated.

Man is able *to change his ideas and behavior,* to the effect of *deliberately* increasing the productive powers of labor. The result is, that mankind is able to sustain people with less land. And, to not only sustain a person with less land required, but to, at the same time, increase significantly the standard of living of the person who's sustained. So, it costs less to maintain a person, but that person has more. They have more in terms of life expectancy. They have more in terms of consumption. They have more in terms of leisure and time for the development of their powers of mind as human beings, and so forth and so on. Child labor is abandoned, and children get into labor at a later point in life, and thus have more time to develop, and develop more richly.

Only the human species can do this. No animal species

can do this, no machine can do it. No computer, no matter how articulate the machine may be, can do that.

We find this quality of mind is associated, most obviously, with what we call fundamental scientific discoveries. The human mind is able to discover fundamental laws of nature and to correct its understanding of those laws, in a very fundamental way.

It is this creative power of the mind, or creative reason, as opposed to logic, which is the essence of human mind.

Let's look at that just briefly again, to make sure we're absolutely clear.

Most people are acquainted with what they think is high school and college physics, for example. They think of this physics in terms of a kind of mathematics, which is based on arithmetic, in which geometry may enter, but it's only as a helping device, only as a means of illustrating the point. The algebra they're familiar with, is based on arithmetic—not geometry—and they assume that everything that physics says, from the standpoint of experiment, can be said in terms of algebras derived from arithmetic or from deductive logic.

But then, consider the case of any fundamental scientific discovery. By a fundamental scientific discovery, we mean an experiment which overturns, implicitly, the entirety of an existing mathematical physics; which says you have to go back over the entire physics, and change all the so-called underlying assumptions of physics; and correct all of the theorems to allow for this sudden discovery of this correction of the error.

Now, that process of correcting the error cannot be represented deductively. And yet, all science is based on nothing but fundamental discovery. All scientific progress, all improvement in the condition of man, is based on these kinds of discoveries, which cannot be represented deductively, and yet which nonetheless occur, which are efficient, and which are directed, in the sense that mankind, somehow or other, knows

how to seek a discovery which increases man's power, and, if
he does it well, a man can actually do what he sets out to do.
It does not occur by random evolution, by random selection.
It occurs by intent.

Every great inventor discovered things because he *intended* to discover them. He may not have discovered exactly
what he intended, but *he intended to discover something
which would increase man's power over nature;* and he ended
up doing just that, if he was any good at it. He didn't make
a lot of inventions and then throw them out as random experiments, such that the successful ones survived and the others
didn't. No, it was all done by intent. And this kind of process,
of discovery by intent, cannot be represented by any deductive
system.

It is precisely this ability *to intend to discover,* or to
transmit such a discovery, or to assimilate such a discovery for
productive or other practice, which distinguishes the human
being from an animal: which sets the human species as a
whole apart from all animals, and which sets the human
species above all animals.

So, the characteristic of the human species is this quality
of mind, which is associated with creative reason.

Now, it is not our purpose here to go into this aspect of
the matter, but it is necessary to report the fact, that *we can
represent the processes of creative reason, in an intelligible
way.* Not a deductive way, but an intelligible way, fully as
rigorous as one might assume an algebra to be.

We can map this, we can describe this, we can show this.
There are methods in geometry, by which we can do that,
with increasing precision. But, we cannot do that in deduction.

So, the important thing is to know that this power of the
mind exists.

Now, the second thing about this power of the mind is,
that it is *sovereign.* No matter how much social influence and
suggestions and collaborations and so forth, go into enabling
an individual to make a fundamental scientific discovery, in

the final analysis, the actual act of discovery, the creation of the idea, is done entirely inside the head of the person who makes the discovery. There is no outside participation in that process itself. There may be outside stimulation, collaboration, input, and so forth. But, in the process itself, there is no outside intervention. It is directly done inside the person. Therefore, it is a *sovereign process* of the individual, as a necessary person, as an individual.

Now, those are the qualities with which we have to deal. That being the case, it is desirable in society, that that quality of the individual, in every individual case, be developed to the maximum degree possible. You don't get discoveries by the *average* behavior of individuals. You get discoveries by developing the individual *as an individual*.

It's more than just the individual scientist that makes the individual discovery. In order to have a discovery work, in society, it cannot be confined to the mind of the original discoverer alone, or to a few scientists. It must be *transmitted* to teachers and others. These people must be educated and developed to the point they can *assimilate* the discovery, and go through, in a sense, a process of *retracing the discovery,* as made by the original discoverer: through their own mind, their own sovereign powers of creative reasoning, in their own sovereign mind. These people, in turn, must transmit this to others, who receive it, as people who work with these ideas, in machine-tool shops or other ways. These persons, too, must go through the process of retracing, at least to some degree of approximation, the kinds of mental processes represented by the scientist's discovery.

And thus, to have this kind of progress, we must educate *all* people in society, more or less. We must develop the *sovereign creative powers of reason, of each and every child*, and foster that quality in each and every adult.

This will give us the highest possible rate of discovery and the highest rate of improvement, in both the productive powers of labor as such, that is, the ability to produce more for human need, and the ability to produce that more, with less land required to do it. This relationship, this happy result, we call an increase in the potential population-density. I won't go more into that right at this point, because that's covered in a textbook of mine which has been published, which covers some of the complexities of this process. *But, that particular quality of mankind is key.*

Now, let's look at another aspect of this quality.

What does that mean? What does it mean, when mankind as an individual has discovered a law of the universe? What that means is, that in discovering a law of the universe, at least getting to know it less imperfectly, the human mind is converging upon the truthful, actual form of that lawful arrangement in the universe. And, in that degree, the mind of the individual person is converging upon agreement with the mind of the Creator, with the mind of God. And, with the will of God.

Thus, within these creative powers of reason, if they are sufficiently developed in the individual, that individual mind *approximates a map of the lawful organizing of the universe as a whole.* That is the wonderful thing.

Thus, we, the Minimum, the little small thing, the indivisible smallness of the universe, our little intellect, is, in that respect, in a direct relationship with the total organization of the universe, and is, implicitly, potentially, a map of the whole universe.

So, the largest and the smallest are thus unified, sort of projectively, as having one character, through the exercise of creative reason. And, through that faculty, it is possible for man to know the Creator—in that respect, and to that degree, and with those limitations. It is possible for man to say, that everything he or she knows, he or she knows by means of the possibility of intelligible representation, i.e., as by construction of any valid idea.

It is not necessary to assert anything arbitrarily. We can find an intelligible representation, which shows us whether the idea is a truthful one or false one. That is the nature of the situation; and, those are things which are covered with many more things as well, in the two reference works which I've cited.

III: What We Mean by the Superiority of Christian Civilization.

When we say that Christian civilization is the highest form of civilization devised by man, with the references I've given at the outset as to what that means, we are saying, *scientifically,* that Christian civilization affords society the *highest rates of growth* of potential population-density, *the highest rate of development of the human mind,* and *the most concentrated and effective kind of development of that mind.*

The crucial feature of Christianity, in this account, is something which is summed up by an emphasis applied to the Christian Credo, by St. Augustine: what is called in Latin the *Filioque,* that Christ is both the Son of God, and God, such that the Holy Spirit flows from Him, as from the Godhead.

What this signifies, without going through the whole issue, is that through this view of Christ and through this intermediating role of Christ, the individual human being is able to recognize, efficiently, his or her identity as *imago viva Dei:* as a living being in the image of a living God. Not some king, not some arbitrary monarch, but the Creator as the Creator, not some petty tyrant like Zeus, spitting from some mountaintop, playing tricks upon men, but a true loving Creator, in whose image we are.

In what sense are we in God's image?

We are in God's image, by virtue of creative reason, and nothing but creative reason. We are in God's image in terms of that potential creative reason, which makes us the Minimum, in correspondence, efficiently so, with the Maximum: a lawful universe and its ordering, as a whole.

It is that image of individual man, the Christian image of individual man, as being born with the divine spark, this poten-

tial for creative reason, this quality like that of the Deity, like
God, in the image of God, which makes Christian civilization
work, it is the secret of Christian civilization, its power; which
is why Christian civilization is based on reason, creative reason,
rather than arbitrary teaching of revealed, arbitrary dogma.

Thus, Christianity and science go together. Not the kind
of image of science we associate with Newton, or Descartes,
or the deductionists generally, or with Aristotle: not *that*
kind of science. More the kind of science we associate with
Nicolaus of Cusa, if we're familiar with his work, or with
Leonardo da Vinci, or Johannes Kepler, Blaise Pascal, or,
above all, Gottfried Wilhelm Leibniz. That kind of science.

That kind of science was created by Christianity. It really
didn't exist before Christianity, even though there was a por-
tent of it in some of the Greeks, especially through the work
of Socrates and Plato, and, to some degree, Archimedes (287–
12 B.C.), of course, as well.

Of course, the rudiments of science exist in many cultures.
We are much indebted, as a matter of practice, in Europe,
to contributions of other cultures, in this and many other
respects.

But, the idea of a science, a universal knowledge, of the
lawfulness of the universe, in a manner totally subject to
intelligible representation, as I've indicated: that is something
peculiar to Christian European civilization. And, it is peculiar
in its actual development, to what was founded as a scientific
method, during the fifteenth century, or so-called Golden Renaissance, particularly the influence of Cusa and others drawn
around Cusa in that period.

This is the essence of the power, the practical power, of
Christian civilization: its ability to foster productivity; because, mankind, in Christian civilization, is not a tradition-
ist, in terms of economy. Mankind does not accept being like
the brute beast, working in the field as his father and his
father's father before him.

Under Christian civilization, man must use that quality,

which places him in the image of the living God, or the living image of the living God. He must use his reason. His work must flow from reason, not from ox-like, repetitive toil. Not the work of a beast. He must innovate, constantly; and must innovate in a way which corresponds to reason, to lessening the imperfections of his work, to increasing the *power* of his work and the *power* in terms of benefits for mankind. *Power* in terms of benefits, as measured in the development of the minds of his children, and so forth and so on.

That drive for progress so defined, as being necessary to the work of the individual, as rejecting so-called traditionalist forms of labor, in favor of technological and scientific progress, that has a twofold impact on civilization. First of all, it creates the necessary preconditions for the development, to the full extent, of the moral potentialities of the character of the individual. And secondly, it provides the means for solving all of the problems we associate with material want and misery, insofar as these afflict society and lead to great evils.

We in European civilization, have thus acquired a great treasure, which, since it is a gift of the Creator, belongs not to us, but is entrusted to us, to our care, as the common property of all mankind. And, whether mankind in general is willing to come forth, to embrace Christianity on this account or not, makes a difference, but not a difference in this respect: that we hold that in trust. We hold that in trust for all mankind. Whoever knocks at our door, so to speak, and seeks that, must receive it, because it is not ours to withhold. It is only ours in trust, to bequeath. That is our power. And, that is precisely why, from an ecumenical standpoint, my proposition is a true one, that the Vedantist, the Jew, the Buddhist, and the Muslim, must join with us, in defense of Christian civilization, against that bestialist, satanic movement, the attempt to destroy Christianity and Christian civilization, with which, unfortunately, the British Royal Household's Prince Philip has lately associated himself.

Thank you.

| Physics and Natural Law

The characteristic flaws of modern classroom opinion on the subject of science are chiefly, that in virtually all instances, these opinions are *empiricist* and therefore *reductionist,* and in all but a dwindling minority of cases, frankly *radical-positivist.* For purposes of illustration, two summarized cases are chosen as examples here. The first is the fact that, although the formulation of Newtonian gravitation is a simply algebraic parody of Kepler's Third Law, Newtonian physics is subject on this point to a pervasive folly, called the "three-body" paradox.[1] The second is the reductionist's refusal to recognize the most crucial kind of importance of non-algebraic functions in shaping the internal history of mathematical physics.

As viewed today, the first mathematical physics, that of Johannes Kepler, is shaped, in effect, by the corollary of the "hereditary principle," that the existence of a single case of a theorem required by nature, requires a corresponding theorem to be implicit in whatever corresponds to an underlying, integral, indivisible set of axioms and postulates of science (mathematical physics) as a whole.[2] The relevant point permeates Kepler's three principal works,[3] and appears, as a central feature, in the "Snowflake" paper.[4] That is key to understanding the Newtonian blunder underlying the three-body problem, and addresses thus implicitly the objections which might arise to our use of the "book of nature." (See also Appendix V.)

As the success of Kepler's "missing, exploded" planet, illustrates the point (see Appendix VI), the Keplerian laws are derived from the construction in which available planetary orbits are not left undetermined. In the London Royal Society's algebraic manipulation of Kepler's Third Law to derive a formula for Newtonian gravitation, the "hereditary principle" of Keplerian physics is disregarded, thus generating the "three-body" paradox.

That "hereditary principle" is a fact, adduced by Leonardo da Vinci et al., that the universe includes living processes, which are harmonically ordered, in morphology of growth and function, in implicit harmony with the Golden Section. This is the characteristic feature of *action* in Keplerian mathematical physics, which defines action within the universe as reflecting a specific physical space-time curvature of the universe as a whole, in contrast to the simplistic error of Descartes, Newton et al., in presupposing a *linear* space-time-matter manifold.

The relevant, general import of this first example, is that *reason,* whose action here is centrally expressed as the corollary of the "hereditary principle," shows the mere existence of *a crucial phenomenon*—in this illustration, the Leonardo da Vinci-Kepler harmonic topology of all living processes—to be sufficient to define an aspect of universal natural law.[5] This example demonstrates the dangerous absurdity of Isaac Newton's famous motto, "I don't make hypotheses." To reject a hypothesis, is to reject all hypotheses but those embedded in one's own peculiar choice of blundering ignorance. Newton, for example, adopted his own policy in defiance of the available (e.g., Kepler's) evidence, and thus, as a result, stumbled into the "three-body" paradox.

The second example to be considered here, is the proper view of the discovery of *physical least-action* and isochronic qualities located within non-algebraic geometries, as two facets of the same, single concept. The discovery of these qualities by Christiaan Huygens, Leibniz, and Jean Bernoulli, was *explicitly* the foundation of Leibniz's calculus; the rejection of the cycloid-related, non-algebraic functions by the Cartesians and Newtonians is analogous to the nature and consequences of Newton's blundering into the "three-body" paradox.

Look at the physics of the *tautochrone* and *brachistochrone*, as Huygens, Leibniz, and Bernoulli viewed this successively.[6] Here is the germ of not only a large family of constructions based upon multiply-connected circular action, but these constructions are also the determination of a mathematical physics of Leibnizian *physical least-action*. Or to state the corollary point, these physical principles, which are common to all processes of a definitely curved physical space-time, are thus susceptible of a constructible, and therefore *intelligible representation*.

That said, turn attention to the central ontological issues of classical Greek and Hellenistic philosophy, from Pythagoras through Archimedes. As Plato and Archimedes, and Nicolaus of Cusa after Archimedes, exemplified this, the Platonic dialectic treatment of the universalist issues of ontology and form of process of knowing, is best fostered by situating the propositions to be examined in the context of such an effort to supply intelligible representation by means of a proper selection of *transfinite* "hereditary principle" of geometrical construction.

It is the case, that for physics as physics, the generation of the non-algebraic family and its functions is the most appropriate method of such intelligible representation.

Now, to close the circle, to sum up the point immediately at issue. We reject the notion, that the authority of the "book of nature" is extended to the reductionist view of physical science in general or to the modern positivist views in particular. Add the following important observation.

The common characteristic of the practice of law under Adolf Hitler and the U.S. federal courts today, is radical positivism in law. For Nazi Germany, the forerunners are Prof. Friedrich Carl von Savigny and Carl Schmitt. For fascist trends in the U.S. law-practice today, the authors are the British empiricists: Hobbes, Locke, Hume, Adam Smith, Jeremy Bentham, and John Stuart Mill, for example.

The Nazis cried, "All is permitted!" The Anglo-American liberal empiricists propose a global neo-Malthusian mass murder ("population control") vastly more extensive and savage than that of the Harriman-Hitler "eugenics" movement of the 1920s–40s.[7]

The argument to be made against the more obvious objections to our "book of nature" is summarized as follows.

The fundamental laws of our universe are embedded for human reason's knowledge in respect to the principles of hypothesis formation which we bring to observation of crucial empirical evidence. The use of crucial-experimental evidence to explore the question of the validity of the hypothesis-forming functions, is the universalizing aspect of rigorous scientific thinking. This is the classical method of such as, notably, Plato, Cusa, Leonardo da Vinci, and Leibniz, for example. The principle of hypothesis-formation, tested experimentally in this way, is of the form of a higher-order *Cantorian transfinite*. The higher reality under which that latter transfinite is subsumed, is itself a transfinite ordering-principle of yet a higher order. (See Appendix VII.)

By knowing these three levels—the immediate generation of hypothesis, the higher hypothesis, and the hypothesis of the higher hypothesis—we make each and all of these three directly the subject of our conscious reason. We know each of these levels *consciously,* by means of constructing a geometrical or analogous intelligible representation of each, and also of the relationship of each to each and all. To say that, is simply to supply a succinct description of the dialectical method of classical philosophy as Plato, Cusa, and Leibniz exemplify that practice.

In the chapters following the preface to *The Science of*

Christian Economy, the branch of physics which is the Leibnizian science of political-economy is presented in its essentials, from the standpoint of those features which are crucial in such a way as to bear more or less directly upon a constructible form of intelligible representation of *natural law*, upon the ecumenical content of the "book of nature."

Notes

NOTES TO | # In Defense of Common Sense

Foreword

1. The reference is to Friedrich Schiller's poem, "The Cranes of Ibykus" (Die Kraniche des Ibykus), William F. Wertz, Jr., trans., *Friedrich Schiller, Poet of Freedom,* (New York: New Benjamin Franklin House, 1985).

2. Plato, *Theaetetus,* Robin A.H. Waterford, trans., (New York, 1987).

3. "Constructive" or "synthetic" geometry signifies a system of intelligible representation of both abstract-geometric and physical domain by no other means than construction based originally and pervasively upon a single principle of universal action, e.g., Nicolaus of Cusa's "isoperimetric," "Maximum-Minimum" principle (*On Learned Ignorance*), and G. Leibniz's *principle of least-action.*

This is also the definition of a true "non-Euclidean geometry," one which excludes reliance upon a set of axioms and postulates, and prohibits any use of the methods of deductive logic, *except negatively,* throughout.

4. This is a central feature of the work of Nicolaus of Cusa, as carried from Cusa into the central features of the work of G. Leibniz. Nothing is known, unless its existence is shown to be *possible* and *necessary* from the standpoint of a *physical geometry* congruent with a *synthetic geometry.*

Such a constructive representative of a conception is otherwise best termed an *intelligible representation.*

Chapter IV

1. *Cf.* Lyndon H. LaRouche, Jr., *So, You Wish to Learn All About Economics,* (New York: New Benjamin Franklin House, 1984). Here, we need not consider the reasons *relative* population-density must be employed in the concrete professional-application practice of Physical Economy.

2. It is implicitly the case, that any form corresponding to a process existing in the universe, is *ultimately* susceptible of intelligible representation. Until a form is rendered intelligible by means of specification for construction of such a representation, the form is classed as *apparently arbitrary.* Cf. text, *infra.*

Chapter V

1. The use of the term *synthetic geometry* references Bernhard Riemann's geometry instructor, Professor Jacob Steiner. Because Riemann associated his own work explicitly with Professor Steiner's constructive method of *synthetic geometry,* the term *synthetic geometry* is the most meaningful usage today.

Chapter VI

1. Today, the term "non-Euclidean geometry" is popularly associated with what are properly termed "neo-Euclidean" geometries. These are Euclidean geometries, but with one or more postulates altered. In a true "non-Euclidean" geometry, there are no axioms or postulates and the use of deductive methods is not permitted, except for negative functions.

2. Cf. Leibniz-Newton-Clarke correspondence of 1715-1716.

3. John Von Neumann, *The Theory of Parlor Games,* 1928; and Von Neumann and Oskar Morgenstern, *Theory of Games and Economic Behavior,* (Princeton, 1944).

Chapter VII

1. Cf. Johannes Kepler, *The Six-Cornered Snowflake,* Colin Hardie, trans., with the Latin text on facing pages and essays by F.J. Mason and L.L. Whyte, (Oxford: Clarendon Press, 1966).

Chapter VIII

1. Cf. Georg Cantor, *Foundations of a General Theory of Manifolds* (Grundlagen einer Allgemeinen Mannigfaltigkeitslehre), Uwe Parpart, trans., *The Campaigner,* Vol. 9, No. 1-2, Jan.-Feb. 1976. What is stressed most immediately in the present author's contributions to Leibniz's science of Physical Economy, is the Cantorian (implicit) enumerability of the density of discontinuities lying within an arbitrarily small interval.

2. From late 1979 to the close of 1983, the international newsweekly *Executive Intelligence Review* produced a quarterly economic forecast based upon the *LaRouche-Riemann method.* This report was constructed quarterly from, primarily, a GNP-defined data-base, using a set of con-

straints supplied by this author. During this period, that was the only consistently reliable published forecast available from any U.S. source. This forecasting was discontinued during early 1988, at this author's recommendation. The margin of fakery in U.S. government and Federal Reserve System data rendered any report using such data worthless.

Chapter IX

1. Bernhard Riemann, *On the Hypotheses Which Underlie Geometry* (1853), inaugural dissertation, published 1854.

2. *Cf.* Riemann's posthumous notes on Herbart, *inter alia,* in the appendices of his collected works.

3. *Cf.* "Note on So-Called 'Non-Euclidean' Geometries," *Executive Intelligence Review,* Vol. 15, No. 21, May 20, 1988.

4. *Cf.* Eugenio Beltrami, *Hydrodynamic Considerations* (Considerazioni idrodinamiche), 1889.

5. Nearly 200 years before the first discovery of the existence of the asteroids, Kepler's solar astrophysics required both the prior existence and self-destruction of a planet in the asteroid-belt orbit between Mars and Jupiter. Kepler supplied the harmonic-orbital values for this. Carl Gauss showed, that if the asteroids had these Kepler harmonic values, that was conclusive proof that all of Kepler's opponents, including Galileo, Descartes, and Newton had been wrong, that the entirety of their physics was axiomatically wrong, relative to Kepler's. *Cf.* also Appendix VI.

6. An absurd, but popularized *musicologism* lists the famous anti-Romantic, classical composer Johannes Brahms, as in the same category, *Romantic,* as Brahms's adversary, Richard Wagner. The neo-Hegelian and kindred critics believe that the European *Zeitgeist* was converted to the *Romantic* aesthetic persuasion at the 1815 Congress of Vienna and persisted in that custom throughout most of the century. On such premises, classical composers such as Franz Schubert, Friedrich Chopin, Robert Schumann, and Brahms are listed as "Romantic," and, worse, often performed as if they were.

Chapter X

1. E. Beltrami, *The Mathematical Theory of Electrodynamic Solenoids* (La teoria matematica dei solenoidi elettrodinamici) (1871); *Hydrodynamic Considerations; Essay on Interpretation of Non-Euclidean geometry* (Saggio di interpretazione della geometria non-euclidea) (1868); for writings on negative curvature and on negative curvature of singularities in a Riemann surface.

2. *Cf.* Christiaan Huygens, *The Pendulum Clock* (1673), Richard J. Blackwell, trans., (Ames: 1986). *Oeuvres,* Paris, first published 1673.

Project A

Foreword

1. For an exposition on the differences in the law-giving of Lycurgus and Solon, *Cf.* Friedrich Schiller's "The Legislation of Lycurgus and Solon," George Gregory, trans., *Friedrich Schiller, Poet of Freedom*, Vol. II, (Washington, D.C.: Schiller Institute, 1988.)

Chapter I

1. Bostick's sequel, titled "How Superstrings Form the Basis of Nuclear Matter," was published in the Fall 1990 issue of *21st Century Science & Technology* magazine. The paper to which it is a sequel is titled, "The Plasmoid Construction of the Superstring: Morphology of the Photon, Electron, and Neutron," and will be published in the same location. Wells's paper appeared in *IEEE Transactions on Plasma Science*, 17:270 (April 1989). "Noisy Foam in Astrophysical Space" is a reference to research subsequently reported in *New Federalist* newspaper, June 22, 1990, p. 11 ("Is the Universe Cellular in the Large?").

2. The Schiller Institute sponsored a conference of the Martin Luther King, Jr. Human Rights Tribunal on "Democracy Movements and the Fight Against Judicial and Political Repression" on June 2, 1990. At that conference, held in Silver Spring, Maryland, three panels discussed the connection of human rights and natural law. Mr. LaRouche submitted a paper, titled "On the Subject of Human Rights and in Honor of the Late Martyr, the Reverend Martin Luther King, Jr." the text of which was adopted as a resolution by the 500 persons in attendance, founding an international Human Rights Coalition. That resolution was motivated for passage by American civil rights leader Amelia Boynton Robinson, honored in 1990 with the Martin Luther King, Jr. Foundation Freedom Award for her work with Dr. King in Selma, Alabama, to bring about civil and human rights for all Americans, during the civil rights movement of the 1960s.

Chapter IV

1. Cf. Plato, *Parmenides*.

Chapter VI

1. This part of the argument is presented a little early, prompted by a note from Khushro Ghandhi on Christiaan Huygens. Ghandhi mentions the connections between the principles of least-time and least-action—this isochronicity, by the way, has to be looked at a little more carefully—and between least-area (minimal surfaces) and least-perimeter. But here I will comment on his elaboration of the relations of cycloid, epicycloid, and hypocycloid as members of a single family, with the shared characteristic that in every case the involute is identical with the original figure.

Ghandhi proposes to relate the epicycloid to the cycloid by allowing the radius of the circle that does not roll to become infinitely large, such that its circumference constitutes a straight line. The essential thing here, which I have stressed all the way through, is what I've referred to, for pedagogical reasons, as the hereditary principle of a properly ordered constructive geometry; and, in this connection, I have located the ontological actuality of physical space-time, in respect to that hereditary principle, as the primary reflection of ontological reality. Thus, that which unifies all of these figures in a single, shall we say, virtually monotonic expression of this transfinite, this hereditary principle, is the referent for ontological actuality in physical space-time. That's the essential point.

What you're seeing with the circle and the relationship of the spiral to the circle, is the character of an envelope. What must not be forgotten, is that we're also seeing the way in which the discrete is defined, harmonically, by sections of the circle, or sections of circular action, or in respect to sections of circular action as we have, for example, in the case of the Golden Section and its significance. So the relationship of the circle, as an envelope for cycloids (which is what the epicycloid and hypocycloids represent), is the essential thing to be borne in mind in respect to defining the universe as based upon multiply-connected circular action, in respect to the hereditary principle.

Now, a straight line cannot be represented ontologically as a small portion of the perimeter of a very large circle relative to a unit circle. That is fallacious, because a straight line and a circle are ontologically two different things. That is, circular action, the circular perimeter, they're not the same thing: One, the essential definition of the straight line, is *without* curvature; and we have a very simple means, without curvature, because it's defined with respect to both negative curvature and positive curvature, two ways you can define a straight line passing through a circle; on the one side, internally, it is in respect to negative curvature; outside the circle, the same line extended is in respect to positive curvature. It's normal. It's not something that lies upon the perimeter sufficiently extended; it's normal to the perimeter, the perimetric action. It's quite different. So, we have to be careful about that. The straight line is something we derive by construction from multiply-connected circular action; and we can derive it in various ways from multiply-connected circular action, but they all amount to the same way, in the final analysis. The essential thing is, we must derive it together with the notion of a point, within any definition of circular action, within any particular transfinite ordering, to go through the corresponding elaboration of the specific geometry analogous to a constructive version of a Euclidean representation. We must develop this in order to make that particular phase construction at each, shall we say, point, in the transfinite series generated by hereditary action.

2. It is part of their right to have silence, to have some of these characters shut up, so they get a chance to read and think and concentrate, so that they don't have to listen to people yelling that stupid word "motherf–er" over, and over, and over, and over again, as if it were almost the only word in their vocabulary. A right to be free of that word, of hearing that word, is also a right, to put a fine point on the matter.

Chapter XVI

1. Max Planck, *A Survey of Physical Theory* (formerly titled *A Survey of Physics*), R. Jones and D.H. Williams, trans., (New York: Dover Publications, 1961).

Chapter XVII

1. LaRouche, *In Defense of Common Sense,* Foreword.

2. *Cf.* Plato, *Parmenides, passim.*

The Science of
Christian Economy

Preface

1. Pope Leo XIII, *Rerum Novarum (On the Condition of the Working Classes)*, N.C.W.C. translation, (Boston: Daughters of St. Paul, 1942).

2. *Ibid.*, p. 44*n*; St. Thomas Aquinas, *Summa Theologica*, (Chicago: University of Chicago Press, 1952), I-II, Q. 93, Art. 3 ad 2.

3. Pope Leo XIII, *op. cit.*, p. 7.

4. As early as 1975-76, this author had warned of the genocidal policies of the neo-Malthusian faction centered around Henry A. Kissinger and the faction he represented within the United States government. On Nov. 3, 1976, the author, in an election eve broadcast, warned of the genocidal intent of the Paddock Plan, which called for closing the Mexican borders and "let them scream," and the similar policies of George Ball.

Newly declassified National Security Council documents reveal that from 1974-77 Henry Kissinger and Brent Scowcroft (successive national security advisers) outlined a strategic plan to reduce the population of the Third World. The plan was forwarded to then CIA director George Bush, among others, for implementation.

The 1974 Kissinger-supervised National Security Study Memorandum 200 (NSSM 200), "Implications of Worldwide Population Growth for U.S. Security and Overseas Interests," argues that U.S. national security interests demand the imposition of population control or reduction on the LDCs—the lesser developed countries, otherwise termed the Third World. Thirteen of these states are defined as "key countries" in which there is "special U.S. political and strategic interest," which requires special emphasis. The primary reason these states are so defined, is that the effect of their population growth is judged likely to increase their relative political, economic, military, regional, and even world power. These key states are: India, Bangladesh, Pakistan, Nigeria, Mexico, Indonesia, Brazil, the Philippines,

Thailand, Egypt, Turkey, Ethiopia, and Colombia. The countries on this special target list, as well as the LDCs generally, are ones which the author and his associates have fought to defend against precisely those population control policies over the years.

Among Kissinger's biggest fears is that leaders of the lesser developed countries might realize that international population reduction programs are designed to undermine their development potential. As he puts it: "There is also the danger that some LDC leaders will see developed country pressures for family planning as a form of economic or racial imperialism; this could well create a serious backlash." He adds: "It is vital that the effort to develop and strengthen a commitment on the part of the LDC leaders not be seen by them as an industrialized country policy to keep their strength down or to reserve resources for use by 'rich' countries. Development of such a perception could create a serious backlash adverse to the cause of population stability."

Consequently, one of the major concerns of NSSM 200 is to check the spread of ideas which are hostile to population control and which demand economic development as the solution to Third World problems. According to Kissinger's definition, such LaRouche-associated ideas are a threat to U.S. national security.

To highlight the dangerous growth of such ideas, the document presents the case of the World Population Conference in Bucharest in August 1974, where Helga Zepp (now Mrs. LaRouche) intervened to denounce the Club of Rome's population control policies and John D. Rockefeller III in particular. The document complains that the conference's proposed World Population Plan of Action was rejected by many of these states, because of the spread of such anti-Malthusian ideas. The failure of the conference is one of the cited reasons for the drafting of the NSC memorandum.

Referring to this conference, the document states: "There was general consternation, therefore, when at the beginning of the conference, the Plan was subjected to a slashing, five-pronged attack led by Algeria, with the backing of several African countries; Argentina, supported by Uruguay, Brazil, Peru and more limitedly, some other Latin American countries, the Eastern European group (less Romania); the P.R.C.; and the Holy See."

Kissinger reports that the objections to the Plan were based on the idea, that a "New World Economic Order" could be a basis for social and economic development of the former colonial sector. Related NSC memoranda from the period define the "wishful thinking that economic development will solve the problem" generated by supposed overpopulation, as the thinking necessary to eradicate. The reference is clearly to the Zepp intervention at Bucharest and the author's influence more broadly.

Kissinger outlines various formulations to counter these ideas. For example: "The U.S. can help to minimize charges of an imperialist motivation behind its support of population activities by repeatedly asserting that such support derives from a concern with: (a) the right of the individual to determine freely and responsibly their number and spacing of children . . . and (b) the fundamental social and economic development of poor countries."

On Nov. 26, 1975, Brent Scowcroft (who had succeeded Kissinger as national security adviser, while Kissinger remained as secretary of state) issued National Security Decision Memorandum 314, which endorsed NSSM 200, making it the official, if covert, policy of the Ford administration.

In May 1976, the NSC released its "First Annual Report on U.S. International Population Policy," which examined the progress made over the previous year in implementing Kissinger's memorandum. The classified report was forwarded to then-Director of Central Intelligence George Bush for implementation.

Among the findings of the report was, that it was difficult to implement population reduction in Third World states without the appropriate form of draconian government: "Prerequisites for real success are likely to involve three approaches that are interrelated and have proved highly effective, as follows: 1) strong direction from the top; 2) developing community or 'peer' pressures from below. . . .

"With regard to (1), population programs have been particularly successful where leaders have made their positions clear, unequivocal, and public, while maintaining discipline down the line from national to village levels, marshaling government workers (including police and military), doctors, and motivators to see that population policies are well administered and executed. Such direction is the *sine qua non* of an effective program. In some cases, strong direction has involved incentives such as payment to acceptors for sterilizations, or disincentives such as giving low priorities in the allocation of housing and schooling to larger families."

Although relevant population policy documents from the subsequent Carter, Reagan, and Bush administrations remain classified, information in the public domain indicates that the approach outlined in the 1974-77 NSC memoranda remains U.S. government policy.

5. *Cf.* Nicolaus of Cusa, *On the Peace of Faith* (De Pace Fidei), William F. Wertz, Jr., trans., *EIR*, Vol. 18, No. 1, Jan. 4, 1991, pp. 19-39; or Jasper Hopkins, trans., (Minneapolis: A.J. Banning, 1990), pp. 33-70; for the Latin text, *cf. Nicholas of Cusa, Philosophical-Theological Writings, Vol. 3,* (Vienna: John Herder & Co., 1982).

6. *Cf.* Bal Gangadhar Tilak, *The Orion, or Researches into the Antiquity of the Vedas,* (Poona, 1893; repr. Poona: Tilak Brothers, 1972); and *The Arctic Home in the Vedas, Being Also a New Key to the Interpretation of Many Vedic Texts and Legends,* (Poona, 1903; repr. Poona: Tilak Brothers, 1956).

7. Plato distinguishes between the hypothesis of the higher hypothesis and the Good. The hypothesis of the higher hypothesis is typified by the kinds of transfinite orderings referenced within the chapters of the following text. This is the *Becoming,* a *Becoming* which is a notion of a transfinite ordering of *changes* moving toward increasing perfection or decreasing imperfection. The Good, by contrast, is the state of perfection. It efficiently is the *changeless* idea of perfection which governs the process of change in the direction of increasing perfection or lessening imperfection. For one reading Plato, for example, we can say that the Good has the ontological quality of being, as distinct from the quality of *Becoming.*

8. Luke 19:40.

9. *Cf.* LaRouche, *In Defense of Common Sense,* Chapters III-V.

10. Nancy Spannaus and Christopher White, *The Political Economy of the American Revolution,* (New York: Campaigner Publications, 1977), p. 380.

11. *Cf.* Mathew Carey, *Autobiographical Sketches in a Series of Letters Addressed to a Friend Containing a View of the Rise in Progress of the American System, The Efforts Made to Secure its Establishment, The Causes Which Prevented its Complete Success,* (Philadelphia: John Clark, 1829); and Henry C. Carey, *Principles of Political Economy,* Reprints of Economic Classics, (New York: Augustus M. Kelley, 1965); and *The Harmony of Interests: Agricultural Manufactures and Commercial,* (Philadelphia: J.S. Skinner, 1851, reprinted by A.M. Kelley, 1967).

For Friedrich List, *Cf. Outlines of National Economy* (1827), *National System of Political Economy* (1837) (New York: Augustus M. Kelley, 1966). Also: Prof. Dr. Eugen Wendler, *Friedrich List—Politische Wirkungsgeschichte des Vordenkers der Europäischen Integration,* (Germany: Oldenburg-Verlag, 1989). Although this history defends the idea of a single European Market in 1992, Dr. Wendler's work was written to give new stimulus to the discovery of List's life and his ideas of national economy.

12. *Cf.* H. Graham Lowry, *How the Nation Was Won, America's Untold Story, Vol. I, 1630-1754,* (Washington, D.C.: Executive Intelligence Review, 1987), Chapters 5-8.

13. Pope John Paul II, *Sollicitudo Rei Socialis* (On Social Concern), (Boston: Daughters of St. Paul, 1987).

Introduction

1. *Cf.* Lyndon H. LaRouche, Jr., "U.S. Policy toward the Reunification of Germany," *EIR,* Vol. 15, No. 40, Oct. 21, 1988, p. 40, for the full text of LaRouche's Oct. 12, 1988 speech on German reunification; reprinted in *EIR,* Vol. 17, No. 38, Oct. 5, 1990, p. 23.

2. President George Washington commissioned his treasury secretary, Alexander Hamilton, to organize these founding policies for the U.S. government: a Bank of the United States, to counter private usury and to secure credit for private industrial and agricultural investment; national sponsorship for the development of manufacturing, through tariffs, patents, and other means; federal investment in canals, roads, harbors, and other infrastructure. This general program, personally identified with both Washington and Hamilton, was implemented by the nationalist political faction up through the Lincoln and Grant administrations, and accounts for the industrialization of the U.S.A.

3. Irish Catholic rebel leader Mathew Carey (1760-1839), a protégé of Benjamin Franklin in Paris, continued Franklin's humanist political tradition in Philadelphia after the Revolution. Publishing scores of pamphlets, Mathew Carey attacked Adam Smith and the British Empire, revived Hamilton's policies, and instructed Henry Clay and the second generation of American nationalists.

Mathew's son Henry Carey (1793-1879), the most widely known American economist of the nineteenth century, led the Philadelphia nationalist circle which built the main American railroads, iron, steel, and machine industries. Abraham Lincoln learned his economics from Henry Carey's books, as did many republican patriots in Latin America, Europe, and Asia.

4. Henry Clay (1777-1852) was the public spokesman for the "American System" of political economy, and for national sovereignty, in explicit opposition to imperialism. Clay created the pro-defense movement, that rescued the U.S. from British aggression in 1812. Leading the Whig Party against the Southern slavocracy and the Northern Anglophile merchants, he pushed protective tariffs through Congress, which transformed the economy. Abraham Lincoln served as Clay's political lieutenant in Illinois before becoming U.S. President.

5. The Philadelphia-based nationalist American leadership adopted exiled German republican leader Friedrich List (1789-1846) as their tutor in economic science, principally through List's text, *The National System of Political Economy*. After helping form the Carey-Clay party, List returned to Germany, pioneered the railroads, and created the customs union (Zollverein), which led to German national unity and industrial greatness.

6. The concept of just war, principally developed by St. Augustine (354-430), can be summarized as follows: 1) Wrongs are preferably redressed by patiently submitting to the loss of temporal advantage, to produce amendment of the ways of the wicked and to thus overcome evil with good; 2) war should be conducted only as a last resort, since it were better to stay war with a word and to procure peace through peace, than to slay men with the sword and to achieve peace through war; 3) however, the legitimate sovereign of a state has a natural right to conduct war to defend the common weal; 4) such a war requires a just cause in the form of some injury inflicted and the failure to make amends; and 5) in the event of a just cause, war, to

remain just, must have a right intention and be waged mercifully, in the spirit of a peacemaker.

Before the American patriots launched their War of Independence against their evil adversary, King George III and his liberal backers, they sent a number of missions to Great Britain to attempt to reach a compromise with the stubborn monarch and his advisers. It was only after King George III and the British government refused to ameliorate the economic policies which were strangling the American colonies, that the war was fought.

7. *Cf.* Lowry, *op. cit.* By the early eighteenth century, the same Leibniz, sometimes called "the last universal intellect" of history, was at the center of not only the world's leading scientific circles, but was also the principal intellectual figure of Europe-based, global political movements. Among those were included the circles of Cotton Mather, Benjamin Franklin et al., in the future United States. Franklin's far-flung conspiratorial networks of the period from approximately 1766 onward, for example, were principally Franklin's intersection with the still active networks of Leibniz's followers in Europe.

8. The ongoing war against the United States included the 1776-81 War of Independence, the War of 1812, the role of the Duke of Wellington in directing Mexican forces in the war with Mexico, and also the Civil War of 1860-65, which was being orchestrated by Britain, chiefly, with the intent of dividing the United States into several quarreling baronies, which might then be taken over by Britain.

9. Lowry, *op. cit.*

Chapter I

1. *Cf.* Lowry, *op. cit.*, Chapters 4 and 5, and *passim*.

Chapter III

1. *Cf.* "Powered Flight to Mars in Less than Two Days," Heinz Horeis, *EIR,* Vol. 14, No. 12, March 20, 1987, p. 18.

For the case that human space travel occurs at a constant acceleration/deceleration of one Earth gravity in a fusion process using helium-3, a spaceship could carry enough fuel for a round trip approximately as far as the asteroid belt. If we use the matter/anti-matter reaction, it gives a fuel-to-work ratio such that the same ship would be able to go three orders of magnitude greater distance, approximately, which would carry us to the extremes of the solar system in a round trip.

Chapter IV

1. The term "technological attrition" refers to the depreciation or devaluation through relative or marginal obsolescence, of tools, equipment, and so forth, rendered less competitive in quality through being superseded by more technologically advanced means. This is associated with a relative lowering of the value of labor using the older equipment, relative to labor using the new.

2. Consulting the original manuscripts in the Hanover Leibniz Archive establishes not only that Leibniz had completed the work leading to a submission of the first published discovery of the differential calculus, submitted to a Paris publisher in 1676, but that, at that time, prior to that date of publication, he had also made many more advanced discoveries in this connection, discoveries which were attributed ordinarily to decades later in time than Leibniz by others.

This is to be compared with the examination of the newly discovered papers of Newton, during the course of the present, twentieth century, in which it is discovered, that Newton had done no significant work toward any calculus, but had instead concentrated most of his laboratory and related activity on experiments in black magic. *Cf.* Carol White, "Refuting the Second Law," *Fusion,* Vol. 8, No. 1, Jan.-Feb. 1986, p. 63.

3. The hypothetical structure of the atomic nucleus, as developed by the late Dr. Robert J. Moon, professor emeritus at the University of Chicago and veteran of the Manhattan Project, is presented in Laurence Hecht, "The Geometric Basis for the Periodic Table of the Elements," (*21st Century Science & Technology,* Vol. 1, No. 2, May-June 1988). Moon's model is explicitly derived from Keplerian considerations of the structure of space-time and the necessity for expression of the Golden Section, or "Divine Proportion," as he always referred to it.

4. The popularity of both the textbook and of the textbook-based classroom course has tended to distract modern opinion's attention away from the fact, that earlier, prior to the development of the textbook, a superior form of education had been used, in which the student had to re-work original experiments with the guidance of original literary sources, and thus relive as closely as possible the mental experience of the original discovery.

Chapter V

1. Classical philosophy divides beauty between two related forms, "natural," and "artistic." Natural beauty refers to harmonic compositions which are congruent with forms characteristic of healthy living processes as distinct from non-living ones. These natural forms all have harmonic orderings, ultimately congruent with the Golden Section of geometry. (*Cf.* Appendix III.) All such living forms, which have this harmonic ordering,

are physically (thermodynamically) negentropic. The mere imitation of nature is not considered artistic beauty. Artistic beauty consists in creating forms which are products of those creative aspects of the human mental processes otherwise responsible for valid, fundamental discoveries in physical science. The further requirement of artistic beauty is that it be coherent with, although distinct from, natural beauty.

2. Herein, the definition of *creative* used is that supplied, for example, *inter alia,* in the author's *In Defense of Common Sense,* and *Project A.*

Chapter VI

1. G.W. Leibniz, *Monadology,* George Montgomery, trans., (Peru, Ill.: Open Court Publishing Company, 1989), pp. 251-72.

2. In this portion of the step-wise presentation of the relevant conceptions, we are stressing the absence of any linear commensurability of the units of per capita potential population-density among different qualities of potential population-density. It is nevertheless the case, that the numerical value of potential population-density rises with realized technological progress. *Cf.* LaRouche, *In Defense of Common Sense,* Chapter II.

3. *Cf.* LaRouche, *Project A.*

4. Henry C. Carey, *op. cit.*

5. For example, a simple indifference curve can be constructed, comparing the net cost to the economy of moving coal by ship, or inland-canal barge, leafy vegetables by rail or truck, and transistors by air freight over long distances.

Chapter VII

1. For Leibniz's writing on Physical Economy, *Cf.* Gottfried W. Leibniz, "Society and Economy," pp. 12-13, in *EIR,* Vol. 18, No. 1, Jan. 4, 1991. *Cf.* also "On an Academy of Arts and Sciences," *Leibniz Selections,* Philip P. Wiener, ed., (New York: Charles Scribner's Sons, 1951), pp. 594-99.

2. *Cf.* LaRouche, *In Defense of Common Sense,* Chapters III-V.

3. To leave no reasonable margin for scholarly objections from relevant mathematical specialists, the issues represented by the two cited corollaries can also be represented in different terms, differently posed, but equivalent to what we have just said above. We summarize this alternative representation of the paradox of formalism and then leave the reader to resume the argument in progress.

No formal, deductive system of argument or thought could ever escape two devastating formal problems: *ambiguity* and lack of *completeness.* For the purpose of recognizing this twofold formal problem, it is important to adopt the older sort of conventional distinction between *axioms* and *postulates.* In other words, the basis for the entire deductive theorem-lattice system is located within the *integralness of a set of axioms,* and the postulates are other arbitrary, unproven assumptions, which address the problems of ambiguity and completeness inhering in the set of axioms. This should forewarn us, that no deductive mathematics could ever support a durable form of valid mathematical physics.

The author's preference for the two corollaries he supplies above, over the more traditional emphasis on paradoxes of *ambiguity* and *completeness,* is prompted by the fact that the problems of *ambiguity* and *completeness* are not containable within the arbitrary domain of mathematical formalism. They are, in the first instance, reflections of *physics:* It is the physics of such relatively crucial matters of non-algebraic functions as isochronism, harmonic orderings, least-action, and so forth, which forces the overthrow of a deficient, analytical formalism, in favor of a constructive-geometric generation of functions cohering with such crucial physical actualities. In the second degree, physics itself is incompetently defined as subject-matter, as long as the deluded quest for a "non-subjective" form of physical knowledge is tolerated. The two corollaries given above, thus appear a more efficient way of representing these issues of formalism.

4. The distinction, *crucial* experiment, references most directly Riemann's habilitation thesis published in 1854, which in English is "On the Hypotheses Which Underlie Geometry."

5. *Cf.* LaRouche, *In Defense of Common Sense,* Chapters III-V.

6. Ultimately, *being* is a quality which resides in the *Good,* not the Becoming. The Good is *the changeless being of universal change,* the former One, the latter the generation of the *manifoldness* of the universe. *Being* is thus the residence of the highest which is higher than the transfinite ordering of change. The constant Good is this constant cause of change: *matter* is thus generated (created) continuously by the Good (being).

7. One example of this is the eruption of the "three-body" paradox, caused by the Newtonian parody of Kepler's three laws as the Newtonian Gravitation = $(km_1m_2)/r^2$. The paradox, which does not exist in the Keplerian original, is introduced by substituting mechanistic, pair-wise interaction-at-a-distance for the unified Keplerian harmonic ordering principle. (*Cf.* also Appendix V.)

8. The derivation of Newton's inverse square law of gravitation from Kepler's Third Law is as follows: Kepler's Third Law states, that the cube of the mean radius (a) for any planet, divided by the period (T) of the planet squared equals a constant (k).

(1) Kepler's Third Law:

$$a^3/T^2 = k$$

(2) The speed or velocity of a planet in terms of the radius of the circular orbit (assume circular rather than the almost circular elliptical orbit of the planets) and the period is:

$$v = 2\pi r/T$$

where v = speed of planet;
       r = the radius of a circular orbit;
       T = period time for one revolution.

(3) Christiaan Huygens (1629–1695) had shown centrifugal acceleration to be:

$$A = v^2/r$$

where A = acceleration.

(4) Since $F = mA$, $F = mv^2/r$.

where F = force;
       m = mass;
       A = acceleration.

From equation (2):

$$v=2\pi r/T \text{ thus } v^2=4\pi^2 r^2/T^2.$$

From equation (1):

$$a^3/T^2=k \text{ or } T^2=a^3/k.$$

Since the orbits of the planets are nearly circular, assume $a=r$, so

$$T^2 = r^3/k.$$

By substituting this value for T^2 into the value for v^2—i.e.,

$$v^2=4\pi^2 r^2/T^2$$

the result is

$$v^2=4k\pi^2/r.$$

Now take this value for v^2 and substitute it in the value for $F=mv^2/r$. The result of this final substitution gives you the inverse square relationship in Newton's law of gravitation. So

$$F=4\pi^2 km/r^2.$$

The preceding derivation shows how the inverse square laws and Newton's law for Universal Gravitation can be derived from Kepler's Third Law, assuming the orbits of the planets are circular, rather than the nearly circular ellipses that they are. It is generally accepted that Kepler's Second Law of constant real velocity had indicated to Newton, that whatever forces were acting on the planets were directed toward the Sun instead of tangentially to their paths. (Also *cf.* Appendix V.)

Furthermore, Kepler had proposed this proportionality of masses and a force relationship in his 1609 *Astronomia Nova*. Kepler writes, "If two stones were removed to any part of the world, near each other but outside the field of force of a third related body, then the two stones, like two magnetic bodies, would come together at some intermediate place, each approaching the other through a distance proportional to the mass [moles] of the other." E. Hoppe claims the concept of mass, not for Newton, but for Kepler, who designates it by the word *moles*.

9. In the summer of 1801 the great mathematician Carl Friedrich Gauss became acquainted with the astronomical discovery of the small planet Ceres. After Ceres, coming too close to the Sun, became invisible, Gauss developed new methods for calculating the orbit of Ceres. Gauss applied his new methods to the discoveries of other small planets, Pallas in 1802, and Vesta in 1807. Gauss, recognizing that there are no two-body problems in the solar system, but n-body problems where other planets attract a given planet and perturb the elliptical orbit, applied his mathematical genius to the problems. In 1818, Gauss published a paper on the theory of perturbations. Gauss determined the distribution of mass on the circumference of the ellipse by assuming a distribution of the mass according to Kepler's Second Law, the law of constant real velocity: Equal amounts of mass will be distributed on that length of the ellipse that requires equal times.

Pair-wise interactions and point masses—Newton's method—did not and could not predict the existence of Ceres and Pallas, as Kepler's method did. For Newton, mass is primary and the size of the two masses determines the orbit of the planet. For Kepler, the mass is determined by the orbit and the orbit is determined by the curvature of physical space-time. For Kepler musical harmonies and the uniqueness of the five Platonic solids were the keys to determining why certain orbits were permissible, and others not. From these conceptions, Kepler developed his three planetary laws. God's universe flows from a principle of sufficient reason, which has manifested itself in the universe through the musical harmonies and the Platonic solids. From this standpoint, Kepler developed his three laws by exploring least-action/least-time/isoperimetrical qualities that the Creator has built into His creation. Kepler's solutions work for the multi-body problem presented by our planetary system. (*Cf.* Appendices V and VI.)

Newton's point-mass/pair-wise interaction approach falls apart as soon as the three-body problem appears. It also falls apart if we think of the simple problems presented by modern spectroscopy. Electrons revolving around a nucleus of an element in the gaseous state emit and absorb light at definite frequencies, which are characteristic of the element making up the gas. This is how spectroscopy can identify the gaseous elements of which planets are composed. There are many possible orbits around a nucleus. In some orbits electrons circle faster than in others. Why do electrons only circle in those orbits which have the assigned frequencies, and why do orbits only have an assigned frequency? Furthermore, the collisions of atoms in a gas are occurring at 10^{12} times per second, which creates a large amount of heat energy. The impacts are powerful and should change the orbits of the electron completely, in respect to size, shape, and frequency, were Newton's force prescriptions to apply. This does not occur, because the orbits are determined by a more fundamental process, the curvature of physical space-time, and not some simplistic notion of pair-wise interactions of point masses.

10. The fact that 90-odd elements were known to exist in our solar system seemed impossible, if the elements had been generated by the kind of simple thermonuclear fusion which was usually thought to have occurred within our Sun, and that process was thought to have been the source of the material for these planets with 90-odd elements. However, if we assign the process of fusion generally, less to the interior of the Sun, and rather mostly to polarized-fusion occurring in the indicated plasma envelope around the Sun, under the indicated early conditions, 90-odd elements are implicitly accounted for by action within our solar system.

11. Adam Smith, *Theory of the Moral Sentiments* (1759), (Glasgow: Liberty Classics, 1984). Man, according to this Calvinist's argument, is not morally responsible for the consequences of his actions for humanity in general. If his blind indifferentism to morality, in following nothing but his hedonistic impulses, causes cruelty and other great harm to large numbers of humanity, then God is to be blamed for having provided such a Calvinist with his hedonistic instincts.

12. The Smith family biography documents transactions between the Second Earl of Shelburne and Adam Smith, during a 1763 carriage journey. Shelburne coopted Smith as his personal agent and instructed Smith on the axioms of a program for destroying the economy and semi-autonomous governments of the English colonies in North America. To train Smith for this activity, he was sent by David Hume to Switzerland and France for education in political-economy by such Swiss bankers' assets as Quesnay and the circles of Voltaire. The content of Smith's anti-American *The Wealth of Nations,* chiefly a plagiarism of A.M. Turgot, reflects the anti-Colbert physiocratic elements of his indoctrination by Hume's cronies in Switzerland and France.

Cf. also Lyndon LaRouche and David P. Goldman, *The Ugly Truth About Milton Friedman,* (New York: New Benjamin Franklin House, 1980); and *EIR Special Report* "The Trilateral Conspiracy Against the U.S. Constitution: Fact or Fiction?" Sept. 30, 1985.

13. Adam Smith's *The Wealth of Nations* was considered by many (including eighteenth-century economist Pierre du Pont de Nemours) to be a rehash of French physiocrat Turgot's *Reflections on the Formation and Distribution of Wealth.* However, "Everything added by Smith is inaccurate," stated Du Pont de Nemours.

14. When Karl Marx was offered free access to the British Museum, its director, David Urquhart, fed convenient documentation to Karl Marx. Urquhart is best described as the "St. John Philby of his day." Carol White, *The New Dark Ages Conspiracy,* (New York: New Benjamin Franklin House, 1980), pp. 326-27.

15. During the eighteenth century, the influence of Leibniz's economic science was strong in many parts of Europe and spilled into circles around Benjamin Franklin in America. Over the period from 1791 through about

1830, Leibniz's economic science became identified worldwide as the *American System of political-economy*. This name was coined by U.S. Treasury Secretary Alexander Hamilton, in 1791, within a U.S. policy document submitted to Congress entitled, "A Report on the Subject of Manufactures."

16. Friedrich Schiller, "The Legislation of Lycurgus and Solon," George Gregory, trans., *Friedrich Schiller, Poet of Freedom, Vol. II*, (Washington: The Schiller Institute, 1988), pp. 273-307.

17. Cf. St. John's "Book of Revelation," or "the Apocalypse," in the Bible. If the Apocalypse is read in terms of the concrete realities of the century when it was written, there is no part of that book which is either allegorical or symbolic. The same kinds of forces which St. John identified as arrayed in support of the "Whore of Babylon" then, are the concrete forces of Evil in the world of today.

The personality of Evil is clearly and concretely identified as "the Whore of Babylon." This is no symbolism; it is the name of a very specific mother-goddess, whose priestesses practiced prostitution as part of religious ritual, to such effect that the names of Ishtar, Astarte, Isis, and Venus, are venerated as the goddesses of the lesbian's and the whore's professions into modern times. The source of these whore-goddesses' cults, in Mesopotamia, in Sheba-Ethiopia, in Egypt, in Palestine, and among the Phrygians, is the worship of the whore-goddess Shakti by the ancient "Harappan" culture of the Indian subcontinent, introduced to Mesopotamia through the "Harappan" colony at Sumer. The Satan-figures of ancient Mesopotamia, Sheba in Ethiopia, and of Osiris, Apollo, and Lucifer, are, like the Phrygian Dionysos, derivatives of the "Harappan" Siva. The most powerful form of this satanic cult then, was the Syrian magicians' cult of Mithra, which had been established as the leading cult of the Roman imperial legions, through an agreement reached between Augustus and Syrian magicians at the Isle of Capri.

In the Apocalypse, St. John attacks the question: To what consequence must the war between Good and Evil ultimately lead? Evil's persistence must bring the very existence of mankind into jeopardy, through such features as wars and pestilences. Evil must reach such a state, that it prizes its gains in power so much, that it would prefer to allow humanity to be destroyed, rather than compromise the policies promoting such apocalyptic destruction. The men and women who then adopt the cause of Evil and support its policies, by that adoption adorn themselves with the designating mark of the beast, and seek to exterminate the men and women who are resistant to the policies of Evil.

This conjunction of the struggle between Good and Evil must emerge, because the essence of Evil leads it to no other result than this one. Evil cannot possibly be a permanent condition within the human species. The increase of power at the disposal of Evil, will, by itself, cause Evil to reach the point that it becomes, immediately, the cause of threatened extermination of the human species. That is Armageddon and the Apocalypse.

18. Contrary to that slovenly pack of British "biblical archaeology" gophers, who treated the tablet collections which they uncovered in the most immorally reckless fashion, civilization first came to Mesopotamia very late in ancient times, probably first introduced by the "black-headed people" (i.e., Dravidians) who established their maritime colonies at Sumer, Shebe, Ethiopia, and (according to Herodotus) Canaan. Furthermore, the practice of lunacy in ancient Mesopotamia shows the early civilized Semites of that region to have risen to a much lower cultural level than the Vedic peoples of Central Asia had reached, those responsible for the solar astronomical calendars referenced by Tilak from the Orion Period interval of 6,000-4,000 B.C.

19. For the axiomatically pantheistic feature of British liberalism, *cf.* White, *The New Dark Ages Conspiracy, op. cit.,* Chapter 8, "The Roots of British Radicalism," and *passim.*

20. For the truth about the sponsorship of Hitler and Mussolini, *cf.* Carol White, *The New Dark Ages Conspiracy, op. cit. Cf.* also Schiller Institute, *The Hitler Book,* Helga Zepp-LaRouche, ed., (New York: New Benjamin Franklin House, 1984); and *EIR Special Report* "Project Democracy: The 'Parallel Government' behind the Iran-Contra Affair," April 1987.

21. On the racial purification dogma of the Nazi Party and the agreement of Averell Harriman with that dogma, *cf.* Anton Chaitkin, *Treason in America, From Aaron Burr to Averell Harriman,* second edition, (New York: New Benjamin Franklin House, 1984), Chapter 19.

22. *Cf.* White, *The New Dark Ages Conspiracy, op. cit.,* Chapters 1-3 and *passim.*

23. Allen and Rachel Douglas, "The Roots of the Trust," unpublished MS., 1987. The standard accounts of pre-1917 Russian history, in which a dreaded Czarist secret police (from 1826 the Third Section, then after 1881 the Okhrana) struggled heroically against bands of "proletarian" terrorists to defend the Czar and state, are entirely mythological. The Okhrana was controlled by 102 highly intermarried Russian noble families, known historically as the *boyars,* whose bitter opposition to the Czar and Russian state dated back to the establishment of a Western-style state by Peter the Great in the early eighteenth century. These families deployed the terrorist bands, such as the Socialist Revolutionaries and the Bolsheviks, virtually all of whose leaders (e.g., Stalin) were Okhrana agents, to overthrow that state. The *Raskolniki* ("Old Believers"), a seventeenth-century splitoff of the Russian Orthodox Church, were also fanatically opposed to Peter's Westernizing reforms, and in the late nineteenth and early twentieth centuries, financed the Bolsheviks.

The noble families and *Raskolniki* collaborated with Western oligarchical interests, first to overthrow Westernizing institutions in Russia and then, after 1917, to use the Soviet state as a battering ram against Western civilization more generally. A Western command center for these efforts was the Equitable Life Assurance building at 120 Broadway in Lower Manhattan, dominated by the Harriman and J.P. Morgan interests and home of the notorious Sidney Reilly. Reilly was an Okhrana agent pre-1917, a Soviet intelligence agent afterwards, and an agent of the British Secret Intelligence Service throughout.

24. *Cf.* LaRouche, *In Defense of Common Sense*, Chapter III.

25. The nuclear family is obviously one of those essential institutions.

26. William Manchester, *The Arms of Krupp, 1587-1968*, (Boston: Little, Brown and Company, 1967), p. 455 and p. 639 make this point. This is contrary to the usual nonsense in the U.S., that the U.S. built up German industrial potential after the war.

27. "Lunatic" in this case refers to those societies which base themselves on the lunar calendar, as opposed to the solar calendar.

28. Karl Marx, *Capital*, Vol. III (New York: International Publishers, 1984).

29. Adam Smith, *Theory of the Moral Sentiments, op. cit.*

30. *Cf.* White, *The New Dark Ages Conspiracy, op. cit.*

31. Leroy E. Loemker, editor, *Gottfried Wilhelm Leibniz Philosophical Papers and Letters, Vol. II,* "The Controversy between Leibniz and Clarke," (Chicago: University of Chicago Press, 1956), pp. 1095-1169.

32. Philo, *On the Creation*, (Cambridge: Harvard University Press, 1981), pp. 134-137.

33. The Emperor Constantine's reforms, "legalizing" Christianity under the traditional authority of the emperor as "pope" (*Pontifex Maximus*) of all legalized religious bodies, reflected the failure of the Roman emperors since Nero and Tiberius, to crush Christianity by crude gestapo methods of mass-murder, and reflected most emphatically the strength of Apostolic Christianity among heirs of Plato's tradition within the Greek-speaking population. This "Constantinian reform" meant: "Let them worship the name of Christ with as much devotion to that name as they may choose; we will control what they believe about Christ." Bishops appointed by the emperors of the Eastern Empire (Byzantium), used their authority in matters of doctrine and liturgy, to introduce into Christianity the doctrines of the "Roman mystery religions," and even the priests were assigned to wear costumes of the Ptolemaic cult of Isis. This evil practice is often termed, euphemistically, "syncretism"—the fusion of Christianity with elements of pagan cults, like the Jesuits' "Liberation Theology" and "Christian-Marxist dialogue" of today. The practice is better described as "gnosticism," the transformation of Christian doctrine by saturating taught doctrine with the cult-beliefs of the Roman "mystery" cults, "gnosis."

The gnostics degrade persons, from creatures in the image of the living God, to children of the soil, creatures of immediate and original hedonistic instincts. Man exists, therefore, for the pleasure of other men, the ruled for the pleasure of those who rule, and the people of one race or nation for the pleasure of the rulers of another.

34. The calculus of Leibniz consists of analytical methods for the solution of problems about curves, using variable geometric quantities as they occur in such problems. The starting point of curves for Leibniz can be seen in his theory of envelopes, where curves are viewed as a locus of tangents. Leibniz's "characteristic triangle," which he uses in the transformation of quadratures, came out of his study of Pascal's work on the cycloid. The characteristic triangle, generated by ordinate, tangent, and sub-tangent, or ordinate, normal, and sub-normal, applied generally, gave Leibniz the ability to find relations between quadratures of curves and other quantities, such as moments and centers of gravity. The importance of the involute-evolute relationships in the theory of envelopes, together with the study of the cycloid and caustics, placed the non-algebraic higher curves at the center of the calculus.

Karl Bernoulli's (1834-1878) method of integrals used the "inverse method of tangents," where a curve is determined from a given property of its tangents. Bernoulli teaches, that the property of the tangent has to be expressed as a differential equation. The method of integrals applied to this differential equation will yield the curve itself. So once again, the curves are seen from the standpoint of the theory of envelopes.

Bernoulli applied himself to arc length and quadrature problems involving caustics, cycloids, the catenary, logarithmic spirals, and the form of sails blown by the wind. The brachistochrone-tautochrone properties of the cycloid made it rich in least-action, least-time qualities of self-organization, as did the other higher curves, which made them the appropriate foundation for examining the calculus. They combined geometrical and physical principles. (*Cf.* H.J.M. Bos, "From the Calculus to Set Theory 1630-1910," in *Newton, Leibniz, and the Leibnizian Tradition,* I. Grattan-Guinness, ed., [Duckworth: London, 1980.])

Huygens, in exploring the isochronic property of the cycloid, and the fact that the evolute of a cycloid is another cycloid, discovered that he could design a pendulum clock that wrapped around sheaths in the shape of a cycloid, which would be perfectly isochronic and therefore keep accurate time.

The proof that the path of quickest descent is the cycloid, was a *tour de force* for the Leibniz-Huygens-Bernoulli faction against the Newtonians and Cartesians. Jean Bernoulli's solution combined three different areas—the motion of light, the laws of free fall, and the mechanical laws for a rolling circle. By looking at the laws of refraction for light, and shining a light through a changing medium, Bernoulli was able to come up with a curve. Since light takes the fastest possible path of optical time, and since light changes its speed as it travels through media of varying density, Bernoulli changed the speed by varying the density according to the laws of free fall. Each of these light rays, changing direction as the media changes its density, is tangent to a curve. The curve is the envelope of these tangents. The curve that Bernoulli gets is the cycloid. The use of light traveling through a non-homogeneous medium demonstrates, that gravitational pathways do not have to be determined by an innate quality of mass, but, in fact, can be a reflection of the curvature of the physical universe, which defines least-action pathways throughout nature.

Chapter VIII

1. Percy Bysshe Shelley, "In Defence of Poetry," in *Shelley: Political Writings*, Roland A. Duerksen, ed. (New York: Crofts Classics, 1970), pp. 164-97.

2. It is to be stressed, that Grotius and John Locke represent typically a standpoint wholly antagonistic to the Christian conception of natural law.

3. Charles de Gaulle, *Memoirs of Hope: Renewal and Endeavor,* (New York: Simon and Schuster, 1971), p. 269.

In 1970, Charles de Gaulle wrote: "Thus, from every part of the world, people's attentions and preoccupations were now directed towards us. At the same time, on the Continent, the initiatives and actions that might lead towards unity emanated from us: Franco-German solidarity, the plan for an exclusively European grouping of the Six, the beginnings of cooperation with the Soviet Union. Besides this, when the peace of the world was at stake, it was to our country that the leaders of East and West came to thrash things out. Our independence responded not only to the aspirations and the self-respect of our own people, but also to what the whole world expected of us. From France, it brought with it powerful reasons for pride and at the same time a heavy burden of obligations. But is that not her destiny? For me, it offered the attraction, and also the strain, of an onerous responsibility. But what else was I there for?"

4. As elaborated by Nicolaus of Cusa in 1433 in the Latin treatise, *De Concordantia Catholica.* An English edition is in preparation for 1992 publication: *The Catholic Concordance,* by Nicholas of Cusa, Paul E. Sigmund, trans. and ed., (Cambridge: Cambridge University Press).

5. Other forms of music are "language," but more or less brutish or brutalized degrees of musical illiteracy.

6. For Plato's references on constructive geometry, cf. "Plato's *Timaeus:* The Basis of Modern Science," *The Campaigner,* Vol. 13., No. 1, February 1980; or, for Greek with English translation, cf. Rev. R.G. Bury, ed., (Cambridge: Harvard University Press, 1966) and *Meno,* W.R.M. Lamb, ed. (Cambridge, Mass.: Harvard University Press, 1966).

7. Cf. LaRouche, *In Defense of Common Sense,* Chapters II and III.

8. The concept *Brotgelehrte* (bread-fed scholar) is developed in Friedrich Schiller, "What Is, and To What End Do We Study, Universal History?" Caroline Stephan and Robert Trout, trans., *Friedrich Schiller, Poet of Freedom, Volume II, op. cit.,* pp. 253-272.

9. *Cf.* Carol White and Carol Cleary, *EIR Special Report,* "The Libertarian Conspiracy To Destroy America's Schools," April 30, 1986. *Cf.* also Herbert Kohl, *Basic Skills: A Plan For Your Child, A Program for All Children,* (Boston: Little, Brown and Company, 1982) and E.D. Hirsch, Jr., *Cultural Literacy: What Every American Needs to Know* (Boston: Houghton Mifflin Co., 1987).

10. One among the proud founders of the Malthusian Club of Rome, former director of the Organization for Economic Cooperation and Development (OECD), Dr. Alexander King, provides a real-life example. Dr. King volunteered that his motive had been to rid the world of what he considered an excessive number of darker-skinned races. Bertrand Russell, like King, revealed his racialist motives in books he wrote and caused to be published himself. Russell, like King, was spiritually a follower of the racialism of Cecil Rhodes and Charles Dilke.

In essence, King agreed with Russell's 1921 statement in *Problems of China,* that "the less prolific races will have to defend themselves against the more prolific by methods which are disgusting even if they are necessary."

The motive of the "sincere Malthusians," according to their own repeatedly stated account of the matter, is the practice for which we hanged Nazis at Nuremberg.

11. *Cf. EIR,* Vol. 8, No. 25, June 23, 1981, "Club of Rome Founder Alexander King Discusses His Goals and Operations." On May 26, 1981, in an interview with *EIR,* Dr. Alexander King, Commander of the British Empire and of the Order of St. Michael and St. George, who in 1968 was the director general for the Scientific Affairs Section of the OECD, an apparatus considered a subordinate feature of NATO but which is actually its policy controller, described the role of his office in helping to create the New Math, and shift students' focus away from problem-solving and into a more "practical" approach.

"We invented the whole question of curriculum reform, trying to teach mathematics and chemistry, etc., in new ways," said Dr. King. "We were very much criticized for this. The ministries of education were all culturally based. Education was something that passed down the riches of posterity to new generations, in their view. To tie education to the economic wagon seemed terrible."

12. Sol H. Pelavin and Michael Kane, *Changing the Odds: Factors Increasing Access to College,* (New York: College Entrance Examination Board, 1990). The study indicates, that black and Hispanic students who take at least one year of high school geometry vastly improve their chances of getting into college and receiving a bachelor's degree. The study of almost

160,000 students found that the gaps between college-going rates of whites and minorities virtually disappeared among those who had taken a year or more of geometry. Author Sol Pelavin commented in the Sept. 24, 1990 *Washington Post,* "I think we're looking at something that is more basic than those other courses," and attributed the findings to the "logical-thinking skills taught in algebra and geometry."

13. God is a far more capable mathematician than such as the late Professors Norbert Wiener and John Von Neumann.

14. Winston Bostick, "The Pinch Effect Revisited," in *EIR,* Feb. 8, 15, and 22, 1991, Vol. 18, Nos. 6, 7, and 8; (reprinted from *International Journal of Fusion Energy,* Vol. 1, No. 1, March 1977).

15. Arthur R. Jensen et al., "Environment, Heredity and Intelligence," reprint from *Harvard Education Review* (Cambridge, Mass.: Reprint Series No. 2).

16. In 1939, while working at Bell Telephone Laboratories, William Shockley began to study semiconductors as amplifiers. That work led eventually to the development of the transistor. Between 1942 and 1945, he did antisubmarine research. For their investigations on semiconductors and the discovery of the transistor effect, Shockley, J. Bardeen, and W.H. Brattain shared the 1956 Nobel Prize.

17. Then-Congressman George Bush invited William Shockley and his co-thinker, Arthur Jensen, to testify about their contention, that blacks are genetically inferior to whites, before the Republican Task Force on Earth Resources and Population, on Aug. 5, 1969. In a statement published in the Sept. 5, 1969 *Congressional Record,* Bush reported on Shockley and Jensen's testimony, noting that the Aug. 5 hearings had focused on "the hereditary aspects of human quality" and "the environmental problems created by our rapid rate of population growth." Summarizing the testimony, Bush said: "Dr. Shockley stated that he feels the National Academy of Sciences has an intellectual obligation to make a clear and relevant presentation of the facts about hereditary aspects of human quality. Furthermore, he claimed our well-intentioned social welfare programs may be unwittingly producing a down-breeding of the quality of the U.S. population." During his congressional career (1967-70), Bush was in the vanguard of the drive to institutionalize population control as a key component of U.S. domestic and foreign policy, and personally sponsored the most important initial "family-planning" measures, including the Family Planning Services and Population Research Act of 1970, which sought to reduce the number of people on welfare by funneling taxpayers' money into Planned Parenthood clinics in poor areas.

18. Cabalism is a form of Jewish mysticism and occultism first brought over into Christian culture by the Renaissance scholar Giovanni Pico della Mirandola, who adopted and propagated the belief, that the Old Testament scriptures would disclose deep secrets if interpreted according to the Jewish cabala.

The almost shamanistic spirit of cabalism is more or less accurately reflected in the English word *cabal,* derived from *cabala,* which entered the English language before 1650, meaning "a secret or private intrigue of a sinister character formed by a small body of persons" *(Oxford English Dictionary).*

Some of the prominent Englishmen involved in cabalism in the sixteenth and seventeenth centuries were Robert Fludd (1574-1637), physician, mystic, and Rosicrucian, who entered into controversy with Kepler; Henry More (1614-87), theologian, leader of the so-called "Cambridge Platonists," who twice refused appointment as a bishop; Elias Ashmole (1617-92), antiquary and astrologer, who authored or edited Rosicrucian works, and whose collection of curiosities is preserved in the Ashmolean Museum at Oxford University; and Sir Isaac Newton (1642-1727), according to "Newton and the Wisdom of the Ancients" by Piyo Rattansi in *Let Newton Be!* (*Cf.* note 19 below). Further clues to the employment of cabalism as a medium of oligarchic thought can be gleaned from *The Discovery of Hebrew in Tudor England—A Third Language,* by G. Lloyd Jones (Manchester, U.K.: University of Manchester Press, 1983). Some of the specifics of cabalistic numerology are explained in *The Most Ancient Testimony—Sixteenth-Century Christian-Hebraica in the Age of Renaissance Nostalgia,* by Jerome Friedman (Athens, Ohio: Ohio University Press, 1983), Chapter 4.

19. John Maynard Keynes, the economist, identified Newton as "the last of the magicians, the last of the Babylonians and Sumerians," whose alchemy was "wholly devoid of scientific value." Keynes had purchased at auction a chest of Newton's papers and reported on their contents in "Newton the Man" in the Royal Society's *Newton Tercentenary Celebrations* (Cambridge: Cambridge University Press, 1947), pp. 27-34. It had been hoped by Newton's admirers, that the chest would disclose evidence that Newton actually developed the calculus. This hope was dashed and Keynes was instead shocked by the mumbo jumbo he found there. A new assessment of Newton in light of his obsession with magic and alchemy is *Let Newton Be!,* John Fauvel et al., eds., (New York: Oxford University Press, 1988). Unlike Keynes, the authors are not shocked by Newton's occult interests, and argue the thesis—as familiar as it is false—that science emerges from magic.

20. Sir Isaac Newton, in his *The Mathematical Principles of Natural Philosophy,* (New York: The New York Philosophical Society, 1964), stated, *"hypotheses non fingo"* (I don't make hypotheses), and explained his reasons for this on grounds of *induction* versus hypothesis.

Newton wrote, in part: "In the preceding books I have laid down the principles of philosophy; principles not philosophical, but mathematical. . . . It remains that, from the same principles, I now demonstrate the frame of the System of the World. . . . For since the qualities of bodies are only known to us by experiments, we are to hold for universal all such as are not liable to diminution, can never be quite taken away. We are certainly not to relinquish the evidence of experiments for the sake of dreams and vain fictions of our own devising; nor are we to recede from the analogy of Nature, which uses to be simple, and always consonant to itself. We no other way know the extension of bodies than by our senses, nor do these reach it in all bodies; but because we perceive extension in all that are sensible, therefore we ascribe it universally to all others also. That abundance of bodies are hard, we learn by experience; and because the hardness of the whole arises from the hardness of the parts, we therefore justly infer the hardness of the undivided particles not only of the bodies we feel but of all others. That all bodies are impenetrable, we gather not from reason, but from sensation."

21. The writings of the late Bertrand Russell are models of one Oxbridgean style of laying on the rhetorical "lard." Witness Russell's success in recruiting so many avid admirers among those Indians and other "Third World" intellectuals of nations which Russell plainly proposed virtually to exterminate, by means of famine and fostered epidemic disease.

22. There is more than a hint of *Les Bougres*—the Cathars-Bogomils, of the Manichean, and perhaps Templar Baphomet worshippers, too—in Cartesian formalism's gnosticism on the subject of matters relating to this topic of *deus ex machina*.

23. Lowry, *op. cit.*, Chapter 4.

24. Hume's reported insanity was the reason for his family throwing him out of Scotland, for the sake of appearances before the neighbors, into France, from whence he returned with the first version of his book.

25. Adam Smith, *op. cit.*, and Adam Smith, *The Wealth of Nations: An Inquiry into the Nature and Causes,* Edwin Cannan, ed., (Chicago: University of Chicago Press, 1977). *Cf.* also Goldman and LaRouche, *op. cit.*

26. Adam Smith, *Theory of Moral Sentiments, op. cit.*

27. Jeremy Bentham, *The Works of Jeremy Bentham,* John Bowering, ed. (Edinburgh: William Tait, 1843).

28. *Cf.* LaRouche, *In Defense of Common Sense,* Chapter III.

29. *Cf.* Chapter VI of *The Science of Christian Economy*.

30. For a fuller discussion of the strategic implications of the Diocletian decrees, *Cf. EIR Special Report* "Global Showdown: The Russian Imperial War Plan for 1988," July 24, 1985.

Diocletian's reforms created an oriental despotism of the most pervasive type, in which all aspects of life were most minutely controlled by the state. This was most evident in economic matters. The Codex Theodosianus of Roman and Byzantine law documents the obligation of every citizen to provide compulsory public service in the guild or corporation in which his father served. This was a class society, in which class status was inherited and enforced by administrative sanctions: No one was allowed to change his station or way of making a living. At the same time, the practice of each corporation or guild was rigidly fixed, also by imperial decree, according to "ancient custom." The affairs of shipmasters, breadmakers, charioteers, cattle and swine shepherds, limeburners, wood transporters, and others were prescribed in adamant detail. This amounted in practice to an outlawing of any form of technological innovation, which would have interfered with the stability of the guilds and the value of their property, which could not be transferred or otherwise changed.

The case of George Gemisthos (Plethon)'s economic-policy counsel to the Paleologue dynasty highlights the point, that the early fifteenth-century, onrushing doom of dwindled Byzantium, reflected accumulated centuries of the de facto Malthusian "decay," echoing the earlier demographic collapse of Rome and the West, and echoing also the "socialist, Malthusian" characteristics of Diocletian's code.

31. For a fuller discussion of the implications of Skull and Bones' "old boy" network for U.S. policy-making, *cf. American Leviathan: Administrative Fascism under the Bush Regime,* (Executive Intelligence Review Nachrichtenagentur GmbH, Wiesbaden, Germany, 1990).

The political power associated with Yale is associated with the infamous secret freemasonic lodge called Skull and Bones or the Russell Trust. Among the 15 graduating seniors "tapped" each year for Skull and Bones, we find such key Establishment figures as Col. Henry Stimson, a member of the Republican administrations of the 1920s, and later selected by Franklin D. Roosevelt as secretary of war in the bipartisan national unity cabinet that waged World War II. We find Averell Harriman; several Tafts, including William Howard, the man who became U.S. President in 1908; and former national security adviser, architect of the Vietnam War, Stimson biographer, and former chief Establishment spokesman, McGeorge Bundy, of the Lowell clan of Boston. It is clear that Skull and Bones constitutes one of the most important avenues of advancement toward positions of power in the State Department and, after 1947, in the Central Intelligence Agency. The rituals and ceremonies of Skull and Bones remain secret, although it is well established that they involve the use of human remains.

Skull and Bones has recently fallen on hard times due to its "males-only" policy. In 1991, the club was suspended *by its own board of alumni* for a year, rather than admit women into its ranks, which it subsequently agreed to do.

32. *Cf. American Leviathan, op. cit.* The Population Crisis Committee/ Draper Fund believes that population growth, particularly of non-white races, is a national security issue for the United States, and has promoted "population war," or the use of warfare to reduce population in the developing sector, as a national policy of the United States. Both William Draper, Jr. and William Draper III have had long "public service" careers and their

policies have been promoted by George Bush since his first years as a congressman.

33. Wags may say, this may account for tendencies for sodomy among some British social strata.

34. Loemker, *op. cit.,* pp. 1095-1169

35. *Cf.* Alfred O'Rahilly, *Electromagnetic Theory, A Critical Examination of Fundamentals, Vols. I and II* (New York: Dover Publications, 1965), republished from the original 1938 title, *Electromagnetics,* for documentation of Maxwell's falsifications with regard to the Weber-Gauss-Riemann electrodynamics and Ampère's famous experiments (pp. 110-13, for example).

A more recent work detailing Maxwell's falsifications in this regard and reviewing experimental evidence which demonstrates this is Peter Graneau's *Ampère-Neumann Electrodynamics of Metals,* (Nonantum, Mass.: Hadronic Press, Inc., 1985). Possible major implications of this Maxwell falsification, in terms of frontier scientific work, is exemplified by the recent, controversial "cold fusion" experiments as seen, for example, in the recent paper, "Nuclear Energy Release in Metals," by F.J. Mayer and J.R. Reitz, *Fusion Technology,* Vol. 19, May 1991, pp. 552-57, with the report of the formation of virtual neutrons through the condensation of electrons on protons. According to the Maxwell falsification, condensation of electrons onto protons to form virtual neutrons (hydrons) is impossible, while from the standpoint of the Ampère-Weber-Gauss electrodynamics, and according to the detailed calculations of the late Dr. Robert J. Moon of the University of Chicago, it is possible.

36. *Cf.* White, *The New Dark Ages Conspiracy, op. cit.,* pp. 206-7.

37. *Cf.* "Plato's *Timaeus:* The Basis of Modern Science," *The Campaigner,* Vol. 13, No. 1, February 1980.

38. We hear of the Bogomils for the first time in the tenth century A.D. in Bulgaria. In Bulgarian, *Bogomil* means "beloved of God," and it may be that their founder took this name. Among their beliefs is the characteristically gnostic one, that the Father of Jesus Christ was not the creator of the world. For the Bogomils and later the Cathars, the power of the devil worked through the nature and constraints of the material world. Since God the Father, it was believed, could not have created such an evil instrument (the world, that is), it was logical to suppose that the devil (Satanael) not only frustrated the intentions of God the Father, but had constructed the stage of the world for that very purpose. It was indeed a wicked world. To be bound to the world, then, was evil, and the realization of the source of evil, coupled with the fervent desire to extricate oneself from it by virtuous practice in a religion of love and goodness, was salvation. One was redeemed to Heaven by knowledge of the Good God. In short, matter and spirit were never meant to cohabit. This division and its corresponding principles of good and evil, light and darkness, is broadly called dualism—

the doctrine of two opposing principles between which Man is pulled. *Cf.* also Tobias Churton, *The Gnostics* (London: George Weidenfeld and Nicolson Ltd., 1987).

The cult was known in France as the Bulgarian cult, or "Les Bougres," which translated into English as "the Buggers." Because of the cult's peculiar sexual pervision—that is, the belief that a man putting semen into a woman to impregnate her, was propagating the flesh, and that was evil—it resorted to various other kinds of sexual activity and thus the name "Bugger" became associated in English with homosexuality.

What the Bogomils and their followers, the Rosicrucians and empiricists, did, in separating the human spirit from those things which involve the human flesh, led directly to the doctrine of the Enlightenment—the separation of *Naturwissenschaft* and *Geisteswissenschaft*.

Although Catharism spread across southern France and northern Italy, it was especially prevalent in Languedoc, to the extent that the condemnation of heretics by the Council, held in the town of Albi in 1176, led to their being generally known as Albigensians. The heresy had its roots in much older religious movements, but no precise date can be assigned to its first appearance in Languedoc; its end, however, was another matter. In 1244 Catharism and all it stood for came to a violent and catastrophic end with the fall of Montsegur. On March 16, 1244, more than 200 Cathar "Perfects"—heretics in the eyes of the Catholic Church—were taken from the castle of Montsegur in the foothills of the Pyrenees and burned alive in the fields below.

Cf. also Walter Birks and R. A. Gilbert, *The Treasure of Montsegur: A Study of the Cathar Heresy and the Nature of the Cathar Secret,* (The Aquarian Press, 1987.)

Both Cathars and Albigensians were basically followers of the religion of Manicheanism, which began in Bulgaria and found its way into northern Italy and the southern part of France. Their chief was Manes. He was born about the year A.D. 216 and was crucified and flayed alive by the Persian magi under Bahram I in the year A.D. 277. His Persian name was Shuraik. *Cf.* Lady Queenborough (Edith Starr Miller), *Occult Theocracy*, (California: The Christian Book Club of America, 1933). Attracted in his youth to the Manichean cult, St. Augustine condemned it after his conversion to Christianity in A.D. 386.

39. This is a pun on the names of East Germany's former dictatorship. Erich Honecker (Honi) is the former East German chairman of the ruling Socialist Unity Party (SED), who is now in exile in the Soviet Union. Gen. Erich Mielke is the former Minister for State Security in the SED regime, and, as such, head of the feared Stasi (secret police).

40. Lyndon H. LaRouche, Jr. "Presidential Campaign Paper Number 5: Military Policy of the LaRouche Administration," published in *New Solidarity*, Aug. 18, 1979.

In February 1982, at a two-day conference sponsored by *Executive Intelligence Review,* this author proposed that the United States and Russia agree, that each would proceed with the most rapid possible development of space-based relativistic beam weapons capable of destroying the proverbial 99 percent of all nuclear-armed ballistic missiles in flight; and further agree that such weapons would be employed as part of a policy commitment to thus destroy nuclear weapons fired anywhere in the world by any nation. "EIR Conference Bursts Intelligence Myths," *EIR,* Vol. 9, No. 9, March 9, 1982. *Cf.* also Lyndon H. LaRouche, Jr., "Only Beam Weapons Could Bring to an End the Kissingerian Age of Mutual Thermonuclear Terror," Policy Discussion Memorandum, (National Democratic Policy Committee, 1982).

41. For the Soviet rejection of President Ronald Reagan's March 23, 1983 proposal to make "nuclear weapons impotent and obsolete" through a U.S.-Soviet sharing of beam defense technologies, *Cf.* "World Council of Churches Conclave: A First-hand Report," and "The Two Military Faces of Yuri Andropov," *EIR,* Vol. 10, No. 33, Aug. 30, 1983; "Beam-Weapons Strategy Relaunched at Erice Conference"; "The Soviet Union Threatens Pre-emptive Nuclear War"; and "Open Letter to Yuri Andropov: You Have Chosen to Plunge the World into War," *EIR,* Vol. 10, No. 35, Sept. 13, 1983.

The final rejection of President Reagan's offer came, of course, in the form of the shooting down of the civilian plane KAL-007 by the Soviets on Sept. 1, 1983. *Cf.* "Moscow Goes on a Global Rampage," and "U.S. Policy toward Moscow after the KAL Incident," in *EIR,* Vol. 10, No. 36, Sept. 20, 1983.

42. On April 9, 1977, Maj. Gen. George J. Keegan, Jr., speaking under the auspices of the American Security Council, gave his honest professional assessment of the present strategic situation: "The Soviets on a war-winning philosophy . . . are 20 years ahead of the United States in its development of a technology which they believe will soon neutralize the ballistic missile weapon. . . . They are now testing this technology.

"The intelligence community was consistently wrong in its estimate of the development of broad-based Soviet science," Keegan continued. "When people talk about technological superiority in this country, they are talking about potential and futures that have not yet been bought and paid for, distributed and manufactured and deployed to our forces. . . . I object to the failure to observe the normal checks and balances, of letting the public know, letting the leaders know, letting the press know, and letting the full range of uncertainties be in the open—lest we make the kind of mistakes that have gotten us into every war this country has ever been in."

Cf. Aviation Week, March 28, 1977 and *New Solidarity,* April 12, 1977, "Air Force General Admits: Soviet Technology '20 Years Ahead of U.S.' " In the fall of that year, LaRouche commissioned the publication of a report from the Fusion Energy Foundation, "Sputnik of the 70s: The Science Behind the Soviets' Beam Weapon."

Cf. also White, *The New Dark Ages Conspiracy, op. cit.,* Chapter 2; and Lyndon H. LaRouche, Jr., "Only Beam Weapons Could Bring to an End the Kissingerian Age of Mutual Thermonuclear Terror"; "The LaRouche Doctrine: Draft Memorandum of Agreement between the United States and the U.S.S.R.," *EIR,* Vol. 11, No. 15, April 17, 1984; and *EIR Special Report* "Global Showdown," July 24, 1985.

43. For a list of the relevant works by Bertrand Russell, *Cf.* White, *The New Dark Ages Conspiracy,* pp. 365-390, and *EIR Special Report* "The Trilateral Conspiracy Against the Constitution: Fact or Fiction?" 1985.

44. In October 1946, Bertrand Russell, father of the so-called peace movement, wrote an article in the *Bulletin of the Atomic Scientists,* advocating the creation of a totalitarian world government "to preserve peace":

"When I speak of an international government, I mean one that really governs, not an amiable façade like the League of Nations or a pretentious sham like the United Nations under its present constitution. An international government . . . must have the only atomic bombs, the only plant for producing them, the only air force, the only battleships, and, generally, whatever is necessary to make it irresistible. . . .

"The monopoly of armed force is the most necessary attribute of the international government, but it will, of course, have to exercise various governmental functions . . . to decide all disputes between different nations, and will have to possess the right to revise treaties. It will have to be bound by its constitution to intervene by force of arms against any nation that refuses to submit to arbitration."

45. Russell, in an article titled "Humanity's Last Chance" (*Cavalcade,* Oct. 20, 1945), called for the creation of a world confederation under American tutelage, and in sole possession of nuclear weapons. The Soviet Union would be offered a place in the confederation, but "if the U.S.S.R. did not give way and join the confederation . . . the conditions for a justified war would be fulfilled. A *casus belli* would not be difficult to find." *Cf.* also White, *The New Dark Ages Conspiracy, op. cit.,* pp. 72-3.

46. The "fulcrum" used to establish the Pugwash Conference as a "back-channel" for negotiations, designed by British and Soviet agencies involved to rope influential U.S. accomplices into complicity, was the World Association of Parliamentarians for World Government, or WAPWG.

In response to persisting offers from Russell and Leo Szilard, four official Soviet delegates were sent to the 1955 London conference of WAPWG. This event set into motion the Fabians' launching of the Pugwash Conference series and the adoption of Russell's proposed nuclear deterrence agreements by the New York Council on Foreign Relations, the launching-point for Kissinger's career in diplomacy.

Cf. also Lyndon H. LaRouche, Jr., "How Kissinger Tricked President Nixon on Soviet Beam Weapons," and Lex Talionis, "The Pugwash Papers: Kissinger Imperiled U.S. National Security: Suppressed Evidence on Soviet E-beam Program," *EIR,* Vol. 10, No. 22, June 7, 1983.

47. For Dr. Leo Szilard's proposed arms control arrangements preparatory to world-federalist government at the second, Quebec Pugwash Conference of 1958, *cf. EIR Special Report* "Global Showdown," Appendix, "Leo Szilard's 'Pax Russo-Americana.' "

48. For the text of Henry Kissinger's May 10, 1982 address, titled, "Reflections on a Partnership: British and American Attitudes to Postwar Foreign Policy," before the Royal Institute of International Affairs, *cf. EIR,* June 1, 1982, Vol. 9, No. 21.

49. As Kissinger bragged later, in his May 10, 1982 Chatham House address, during his time in the Nixon and Ford administrations, Kissinger was in fact operating often behind the President's back, as an agent of influence of the British foreign intelligence establishment.

In that May 10 address, Kissinger said, "The ease and informality of the Anglo-American partnership has been a source of wonder—and no little resentment—to third countries. Our postwar diplomatic history is littered with Anglo-American 'arrangements' and 'understandings,' sometimes on crucial issues, never put into formal documents. . . . The British were so matter-of-factly helpful that they become a participant in internal American deliberations, to a degree probably never before practiced between sovereign nations. In my period in office, the British played a seminal part in certain American bilateral negotiations with the Soviet Union—indeed, they helped draft the key document. In my White House incarnation then, I kept the British Foreign Office better informed and more closely engaged than I did the American State Department—a practice which, with all affection for things British, I would not recommend be made permanent. But it was symptomatic. . . . In my negotiations over Rhodesia I worked from a British draft with British spelling even when I did not fully grasp the distinction between a working paper and a Cabinet-approved document."

50. The fictional "Dr. Strangelove," played by Peter Sellers in the famous film, was modeled principally on Szilard's address to the second Pugwash Conference of 1958.

51. Henry A. Kissinger, *A World Restored: Metternich, Castlereagh and the Problems of Peace, 1812-1822,* (Boston: Houghton Mifflin, 1973).

Stanza II of "The Masque of Anarchy: Written on the Occasion of the Massacre at Manchester," reads:

"I met Murder on the way—
He had a mask like Castlereagh.
Very smooth he looked, yet grim;
Seven blood-hounds followed him;"

Top Shelley Poetical Works, Thomas Hutchinson, ed., (London: Oxford University Press, 1970).

52. Henry A. Kissinger, *Nuclear Weapons and Foreign Policy,* Philip Quigg, ed., (New York: W.W. Norton & Co., 1969).

53. The two translations of the first edition are: *Military Strategy,* first edition, with an introduction by Raymond L. Garthoff (New York: Praeger, 1963; London, Pall Mall Press, 1963); and *Soviet Military Strategy,* first edition, trans. and with an analytical introduction, annotations, and supplementary material by Herbert S. Dinerstein, Leon Gouré, and Thomas W. Wolfe, (Englewood Cliffs, N.J.: Prentice-Hall, 1963).

Soviet Military Strategy, third edition, V.D. Sokolovskii, ed.; trans., ed., and with an analysis and commentary by Harriet Fast Scott (Moscow: 1968; Stanford: Stanford Research Institute, 1975), p. 298.

Whereas the first edition had contained numerous references to beam-related weapons, the third edition deleted all such references, which may explain why the Soviets delayed making the third edition publicly available by as much as 16 months. At that time, there were ongoing efforts by the United States to have defensive missile systems included in any future arms reduction talks. Moscow, most probably, had received assurances from its allies among the U.S. presidential advisory community, that the White House was hooked on the fraud of the ABM Treaty and would not be informed of Soviet efforts in the field of directed-beam weapons systems.

54. *Ibid.*

55. "EIR Conference Bursts Intelligence Myths," *EIR,* Vol. 9, No. 9, March 9, 1982.

56. *Ibid., cf.* also *A Program For America,* (The LaRouche Democratic Campaign, 1985), p. 130.

57. Lyndon H. LaRouche, Jr., "The Difference between LaRouche's and Teller's Role in Creating SDI," *EIR,* Vol. 13, No. 38, Dec. 5, 1986.

58. In that April 24, 1983 interview in *Der Spiegel,* Andropov's first widely publicized interview with a Western publication, then-Soviet Communist Party General Secretary Yuri Andropov reiterated his full-scale rejection of defensive beam weapons.

59. Proposed in 1982 were four successive upgradings of a global strategic ballistic missile defense, the deployment of each separated from the other by an estimated three to five years. For a summary of this proposal, *cf.* "How Beam Weapons Would Spur Recovery," in *EIR,* Dec. 28, 1982, Vol. 9, No. 50; and Lyndon H. LaRouche, Jr., *The Power of Reason: 1988* (Washington: Executive Intelligence Review, 1987), pp. 239-240. For a summary of the potential "spill-over effects" of this proposed program, *cf. EIR Quarterly Economic Report, The Recovery That Never Was,* April 15, 1985.

Mark I, estimated at 1982 dollars $200 billion, would be the use of systems based upon new physical principles to provide a margin of strategic defense, acting, in effect, as enhanced strategic deterrence without increasing the "hair trigger" factor; *Mark II* would be the deployment of supplementing elements of strategic defense, developed at the same rate of investment as Mark I; then *Mark III;* then *Mark IV. Mark IV,* deployed about the end of the twentieth century or slightly later, would be a full-blown global strategic defense. The "payback," via the federal tax-revenue base's increase, from economic "spill-overs" into the civilian sector, should hold the total cost of Mark I-IV to not more than the initial 1982 dollars $200 billion outlay or investment.

60. An "SDI" based upon "kinetic-energy systems," such as the Lt. Gen. Daniel Graham's proposed "High Frontier," is not a workable system, physically or economically.

61. *E.g.,* a proposal for a Paris to Vladivostok railway.

62. "Negative" is used here in the sense "negation" is central to Kant's dialectic of "practical reason" (as in the second part of his *Critique of Practical Reason*). This Kantian negativity of the term "peace" is rightly projected also upon all uses of the term, such as "peace agreements," which are consistent with the term "social contract."

63. The reference to "Tavistockian" is to British Intelligence's psychological warfare section's London Tavistock Clinic and Tavistock Institute. The clinic, which was founded and built up in the pre-World War II decade, under leadership of Brig. Gen. Dr. John Rawlings Rees, Dr. Eric Trist et al., is among the principal coordinating centers for "New Age" attacks upon Christian civilization, especially since the 1963 launching of mass recruiting for the drug-sex-rock and neo-Malthusian counterculture inside the United States of America. "Cultural paradigm-shift" was used among such professional social-planners' circles to describe inducing of deep changes in belief, induced in populations, to the purpose of shifting apparently "instinctive" popular values, away from a Christian, to a Dionysian world-outlook of practice.

64. Anton Chaitkin, *Treason in America,* second edition (New York: New Benjamin Franklin House, 1984), Part II, "The True Story of the Civil War."

65. Konstantin George, "The U.S.-Russian Entente That Saved the Union," *The Campaigner,* No. 2, 1978, pp. 5-33.

66. *Ibid.*

67. Chaitkin, *Treason in America, op. cit.,* pp. 256-59, and Paul Kreingold, "Grant and Mexico: When the U.S. Had a Republican Military Policy," March 23, 1990, *New Federalist* newspaper.

68. Allen Salisbury, *The Civil War and the American System: America's Battle with Britain, 1860-1876,* (New York: Campaigner Publications, 1978), pp. 247-51.

69. For a full account of the shift in French foreign policy, *cf.* White, *The New Dark Ages Conspiracy, op. cit.*, pp. 36-79, and Georges Michon, *The Franco-Russian Alliance: 1891-1917,* (New York: Howard Fertig, Inc., 1969).

By way of explanation, the events of 1898-1904 are the relevant events in France and in French-English relations, so we say "circa 1900."

In June 1898, French Foreign Minister Gabriel Hanotaux was replaced by Théophile Delcassé, who had consistently worked to isolate Hanotaux in the cabinet, and had set up the forced French backdown before Britain in Fashoda, Egypt. Delcassé used the ironical end to the Dreyfus Affair to destroy the last remnants of his predecessor's policy.

Indeed, after first initiating the ill-fated expedition of Captain Marchand to Fashoda in Egypt, Delcassé forced France into a humiliating withdrawal in front of advancing British troops. By 1899, Delcassé had accepted a treaty with the British, establishing "spheres of influence" which totally excluded France from the Nile Valley. As part of the package, Delcassé reinterpreted Hanotaux's "Dual Alliance" with Russia into a policy of aggressive encirclement of Germany. The shift was completed with Delcassé's signing of the secret "Entente Cordiale" with Britain in 1904.

70. *Cf.* White, *The New Dark Ages Conspiracy, op. cit.*, Chapters 1-3.

71. Salisbury, *op. cit.*, p. 248. On April 11, 1865, in his last public address, on the subject of Louisiana's re-entry into the Union, Lincoln said, "Some twelve thousand voters in the heretofore slave-state of Louisiana have sworn allegiance to the Union, assumed to be the rightful political power of the State, held elections, organized a State government, adopted a free-state constitution, giving the benefit of public schools equally to black and white, and empowering the Legislature to confer the elective franchise upon the colored man. Their Legislature has already voted to ratify the constitutional amendment recently passed by Congress, abolishing slavery throughout the nation. These twelve thousand persons are thus fully committed to the Union and to perpetual freedom in the state."

72. In 1902, Germany, Great Britain, and Italy surrounded and launched a naval bombardment of Venezuela, followed by a blockade to collect their debts. Roosevelt's administration publicly acquiesced to this action and only complained in order to turn the incident into anti-German propaganda.

Roosevelt perverted the original anti-imperialist intent of John Quincy Adams's Monroe Doctrine with his infamous Roosevelt Corollary, which attempted to arrogate an international police power to the United States. This police power was then repeatedly used for purposes of debt collection in the service or Anglo-American and other international bankers, with a typical script including the seizure of the customs-houses of the country in arrears and the use of import duties to pay the international creditors.

73. In the presidential election of 1912, Theodore "Teddy" Roosevelt ran a third-party presidential campaign in the Bullmoose Party, which split the Republican vote and thereby ensured that Woodrow Wilson would be elected over Republican incumbent William Howard Taft. Much as the Liberty Party had been created around the issue of anti-slavery in 1844, solely for the purpose of denying the presidency to Henry Clay, Roosevelt's Bullmoose or Progressive Party effort, centered around Roosevelt's "new nationalism," an anti-monopoly, anti-corruption corporatism, was a diversionary effort to throw the election to the Harriman-controlled Wilson.

NOTES TO | **Appendices**

Appendix V

1. In his work, "Epitome of Copernican Astronomy," Kepler states: "These properties of light have been demonstrated in optics. The same things are proved by analogy concerning the motor power of the sun, keeping the difference between the works of illumination and movement and between the objects of each. . . . But if the light is attenuated in the ratio of the square of the intervals, . . . why therefore does not the motor virtue too become weaker in the ratio of the squares rather than in the simple?" His answer is essentially that the planets lie on the plane of the ecliptic, which is a two-dimensional circle, rather than a sperical volume. *Epitome of Copernican Astronomy* (1616-21), Book Four, Part II, "3. On the Revolution of the Solar Body Arounds its Axis and Its Effect in the Movement of the Planets," *Great Books of the Western World, 16: Ptolemy, Copernicus, Kepler,* Robert Maynard Hutchins, ed., (Chicago: Encyclopedia Britannica, Inc., 1982), pp. 894-905.

Appendix XI

1. The subject is axiomatics of nonlinearity. I decided to attack some of the problems of conceptual nonlinearity, as against the linear, methods in mathematical physics, from the most elementary, i.e., axiomatic, critical axiomatic standpoint possible. In that respect, some of the sources available through David E. Smith's *A Source Book in Mathematics* (New York: Dover, 1959) and editor Dirk J. Struik's *Source Book in Mathematics: 1200-1800* (Cambridge: Harvard University Press, 1969) are quite useful, as well as some of the other few collateral sources such as Hilbert (see D. Hilbert and S. Cohn-Vossen, *Geometry and the Imagination* [New York: Chelsea, 1952]). I am looking at these, my dear friend Huygens, a few Leibniz things, the Smith and Struik sources, to take some of the most

494

obvious, simple, elementary cases, where the complexities have the greatest relative dependency on the immediate point at issue.

Let's take, just as a point of illustration of what I am doing and what I am thinking about, pages 312 through 316 of Struik, on the L'Hôpital, excerpts.

On pages 312, 313, and 314, we find a development-elaboration of the ground, the basis for two propositions there, and in the following pages, further excerpts from the same source, which give us propositions 163 and 164.

Now if we take that little diagram, as described on pages 313 and 314, pertinent to proposition 1 **Figure 20**, we have there a simple closed curve, which leads to the proposition that the infinitesimal assumption can be added to make, shall we say, the triangle APM equivalent to the triangle *Apm,* in terms of all the functions associated with that.

It's very simple to show the fallacy of that. If the curve is not a simple closed, a simple positive curve, but a hyperbola, then we take in the vicinity of the rapidly ascending slope of the hyperbola, we try to make the same construction and that assumption is no longer even approximately true: that, roughly speaking, an apparently infinitesimal difference, even a relatively small difference, is sufficient to throw the whole thing out of whack, and therefore the infinitesimal assumptions cannot be made.

The same thing applies to postulate 2, which begins on the same page, and the same approach applies obviously hereditarily to postulate 163, 164 in the second selection, which Struik cites from that source.

So, although I think, while this is very simple, what we must do for pedagogical purposes, is look back at the axiomatic assumptions, which we have with Roberval. These axiomatic assumptions in Roberval, the same kind of mathematical assumptions, turn up hereditarily in the case of the

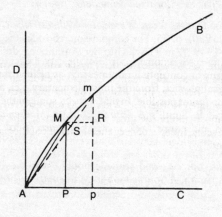

L'Hôpital reflection on the work of the Bernoullis. This shows up in the problems of Euler.

So if we look at this problem of infinitesimals, as defined in these two ways, we find the fallacy of the notion of the infinitesimal wherever discontinuities are generated, as in a Weierstrass function, or this much simpler case of the simple, single hyperbolic application to this first proposition I cited of L'Hôpital.

It's a lot of fun, it's immediately accessible by people. I just throw that in for a suggestion of how we might approach some of these things, from a pedagogical standpoint, and actually get at the deepest, the most elementary, the most simple axiomatic assumptions which cause propositions in physics and as well as mathematics to go awry.

Appendix XIV

1. *Cf.* LaRouche, *In Defense of Common Sense,* Chapters III-V. *Cf.* also LaRouche, *The Science of Christian Economy,* Chapter VII, notes 7 and 8.

2. LaRouche, *In Defense of Common Sense,* Chapters III-V.

3. Johannes Kepler's three principal works are *Mysterium Cosmographicum, The Secret of the Universe,* A.M. Duncan, trans., (New York: Abaris Books, 1981); *Harmonia Mundi,* (Harmony of the World); and *Astronomia Nova* (The New Astronomy) (Paris: Librairie Scientifique et Technique Albert Blanchard, 1979.) (The frontispiece of the 1609 French translation of *The New Astronomy* describes it as, "explaining the causes of heavenly physics presented by the commentaries on the movements of the planet Mars on the basis of observations of the illustrious Tycho Brahe.")

4. Kepler, *The Six-Cornered Snowflake, op. cit.*

5. *Cf.* Max Planck, *Vom Wesen der Willensfreiheit,* (Frankfurt am Main: Fischer, 1990), p. 35. For comparison, consider the definition from Max Planck's 1949 "Zur Geschichte der Auffindung des physikalischen Wirkensquantum": "Was mich in der Physik von jeher von allen interessierte, waren die grossen allgemeinen Gesetze, die für sämtliche Naturvorgänger Bedeutung besitzen, unabhängig von der Eigenschaften der an den Vorgängen beteiligten Körper." ("What I was interested in in physics from the beginning mostly were the great universal laws, which are important for all natural events, independent of the properties of the bodies participating in this event.")

6. Huygens, *op. cit.;* and Jean Bernoulli, "On the Brachistochrone Problem," D.E. Smith, *A Sourcebook in Mathematics,* (New York: 1959), pp. 644-55.

7. Not only did the U.S.A.'s Skull and Bones lodge brothers Averell Harriman and Prescott Bush perform important collaborating roles in putting Adolf Hitler into power in quasi-occupied Germany in 1932-33, but the Harriman family, with shameless openness, avowed sympathy of the Harriman-led eugenics movement in the U.S.A. with the Nazi Party's "racial purification" policies.

This continues into the George Bush-led "new world order" today, out of President Bush's long association with the racialist ("population control") policies of the Draper Fund.

Index